Auroral Processes

ADVANCES IN EARTH AND PLANETARY SCIENCES

Advances in Earth and Planetary Sciences 4

Special Issue of Journal of Geomagnetism and Geoelectricity

Auroral Processes

Proceedings of IAGA/IAMAP Joint Assembly
August 1977, Seattle, Washington

Edited by
C. T. Russell

Center for Academic Publications Japan
Japan Scientific Societies Press
Tokyo

ISBN 978-94-010-9116-9   ISBN 978-94-010-9114-5 (eBook)
DOI 10.1007/978-94-010-9114-5

© CENTER FOR ACADEMIC PUBLICATIONS JAPAN, 1979
Softcover reprint of the hardcover 1st edition 1979

Published by:
CENTER FOR ACADEMIC PUBLICATIONS JAPAN
JAPAN SCIENTIFIC SOCIETIES PRESS
2-10, Hongo 6-chome, Bunkyo-ku, Tokyo 113, Japan

Sole distributor for the outside Japan:
BUSINESS CENTER FOR ACADEMIC SOCIETIES JAPAN
20-6, Mukogaoka 1-chome, Bunkyo-ku, Tokyo 113, Japan

JSSP No. 01782-1104

Preface

Physical and chemical studies of the earth and planets along with their surroundings are now developing very rapidly. As these studies are of essentially international character, many international conferences, symposia, seminars and workshops are held every year. To publish proceedings of these meetings is of course important for tracing development of various disciplines of earth and planetary sciences though publishing is fast getting to be an expensive business.

It is my pleasure to learn that the Center for Academic Publications Japan and the Japan Scientific Societies Press have agreed to undertake the publication of a series "Advances in Earth and Planetary Sciences" which should certainly become an important medium for conveying achievements of various meetings to the academic as well as non-academic scientific communities. It is planned to publish the series mostly on the basis of proceedings that appear in the Journal of Geomagnetism and Geoelectricity edited by the Society of Terrestrial Magnetism and Electricity of Japan, the Journal of Physics of the Earth by the Seismological Society of Japan and the Volcanological Society of Japan, and the Geochemical Journal by the Geochemical Society of Japan, although occasional volumes of the series will include independent proceedings.

Selection of meetings, of which the proceedings will be included in the series, will be made by the Editorial Committee for which I have the honour to work as the General Editor. I and the members of the Editorial Committee will certainly welcome any suggestions that will promote the series. Whenever the convener of a meeting related to earth and planetary sciences is in a position to have to look for a medium for publishing the proceedings please contact us.

Tsuneji Rikitake
General Editor

Foreword

During the second week of the Third General Scientific Assembly of the International Association of Geomagnetism and Aeronomy, held in Seattle, Washington August 22 to September 3, 1977, four special sessions devoted different aspects of auroral processes were convened. The sessions were entitled:

Timing of Substorm Events;

Electromagnetic and Electrostatic Instabilities on Auroral Field Lines;

Rapid Auroral Fluctuations and Associated Phenomena; and

Mechanisms for the Formation of Auroral Structures.

The sessions were convened by the Division II–III Working Group on the Auroral Oval under the chairmanship of Bengt Hultqvist and myself as vice-chairman. Divisions II and III are the IAGA divisions concerned with the ionosphere and magnetosphere, respectively.

The purpose of the session on the timing of substorm events was to come to grips with one of the key experimental issues of substorm research. Many techniques are available for studying substorm phenomena. Substorms may be identified in auroral photographs. They can be recognized by their production of magnetic pulsations. They can be seen in the records of ground based magnetic observatories as ionospheric and magnetospheric currents build up and decay. They can also be identified in spacecraft records obtained deep in space. These multitude of diagnostics lead to problems because they are not intercalibrated. There is intense controversy over whether the onset of a substorm defined by one technique is the same as another.

The purpose of the second session was to review our understanding of plasma wave instabilities on auroral field lines. The auroral ionosphere is filled with a variety of ELF, VLF, and LF emissions. These wave phenomena are not just curiosities, or secondary by-products of auroral processes, but instead appear to be fundamental in the auroral acceleration process. In fact, the energy flux of the terrestrial kilometric radiation (TKR) is a significant fraction of the energy in auroral beams, and hence its source represents an efficient mechanism for the transformation of energy from particles to waves. Study of these phenomena truly involves plasma physics in space, rather than a study of the physics of the space plasma. Many of the techniques used to study these processes, such as computer simulations, are the same as used in other aspects of plasma physics research, such as, fusion research. Many of the observed phenomena have their analogues in other plasma regimes. Yet the auroral plasma is unique. We can probe it locally with multi-diagnostic techniques and yet not disturb it, for our probes are infinitesimal compared to the plasma scale lengths. On the other hand, we are restricted to local measurements and hence often lack the global picture. Thus, these studies often complement laboratory research.

The purpose of the third session was to review rapid auroral fluctuations and their associated phenomena. One of the most characteristic features of aurora is their dynamic behavior. They are constantly moving, changing. This behavior is sometimes chaotic but often very regular. That the aurora can and do undergo regular periodic changes is a cule to the source processes for the aurora. Other clues lie in the nature of other periodic phenomena that accompany the auroral fluctuations.

The final session was an attempt to review first the structure of aurora and the relationship of aurora to field-aligned currents and particle precipitation, and then the main theories on auroral source processes. This session brought together many of the top experimentalists with some of the best theorists and proved to be very stimulating to both groups.

The proceedings of these sessions are being published as two special issues of JGG. In JGG Vol. 30 No. 3 we present the principal review papers presented during the first two sessions; in Vol. 30 No. 4 we present the papers of the second two. In order to publish seventeen review papers in these two issues we have had to ask the authors to limit the size of their papers. The authors have done so, yet in most cases presented a comprehensive overview of their particular topic. In addition to the invited papers, many authors of the contributed papers have prepared summaries of their presentations, and we have included these papers after the invited talks in each of the sessions.

C.T. Russell
Institute of Geophysics and Planetary Physics
University of California at Los Angeles
Los Angeles, California, U.S.A.

CONTENTS

TIMING OF SUBSTORM EVENTS

Pi 2 Micropulsations as Indicators of Substorm Onsets and Intensifications

Gordon ROSTOKER and John V. OLSON

Institute of Earth and Planetary Physics and Department of Physics,
University of Alberta, Edmonton, Canada

(Received October 17, 1977)

Pi 2 micropulsations are impulsive wave trains which accompany the onset of magnetospheric substorm disturbances. At middle and low latitudes Pi 2's are observed as relatively monochromatic damped wave trains, while at auroral observatories onsets of substorms are marked by impulsive increases in noise across the Pi 1 and Pi 2 bands. Because they are often clearly apparent on normal low latitude magnetograms their presence or absence has been used as an indicator of the presence or absence of substorm activity. In this paper we shall demonstrate the relationship between Pi 2 bursts and substorm onsets and intensifications. We shall show that the degree of spatial localization of the Pi 2 disturbance often makes it unidentifiable at middle and lower latitudes using conventional recording techniques. Thus while the presence of a Pi 2 pulsation may be taken as an indication of substorm occurrence, its absence at a particular station may be due to the location of the station relative to the substorm region and may not necessarily imply the absence of substorm activity. Sonograms from an east-west array of observatories in the auroral oval will be employed to demonstrate the use of Pi signals at auroral zone latitudes to define the times of substorm onsets and intensifications. It is recommended that the present subjective means of identifying Pi 2's be replaced by more objective criteria, based upon careful analysis of Pi 2 signals.

1. Introduction

Since the pioneering work of ANGENHEISTER (1912), a great deal of attention has been paid to the morphology of impulsive damped wave trains recorded using conventional magnetographs (which measure absolute values of the main field) and induction coil magnetographs (which measure the rate of change of the magnetic field). These wave trains were initially categorized as sudden impulses (si) and sudden commencements (ssc) and later as pt whose higher frequency component was labelled spt. Eventually, in accordance with the recommendations of JACOBS *et al.* (1964), impulsive micropulsations were given a classification which characterized only their period range, namely Pi 1 ($1 < T < 40$ sec) and Pi 2 ($40 < T < 150$ sec). More recently a third category Pi 3 ($T > 150$ sec) has been suggested (SAITO, 1976) to cover extremely low frequency oscillations detected in substorm disturbed regions. SAITO and YUMOTO (1978) have also introduced the term Ps 6 to cover very low frequency periodic changes in the auroral zone magnetic field during substorm activity. The boundaries of the frequency bands discussed above were not based upon any physical property of the pulsation other than frequency and there was no distinction made pertaining to generation mechanisms (Jacobs, private communication). Thus there is, at the present time, no reason to distinguish between low frequency Pi 1 and high frequency Pi 2 activity, for example.

For many years there was a tendency to study Pi activity as a subject area distinct from the study of substorms, even though it was realized (Saito, 1961) that each Pi 2 observed at low latitudes signaled the occurrence of a magnetic bay at auroral zone latitudes. It was not until Akasofu (1968) accumulated the work done using various research tools (e.g., magnetometers, riometers, all-sky cameras, etc.) under the umbrella of the magnetospheric substorm that it became customary to use any one of the many tools to infer substorm morphology. Rostoker (1968) was one of the earliest workers to make extensive use of the Pi 2 as an indicator of substorm onsets. He found that there was normally a minimum of two Pi 2 wave trains associated with each substorm disturbance, and that a substantial number of substorms were accompanied by more than two wave trains. He suggested, at that time, that the development of a substorm was a two stage process consisting of a trigger bay (the substorm onset) and the main bay (the development of the westward travelling surge) as shown in Fig. 1. Fukunishi and Hirasawa (1970) also noted that there were often several distinctive Pi 2 wave trains associated with a substorm, and they found the wave forms and power spectra of each wave train to be quite similar to one another.

Subsequently it became apparent that substorms developed as an organized series of impulsive electrojet intensifications, each electrojet element lying to the north or north-west of the pre-existing electrojet elements (Sergeev, 1974; Wiens and Rostoker, 1975). Each electrojet element has, associated with its onset, a Pi 2 disturbance which has a source region confined to the region of that electrojet element (Olson and Rostoker, 1975) suggesting that electrojets elements with a small scale size may have an associated Pi 2 which may only be observable close to the electrojet itself.

Since each electrojet element normally has an azimuthal extent of approximately 1 hr in local time, or a time zone, the observations of the two stage character of substorms by Rostoker (1968) may then be understood. Any middle or low latitude observatory may normally detect Pi 2's which originate in its local time sector, viz. it will normally record visually identifiable Pi 2's for substorm onsets occurring up to one time zone on either side of the meridian of the observing station. Any Pi 2 onsets occurring more than one time zone away will not be identifiable due to low amplitudes. Thus for a sequence of several electrojet intensifications each one being to the northwest of the previous one, a normal mid-latitude observatory will detect, on the average, two Pi 2 onsets depending on the strength of the intensifications. It will generally be unable to detect onsets more than one time zone to the east or to the west under normal circumstances. Thus a given mid-latitude observatory may be unable to detect the actual onset times of a sequence of substorm intensifica-

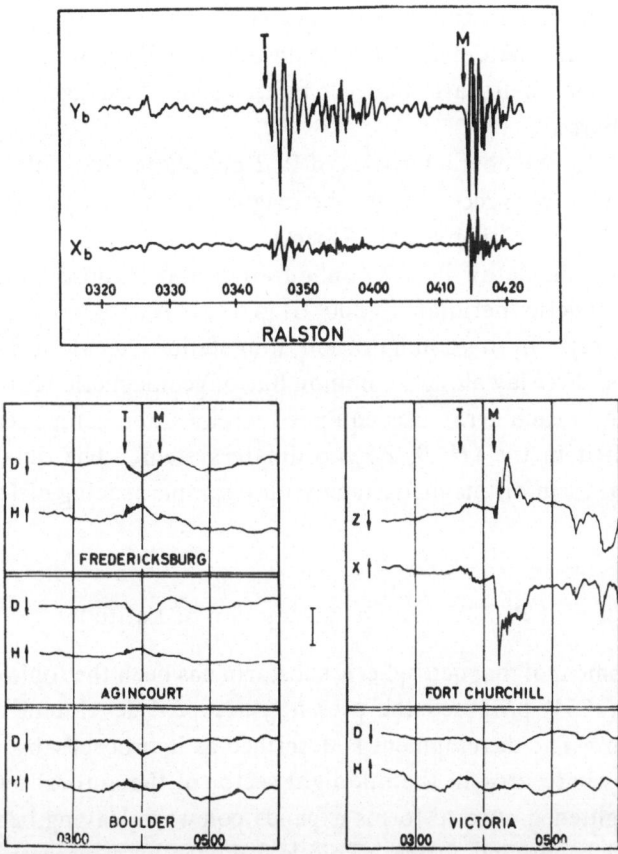

Fig. 1. Example of the structure of geomagnetic bays associated with a polar magnetic substorm and the accompanying Pi 2's recorded on June 6, 1965 (after ROSTOKER, 1968). The top panel shows induction coil magnetograms from the mid-latitude station of Ralston and corresponding mid-latitude (Fredricksburg, Agincourt, Boulder and Victoria) magnetograms and an auroral zone (Fort Churchill) magnetogram are shown in lower panels. The Pi 2's are normally most easily identified in the horizontal perturbation fields. The Pi 2 activity in the auroral zone is masked by the steep ramp of the bay at auroral zone latitudes. The wave form at mid-latitude stations varies from site to site, and weaker Pi 2's may only appear as a thickening of the trace in mid-latitude observatories.

tions if the onset takes place more than one time zone to the east. It may also fail to detect the end of the substorm sequence if the final electrojet element intensifies more than one time zone to the west. In essence, therefore, unless there is an adequate local time coverage of middle and low latitude magnetometers *it may not be possible to identify the actual onset of an episode of substorm activity using Pi 2 pulsations.* Furthermore, an absence of *identifiable* Pi 2 activity at middle and low latitude observatories does not rule out the existence of substorm activity given the present observatory distribution.

In the light of the recent trend towards utilization of Pi 2's as indicators of substorm onsets (SAITO *et al.*, 1976 a, b; SAKURAI and SAITO, 1976; PYTTE *et al.*,

1978) we shall try to demonstrate the characteristics and identifiability of Pi 2 onsets using high latitude magnetometer array data and middle and low latitude normal magnetograms. We shall particularly emphasize the localized character of these micropulsation bursts.

To demonstrate the characteristics of Pi 2 pulsations, we shall utilize two suites of data all of which were recorded by the magnetometer arrays operated in north-western Canada by the University of Alberta:

1)   Meridian line data: In this configuration nine stations were arrayed along a common geomagnetic meridian ($\sim 300°$E) in 1971–72.

2)   Cross array: In this configuration, four stations lay along a common meridian ($\sim 300°$E) and three lay along a common line of geomagnetic latitude ($\sim 67.5°$N).

Each station in each array was equipped with a three-component fluxgate magnetometer oriented in the (H, D, Z) coordinate system. The data were recorded directly in digital form on magnetic tape with a sample spacing of 1.92 sec after low pass filtering.

2.   Development of Pi 2 Activity as a Function of Latitude

The development of magnetospheric substorm has been the topic of intense study since Akasofu (1964) proposed the presently accepted development scheme of the auroral substorm. The development is described as it proceeds by two phases: an *expansion phase* where arcs in the midnight sector of the auroral oval brighten and the region of brightened auroral forms expands poleward leaving behind a region of irregular auroral bands and forms. This is followed by a *recovery phase*, in which the irregular auroral forms gradually reconfigure into east-west aligned arcs. These arcs gradually shift equatorward until the disturbed sector returns to its pre-substorm configuration. In addition the westward edge of the substorm disturbed region, characterized by the *westward travelling surge* expands into the evening sector poleward of the northern edge of the diffuse auroral oval. The substorm disturbed region is generally the site of westward ionospheric current whose magnetic signature is intense negative H-component bays as observed on the ground under the disturbed region.

In recent years, it has become clear that the poleward expansion of the auroral arcs is not steady, but occurs in discrete jumps. Each poleward jump signifies the appearance of discrete auroral forms and a filamentary westward electrojet poleward of the pre-existing substorm disturbed region (Kisabeth and Rostoker, 1971, 1974). This step-wise development of the electrojet is significant in that it suggests that rapid changes in magnetospheric current flow can only involve currents of limited scale size. In addition sudden changes in any electrical circuit will often result in a ringing phenomenon, and Boström (1972) has, in fact, attributed Pi 2 oscillations to the ringing of the substorm current configuration. In this context we may look at Pi 2 development as the substorm electrojet expands poleward, in the manner expounded by Olson and Rostoker (1975). They show an event, for which the magnetograms are shown in Fig. 2 and the filtered micropulsation records in Fig. 3.

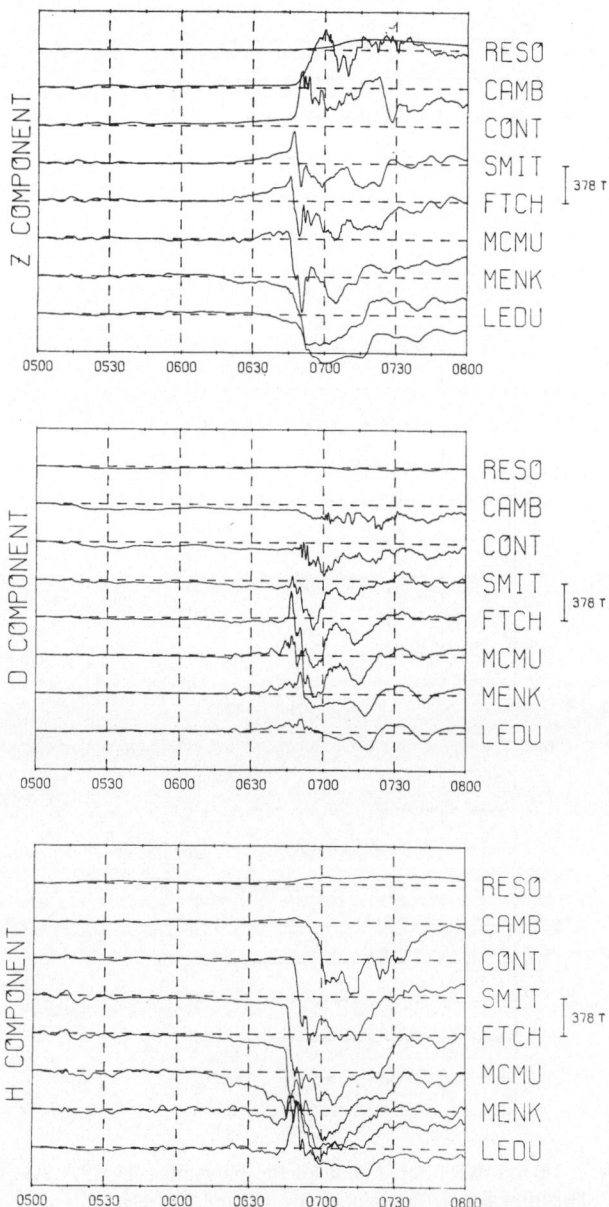

Fig. 2. Magnetograms from the University of Alberta magnetometer chain for the interval 0500–0800 UT on November 23, 1971. The stations are arranged with the northernmost station (RESO, 83°N) at the top of the frame and (LEDU, 60.9°N) at the bottom of the frame. Note the progressive delay in onset times in the high latitude stations.

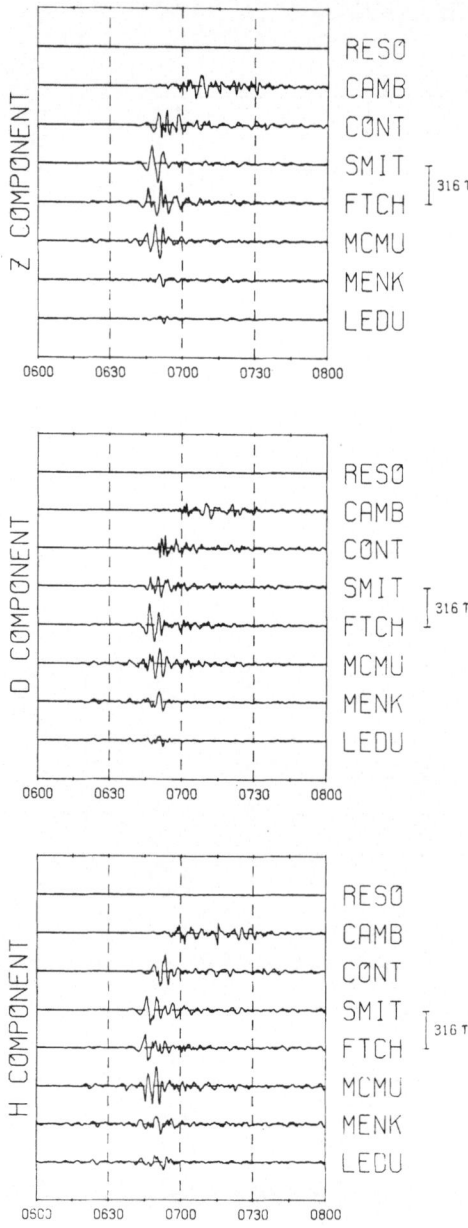

Fig. 3.   Band pass filtered data for the November 23, 1971 event. The filter corner frequencies are 1.0 and 20 mHz.

The Z-component of the micropulsation is a good indicator of the presence of the source of pulsations over the observing stations. Figure 4 shows that the onset of Pi 2 pulsations on the Z-component signals the arrival of the poleward portion of the electrojet over the observing station. Thus, at auroral zone stations, monitoring the Z-component of the Pi 2 allows one to detect when the poleward border of the electrojet has expanded over the observing station.

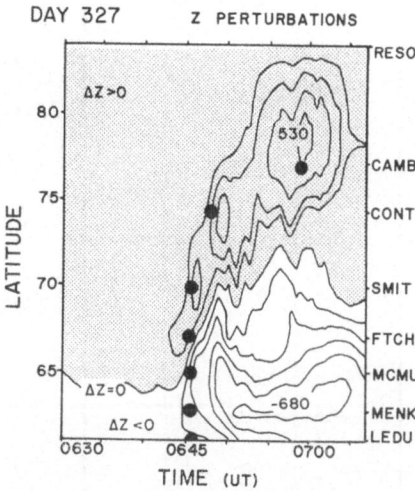

Fig. 4. Contour diagram of the *Z*-component perturbations during the November 23, 1971 event (after OLSON and ROSTOKER, 1975). The diagram shows the intensity of the *Z* perturbations, which indicate the presence of overhead currents, as a function of latitude and local time. Extrema in the *Z* perturbation levels delineate the approximate boundaries of the electrojet at a given instant. The solid dots show the onset times of Pi 2 pulsations at each station in the chain.

From Fig. 3 it can be seen, comparing the SMIT, CONT and CAMB records that the clearly identifiable onsets in all components occur at different times. This is because of the large signal enhancement when the electrojet is overhead compared to when it is not. Thus it is clear that clearly defined Pi 2 enhancement can only be associated with substorm intensifications and not necessarily with the onset of substorm sequences. In addition, low level Pi 2 activity at one location may well (and probably does) indicate strong Pi 2 activity at some distance from the observing point.

3. Development of Pi 2 Activity as a Function of Longitude

In the remaining portion of this paper we shall present a detailed analysis of the magnetic activity which occurred on August 29, 1974. In this analysis the use of low latitude standard magnetograms along with auroral zone records from the University of Alberta magnetometer array will allow us to observe the progress of a series of substorm intensifications and their associated Pi 2 wave trains across North America. Similar analyses have been performed on data from other days with very similar results.

Figure 5, which shows the *H*-component records from various low latitude magnetograms, will allow us to set the scene for the event. The stations are arranged on the figure with the easternmost station (JUAN) at the top and the westernmost at the bottom. We direct your attention to the following features:

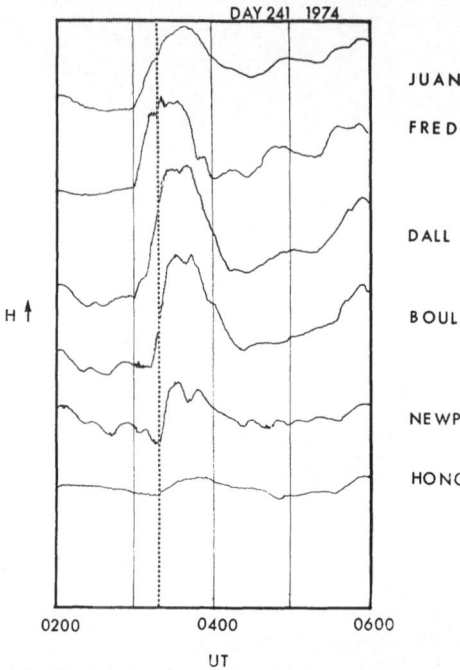

Fig. 5. *H*-component data for August 29, 1974 from low latitude
stations at San Juan, Fredericksburg, Dallas, Boulder, Newport
and Honolulu in the North American sector. The station data
are arranged sequentially from east to west with the easternmost
station data at the top. Substorm onsets may be seen at 0300,
0310, 0315 and 0320 UT. The vertical dashed line marks the
0320 UT event.

1)   A clear substorm onset is observed at 0300 UT from JUAN to DALL. Note
that the associated Pi 2 at BOUL does not show peak amplitude at the moment of
onset. The maximum occurs 3 to 4 min after the hour. The use of this station's Pi 2
would result in an error in timing the moment of substorm onset.

2)   A second onset occurs at 0310 UT. The signature is clear at DALL. The
high latitude data from Fort Churchill (not shown here) showed a sudden negative
excursion in ΔH at this time.

3)   A third onset occurred near 0315 UT. This onset is clearest at BOUL and
an associated Pi 2 is present at FRED. In the auroral oval the onset was masked by
activity associated with the 0310 UT onset.

4)   The fourth onset which signaled the start of major activity in the western
sector of North America occurred at 0320 UT as indicated by the dashed line on
the figure. It is clearly evident at NEWP and shows in the HONO record. The
breadth of this region indicates a large scale event. ΔH variations are positive across
the entire continent and into the mid-Pacific Ocean region.

Although there is evidence for further activity in this substorm sequence we
shall limit our study to these four events, because of the lack of station coverage in
the east Pacific area. Thus in these data we may observe the passage of substorm

activity across North America in four intensifications. We will now focus on the data from our array in the Alberta sector for a detailed view of the activity seen from one sector.

Figure 6 shows the high latitude magnetograms from the University of Alberta magnetometer array. The figure shows three panels which contain the *H*, *D* and *Z* component variations from the array. At the top of each panel is URAN, the easternmost station. SMIT and HAYR lie at the same latitudes as URAN ($\sim$67.5°N) with HAYR some 10° west of URAN. MCMU lies approximately 3° south of SMIT

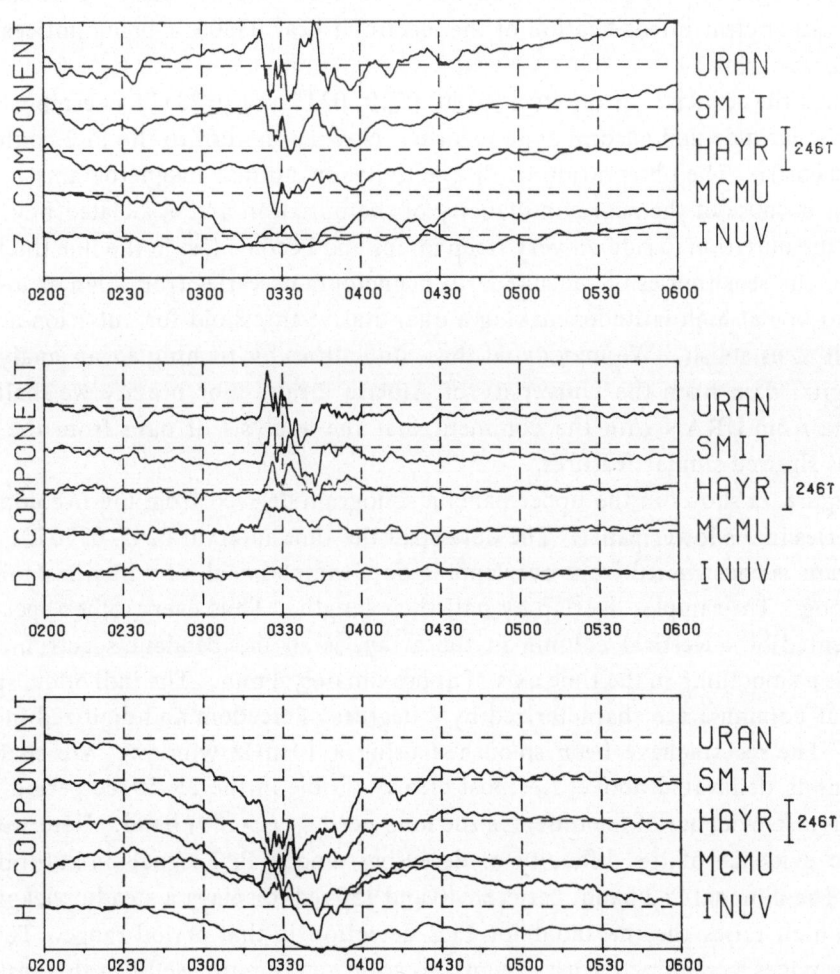

DAY2411974

Fig. 6. Magnetic records from the University of Alberta array showing the high latitude records for the August 29, 1974 event. The principal substorm onset in this sector occurred at 0320 and can be seen in all components. There is also evidence of the 0315 onset which occurred to the east of the Alberta sector.

on the same meridian. INUV lies to the north and far to the west of the other stations.

Recall that we observed onsets at 0300, 0310, 0315 and 0320 UT in the low latitude magnetograms. Here, on this scale, we see no clear evidence of the 0300 UT onset. The first indication of a change in activity occurs at 0315 UT where URAN, at the eastern edge of the array sees a noticeable ΔZ inflection. The other stations do not show this so clearly but they begin to change slightly in accordance with their position relative to the center of activity. The 0320 UT event impacts the entire array and the large $+\Delta D$ signals the arrival of the westward travelling surge over the station array. Upon careful analysis, the ΔD maximum and the onset of the major inflection can be seen to occur at 0321 UT at URAN and 0322 UT and SMIT reflecting the rapid westward expansion (~ 3 km/sec) within the sector disturbed by the 0320 UT onset. As mentioned earlier there is continuing activity after 0320 UT with a rather clear intensification of the electrojet near 0330 UT being noticeable in the Alberta array.

Thus the sequence of onsets at 0300, 0310, 0315 and 0320 UT is established in both low latitude and auroral zone records. Now let us turn to the Pi 2 content of these records. The observation of Pi 2 activity on normal magnetograms is often difficult because of the near simultaneity of the pulsation and associated bay. This causes the pulsation to ride on very steep ramps and become lost in the line thickness. Further the sensitivities of the standard magnetograms varies from high at low latitudes to low at high latitudes making a quantitative threshold for pulsation activity difficult to establish. We may avoid these difficulties by turning to an analysis of the digital data from the University of Alberta array. For brevity we shall treat the data from URAN with the comment that the analysis of data from the other stations showed similar features.

Figure 7a shows in the upper panel a sonogram derived from the H-component data series in the lower panel. The data span the time interval 0245–0350 UT. The sonogram is constructed from overlapping data samples each of which is 4 min and 6 sec long. The samples overlap by half their length. Thus every other spectrum, represented by a vertical column in the array, is an independent spectrum. The result is a smoothing in the time axis of approximately 4 min. The individual spectra (vertical columns) are characterized by 4 degrees of freedom and digitized in 10 db steps. The spectra have been smoothed using a 10 mHz window. We shall find that onsets delineated above are most clearly visible in the 25–50 sec period band. There is also evidence of the onsets in the long period ($T > 100$) band. Here we have graphic evidence of the difficulty of detecting unique Pi 2 signals in auroral zone data. The nominal Pi 2 band, between 40 and 150 sec, displays a steady background level which raises the threshold for Pi 2 detection in that period range. The Pi 1 band provides a greater contrast in signal to noise and is more useful in this instance.

In the first sonogram, data from the H-component at URAN are shown. In this figure the onsets at 0300 and 0320 UT are clearly visible as broad band short duration bursts. The onsets at 0310 and 0315 UT are not resolved in the 25 to 50 sec band due to contamination of the spectra by low frequency noise and the width of the data window. Nevertheless the long period ($T > 100$ sec) portion of the spectrum shows separate enhancements above the background at 0310 and 0315 UT. Note as well the continuing activity after 0320 UT.

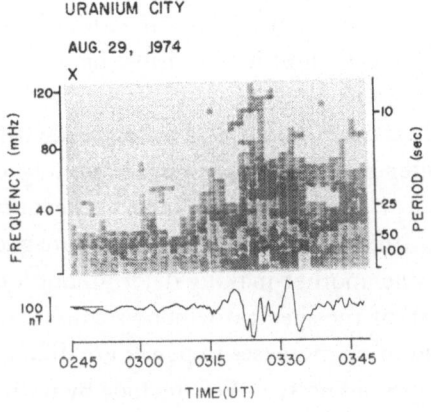

Fig. 7a. A sonogram of the *H*-component data at Uranium City covering 0245–0345 UT on August 29, 1974. The 0300, 0310, 0315 and 0320 UT onsets are most clearly seen in the Pi 1 band where there is good signal contrast. See text for analysis details.

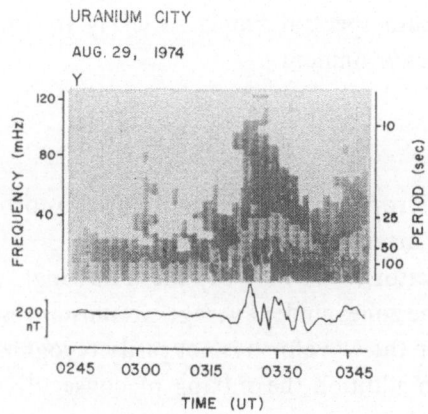

Fig. 7b. A sonogram of the Uranium City *D*-component data covering the interval 0245–0345 UT on August 29, 1974. Onsets at 0300, 0310, 0315 and 0320 UT are clearest again in the Pi 1 band.

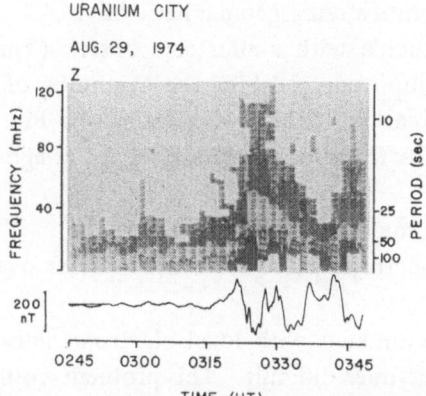

Fig. 7c. A sonogram of the Uranium City *Z*-component data covering the interval 0245–0345 UT on August 29, 1974. Here the 0300, 0315 and 0320 UT onsets may be seen. See text for details.

Figure 7b shows a similar sonogram for the *D*-component at URAN. Again the 0300 and 0320 UT onsets are clearly defined. The 0315 UT onset is also visible in the sonogram even though it is somewhat obscured by background activity. The identification of an onset at 0310 UT from these data alone appears somewhat more tenuous, although there is some suggestion of additional activity near that time.

The final sonogram presented in Fig. 7c shows the *Z*-component activity. While these data are quieter than the *H*- and *D*-component data one can still see the onsets at 0300, 0315 and 0320 UT. The 0310 UT onset is not visible here. It is lost in the general background noise in the Pi 2 band.

We may summarize our results by noting that when we use Pi 2 onsets to identify substorm onsets or intensifications, we encounter basic limitations in the mixing of wave trains (viz. a Pi 2 onset occurs before a prior Pi 2 dies out) and background noise in the Pi 2 range (particularly at high latitude stations). On normal magnetograph recordings the wave trains mix with one another making determination of onsets often difficult. For sonograms the width of the spectral windows causes uncertainty in the estimation of the onset time and often increases in power are difficult to extract from the power associated with background noise. We conclude by noting that the sonogram, when used as we have above, represents a pulsation detector based upon changes of energy density in certain spectral bands. Clearly it is of marginal use for short-lived signals in a noisy environment.

4. Conclusions

We finish our discussion of the use of Pi micropulsations in identifying substorm onsets by summarizing some of the key points made in the foregoing text.

As regards middle and low latitude observatories we may say the following:

i) Pi 2 onsets are observable over ∼1–2 time zones under average circumstances. However, sometimes the amplitudes are low or the waveform is not easily recognizable as a Pi 2 according to existing criteria. In addition the mixing of consecutive wave trains may make some onsets difficult to define.

ii) For normal magnetograms, Pi 2 wave trains may have their structure masked as they ride on a ramp associated with the accompanying geomagnetic bay.

iii) Pi 2 amplitude extrema may not coincide with a substorm onset. (That is, one should not demand an initial large amplitude pulse at the beginning of a wave train as a necessary condition for classification of that wave train as a Pi 2.) This is because the wave train may be shaped by transmission processes at the signal propagates from the source to the field point.

As regards auroral oval observatories, we may say the following:

i) Pi 2 "onsets" occur as the substorm electrojet poleward border passes over the observing station.

ii) The Pi frequency band is often contaminated with local electrojet noise, making the identification of distant onsets sometimes difficult. This problem would be accentuated as the general level of magnetospheric activity increases.

In conclusion, if Pi 2 micropulsations are to be used in the future for the identification of substorm onsets and/or intensifications we recommend that:

a) Agreement be reached on the character of wave trains which uniquely define them as Pi 2.

b) That onsets not be identified solely by the Pi 2 technique. Any individual event should be studied using multistation data and, for each station, all three components of the perturbation field should be considered.

This research was supported by the National Research Council of Canada.

REFERENCES

AKASOFU, S.-I., The development of the auroral substorm, *Planet. Space Sci.*, **12**, 273–282, 1964.

AKASOFU, S.-I., *Polar and Magnetospheric Substorms*, Springer-Verlag, New York, 1968.

ANGENHEISTER, B., Über die Fortpflanzungs—Geschwindigkeit magnetischer Storungen und Pulsationen, Bericht über die Erdmagnetischen Cheltenham und Tsingtau in September 1911, *Goettingen Nachr. Ges. Wiss.*, 1912.

BOSTRÖM, R., Magnetosphere—ionosphere coupling, in *Critical Problems of Magnetospheric Physics*, edited by E.R. Dyer, Jr., pp. 139–156, IUCSTP Secretariat, Washington, D.C., 1972.

JACOBS, J.A., Y. KATO, S. MATSUSHITA, and V.A. TROITSKAYA, Classification of geomagnetic micropulsations, *J. Geophys. Res.*, **69**, 180–181, 1964.

KISABETH, J.L. and G. ROSTOKER, Development of the polar electrojet during polar magnetic substorms, *J. Geophys. Res.*, **76**, 6815–6828, 1971.

KISABETH, J.L. and G. ROSTOKER, The expansive phase of magnetospheric substorms. 1. Development of the auroral electrojets and auroral arc configuration during a substorm, *J. Geophys. Res.*, **79**, 972–984, 1974.

FUKUNISHI, H. and T. HIRASAWA, Progressive change in Pi 2 power spectra with the development of magnetospheric substorms, *Rep. Ionos. Space Res. Japan*, **24**, 45–65, 1970.

OLSON, J.V. and G. ROSTOKER, Pi 2 pulsations and the auroral electrojet, *Planet. Space Sci.*, **73**, 1129–1139, 1975.

PYTTE, T., R.L. MCPHERRON, E.W. HONES, Jr., and H.I. WEST, Jr., Multiple satellite studies of magnetospheric substorms: Absence of substorm-onset signatures during prolonged auroral-zone bay activity, *J. Geophys. Res.*, **83**, 663–679, 1978.

ROSTOKER, G., Macrostructure of geomagnetic bays, *J. Geophys. Res.*, **73**, 4217–4229, 1968.

SAITO, T., Pi 3-type polar magnetic variation due to a fluctuating three-dimensional current system during a magnetospheric substorm, *IAGA Bulletin*, No. 36, 1976.

SAITO, T. and K. YUMOTO, Comparison of the two-snake model with the observed polarization of the substorm-associated magnetic pulsation Ps 6, *J. Geomag. Geoelectr.*, **30**, 39–54, 1978.

SAITO, T., K. YUMOTO, and Y. KOYAMA, Magnetic pulsation Pi 2 as a sensitive indicator of magnetospheric substorm, *Planet. Space Sci.*, **24**, 1025–1029, 1976a.

SAITO, T., T. SAKURAI, and Y. KOYAMA, Mechanism of association between Pi 2 pulsation and magnetospheric substorm, *J. Atmos. Terr. Phys.*, **38**, 1265–1277, 1976b.

SAKURAI, T. and T. SAITO, Magnetic pulsation Pi 2 and substorm onset, *Planet. Space Sci.*, **24**, 573–575, 1976.

SERGEEV, V.A., On the longitudinal localization of the substorm active region and its changes during the substorm, *Planet. Space Sci.*, **22**, 1341–1343, 1974.

WIENS, R.G. and G. ROSTOKER, Characteristics of the development of the westward electrojet during the expansive phase of magnetospheric substorms, *J. Geophys. Res.*, **80**, 2109–2128, 1975.

The Use of Ground Magnetograms to Time the Onset of Magnetospheric Substorms

Robert L. McPherron

Department of Earth and Space Science
and
Institute of Geophysics and Planetary Physics
University of California, Los Angeles, U.S.A.

(Received October 17, 1977)

The original definition of substorm onset was the sudden brightening of an equatorward auroral arc near midnight. Later, other phenomena were correlated with this onset and found to also undergo sudden changes. Because of this systematic relation these phenomena have also been used to time substorm onsets. Two such phenomena are the sudden intensifications of the westward electrojet in the midnight auroral zone and the development of field-aligned currents along field lines connected to the westward electrojet. These two phenomena are observed as characteristic magnetic deviations in ground magnetograms. In the midnight auroral zone it is the sudden beginning of a negative deviation (or bay) while at midlatitudes it is the beginning of a positive bay which has been used as the substorm onset.

Both experimentally and theoretically it has been established that the magnetic signature of substorm onset is spatially variable. In the absence of a perfect distribution of magnetic observatories timing errors may result from incorrect interpretation of limited data. An important question for substorm research is to what extent these errors have contributed to the development of incorrect phenomenological models.

A number of studies have established that many intensifications of the aurora cannot be observed magnetically at midlatitudes. Similarly, some disturbances, particularly events at high latitudes, cannot be observed magnetically by auroral zone observations. Another fundamental question is whether a phenomenological model based on data from the onset of magnetically detectable substorms is a correct basis for physical theories.

A third problem which has been only recently recognized is the frequent occurrence of fine structure within a substorm. Stepping substorms, multiple onset substorms and double substorms are names applied to events in which some of the phenomena normally associated with the classical model of a substorm reoccur with a frequency of 10–20 min, rather than the expected 2–3 hr.

1. Introduction

A fundamental question of magnetospheric research today is, what physical process causes the onset of a magnetospheric substorm? An answer to this question depends on the timing of substorm onset. Only if this time is accurately known can it be determined what physical processes are occurring at different points in space.

For example, PYTTE *et al.* (1976b), NISHIDA and FUJII (1976), and WALKER *et al.* (1977) show that in the midnight sector of the near geomagnetic tail a drop out of energetic electrons precedes substorm onset as defined by Pi 2 bursts and bays on ground magnetograms.

In the original concept of the auroral substorm (AKASOFU, 1964), substorm onset was defined as the sudden brightening of the southernmost auroral arc in the midnight sector. In extending the substorm concept, AKASOFU *et al.* (1965) showed that the polar magnetic substorm accompanied the auroral substorm, and that the magnetic signature of substorm onset was a sudden intensification of the westward electrojet beneath the brightening auroral arc. As pointed out by AKASOFU and SNYDER (1972), a physical variable such as plasma sheet thinning which is systematically correlated with sudden brightening of the aurora can also be used as a signature of substorm onset.

Because of their availability and relative simplicity, ground magnetograms have been used by many investigators to determine substorm onset (ROSTOKER, 1972b, 1974). Results obtained in this fashion have not always agreed with those obtained using auroral data, leading some workers to conclude that the "surest way" to define substorm onset is the one originally made by AKASOFU (AKASOFU and SNYDER, 1972).

In this paper we briefly review the technique of substorm timing using ground magnetograms. We particularly emphasize some of the pitfalls as these have led some workers to err in onset determinations. We also summarize recent developments in the concept of multiple onset substorms. Such substorms appear to account for many of the reported discrepancies in substorm timing. So frequently do such substorms occur that we suggest that present, idealized models of substorm morphology must be modified to include them.

2. Substorm Timing with Auroral Zone Magnetograms

Substorm timing using auroral zone magnetograms is based on the observations reported by AKASOFU *et al.* (1965), that a sudden intensification of the westward electrojet occurs under the same brightening aurora used to define the onset of the auroral substorm. As the auroral substorm develops the location and strength of this electrojet change systematically, expanding northward and westward with the aurora. Near the end of the expansion phase the equivalent ionospheric current system occupies nearly the entire auroral oval as shown in Fig. 1.

The existence of this current system is normally inferred from the magnetic perturbations in the horizontal component measured on the earth's surface. For example, applying the right hand rule to the currents in Fig. 1 at 0100 local time (LT) and 68° latitude gives a large southward ground perturbation or "negative bay in H." Similarly at 2000 LT and 65° latitude one predicts a positive bay in H due to a weaker and more diffuse "eastward electrojet." The location of the electrojet in latitude relative to a ground station is inferred from the sign of the vertical com-

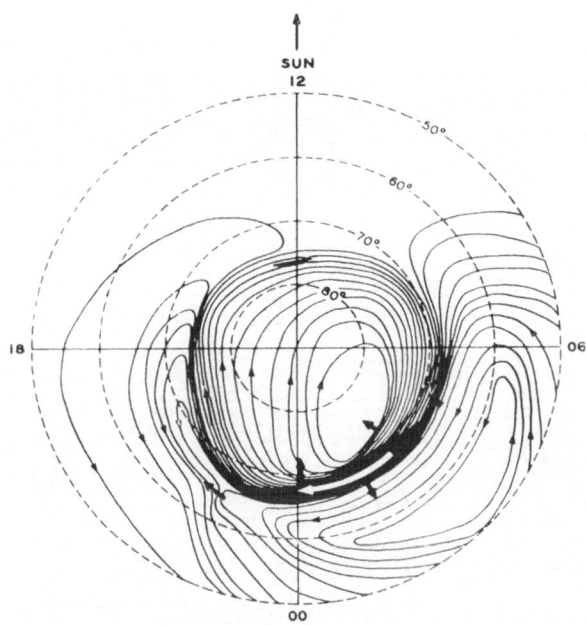

Fig. 1. Equivalent ionospheric current for westward electrojet. This model
assumes the eastward electrojet near dusk is a return current for the westward electrojet (from AKASOFU *et al.*, 1965).

ponent. If the westward electrojet is poleward of the station the vertical component
is out of the ground giving a negative component.

The onset of the substorm is normally inferred from a sudden drop in the H
component at an auroral zone station just before midnight. Two examples are
presented in Fig. 2. In both cases the sudden large drops at Great Whale were used
as the signature of substorm onset (MCPHERRON, 1970).

There are numerous pitfalls in interpreting the magnetic data from a limited
set of magnetic observatories. These pitfalls are primarily a consequence of the
initial localization of the westward electrojet and its subsequent slow spatial development. For example, ROSTOKER (1972a), shows an example (his Fig. 3) in which
the sudden onset at Fort Churchill is delayed 15 min relative to the onset at Great
Whale, ≈600 miles to the east. Similar delays occur in the north-south direction.
As an example (ROSTOKER, 1972), Fig. 13 presents data from a latitude chain illustrating a 20 min delay between two stations separated only a few degrees in latitude.
Other problems arise when the substorm occurs on a contracted auroral oval. In
such cases, as shown by AKASOFU *et al.* (1971, their Fig. 3b) no obvious signature
is seen at standard auroral zone observatories although a station further north records
a clear negative bay.

One tool which has been a great aid in interpreting auroral zone magnetograms
is the latitude profile developed by Rostoker and coworkers. By comparing observed
profiles of all three field components to model profiles it is possible to deduce a

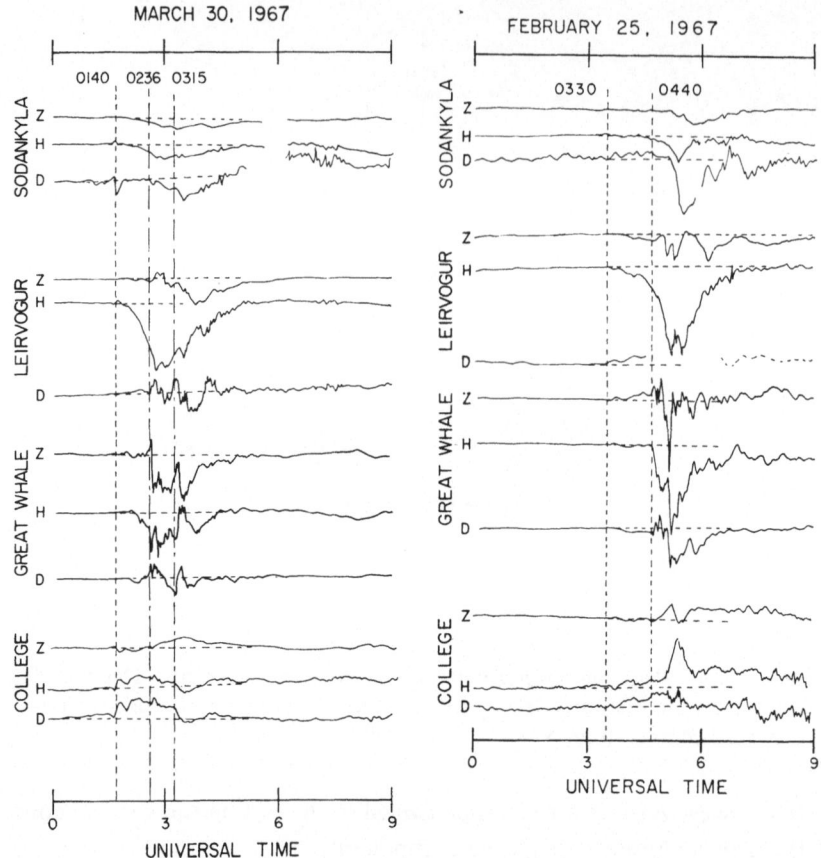

Fig. 2. Magnetograms from an east-west auroral zone chain illustrating the determination
of substorm onset using the sudden development of a negative bay in H (horizontal component) at a premidnight auroral zone magnetic observatory (Great Whale). Note both
panels show double onset structure though only the left panel shows this explicitly (from
McPherron, 1970).

number of characteristics of the electrojet current system. One characteristic which
makes the timing of substorms particularly difficult is illustrated with this technique
in Fig. 3. Three peaks in the perspective plot are separated by about 10 min and
each satisfies the definition of substorm onset as defined above. Such "quasi-periodic
intensifications" or "multiple onsets" are apparent in many auroral zone events
(Kisabeth and Rostoker, 1971, 1973, 1964). Figure 2 discussed above, shows two
other cases in data from widely spaced auroral zone observatories.

The four problems outlined above inevitably lead to errors in the determination
of substorm onset. In most cases the inferred onset will be later than the actual
onset. An upper limit to the magnitude of such errors is of order 20 min, nearly the
duration of a classical expansion phase. Greater timing accuracy can be obtained
by using more auroral zone observatories so that the possible delays due to substorm
expansion are reduced. In addition, as we discuss below, additional types of data
may be used to time substorm onset.

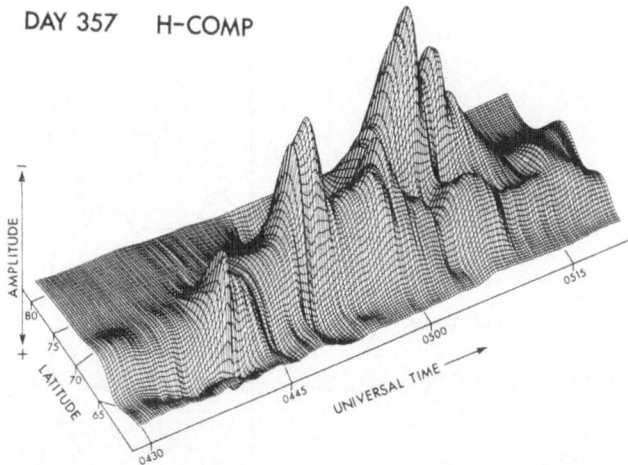

Fig. 3. Dynamic development of the westward electrojet as recorded by a north-south chain of magnetometers. Successive peaks represent quasi-periodic intensification of the westward electrojet with a time scale of about 10 min (from WEINS and ROSTOKER, 1975).

3. Timing Substorm Onset with Midlatitude Magnetograms

Large substorms can be observed in midlatitude magnetograms as well as in auroral zone records. The expected midlatitude magnetic signature and its correlation with auroral zone effects has been well documented by AKASOFU and MENG (1968, 1969) and MENG and AKASOFU (1969). Figure 4 taken from this work shows the pre-midnight variations in H is positive (opposite to the auroral zone change) and about one tenth as large. Originally it was thought these effects were produced by ionospheric return currents of the westward electrojet (cf. Fig. 1). However, synchronous satellite data conclusively show field aligned currents are the primary source as discussed below.

Midlatitude perturbations in the D component depend significantly on latitude and local time (see Fig. 4). On the average in the northern hemisphere positive perturbations (east) are seen before midnight and negative perturbations are seen after midnight. The opposite sign is seen in the southern hemisphere.

The three dimensional current system responsible for both midlatitude and auroral zone substorm variations is shown in Fig. 5. This highly idealized model current system has an equivalent ionospheric current system virtually identical to that shown in Fig. 1 (BONNEVIER and BOSTROM, 1970; FUKUSHIMA and KAMIDE, 1973). Midlatitude magnetic perturbations produced by this current along a longitudinal profile in the northern hemisphere are shown in the bottom half of the figure. It is evident that the D component is very useful in determining the central meridian and width of the field aligned current system responsible for the electrojet.

Substorm timing with midlatitude magnetograms is based on the connection between the field-aligned currents and the westward electrojet. As a sudden intensi-

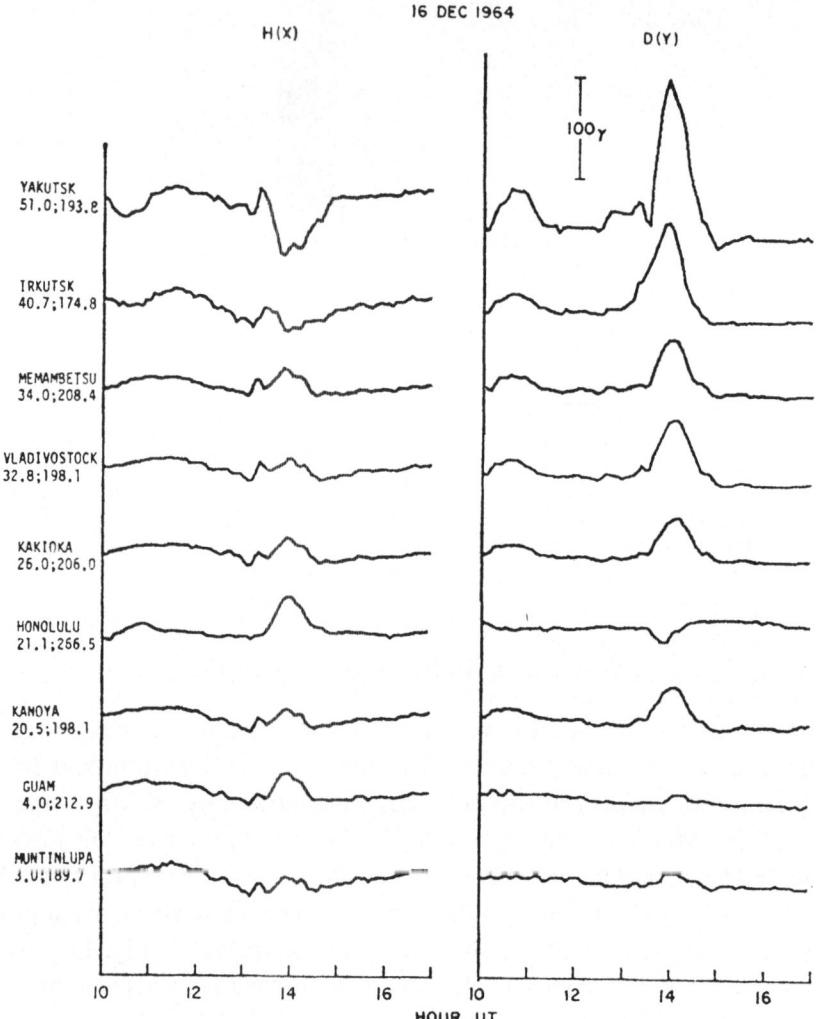

Fig. 4.   The midlatitude signature of a double onset substorm as a function of latitude
          for a number of observatories in the premidnight sector (form AKASOFU and MENG,
          1969).

fication of the electrojet occurs in the brightening aurora, field-aligned currents begin
to flow feeding the electrojet. Magnetic perturbations of these currents are seen at
midlatitudes as positive bays in H for stations inside the current "wedge." Onset is
defined as the beginning of the midlatitude positive H bays.

An example of the use of midlatitude magnetograms in substorm timing is
presented in Fig. 6. Vertical-dashed lines represent times chosen as substorm onset.
As shown by McPHERRON et al. (1973b), the times for this event correspond to the
onset of auroral zone negative bays as well. This is not always the case as auroral
zone magnetic onsets may appear to precede the midlatitude onsets as discussed by
McPherron et al. As discussed below, however, such cases seem to be examples of

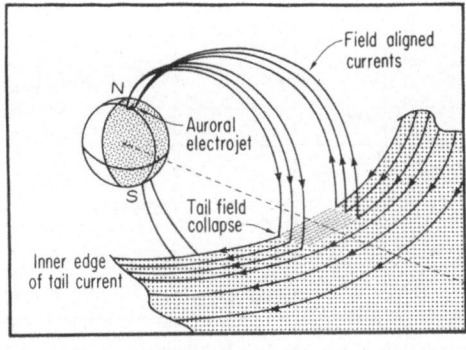

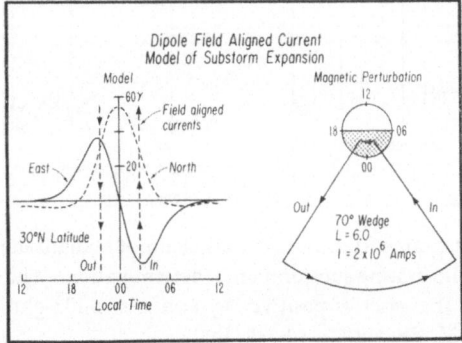

Fig. 5. An idealized model of the field-aligned currents connected to the westward electrojet during a substorm expansion. The bottom panel shows the expected magnetic signature along an east-west midlatitude chain of magnetic observatories. Central meridian, angular extent and strength are the most important parameters of this current system (from CLAUER and McPHERRON, 1974).

multiple onset substorms for which the initial onsets are too weak to be observed at midlatitudes.

There are numerous problems associated with the use of midlatitude magnetograms, especially when only a few stations are used. One of the greatest problems is interference by other current systems. Since the field-aligned currents are far from the midlatitude observatories, perturbations due to other current systems can be of comparable amplitude. These include ionospheric currents driven by solar heating, magnetopause currents responsive to the solar wind dynamic pressure and the ring current created during magnetic storms.

If such effects were absent, midlatitude observatories would still be hampered by lack of sensitivity. Magnetic perturbations of weak substorms are below the noise level of most existing midlatitude observatories.

Another serious problem is caused by the partial ring current. This current is centered near the dusk meridian and opposite in direction to the expansion phase current system of Fig. 5. Unfortunately, these two current systems frequently develop simultaneously. Because the effect of the partial ring current is a negative

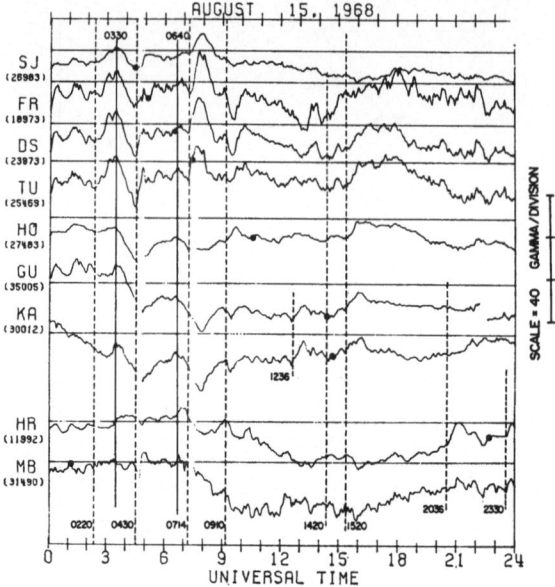

Fig. 6. Magnetograms from an east-west chain of midlatitude magnetic observatories illustrating substorm onset determination. The characteristic
signature is the onset of positive bay at a near midnight station (black
dots) (from McPHERRON *et al.*, 1973b).

bay in *H* it is possible to misidentify the minimum of the negative bay as the onset
of a positive bay if only data from the dusk sector is used. This can be seen in Fig.
6 for the second and third substorms if one uses only the data from Guam (GU)
and Kakioka (KA).

As in the auroral zone, multiple onset substorms are a cause of some difficulty
(CLAUER and McPHERRON, 1974). For example, in the midlatitude data of Fig. 4
two onsets are clearly present in both the *H* and *D* components. Since it is often
the case that one onset is much weaker than the other it is not obvious which onset
to choose. Frequently multiple onsets may be separated by times short compared
to the duration of a typical substorm. In such cases the requirement that substorms
have only one onset has led some investigators to choose as "the onset" the one
after which the largest magnetic perturbations occur. This choice can easily lead
to "errors" as large as 20 min.

4.   Multiple Onset Substorms

The concept of a multiple onset substorm appears to be a considerable departure
from the substorm concept originally described by Akasofu and coworkers or from
the model presented by KAMIDE and MATSUSHITA (1977). As shown above, however,
there is ample evidence in the literature that such substorms are a rather frequent
occurrence. Figure 7 taken from the work of ROSTOKER (1968) illustrates the mutual

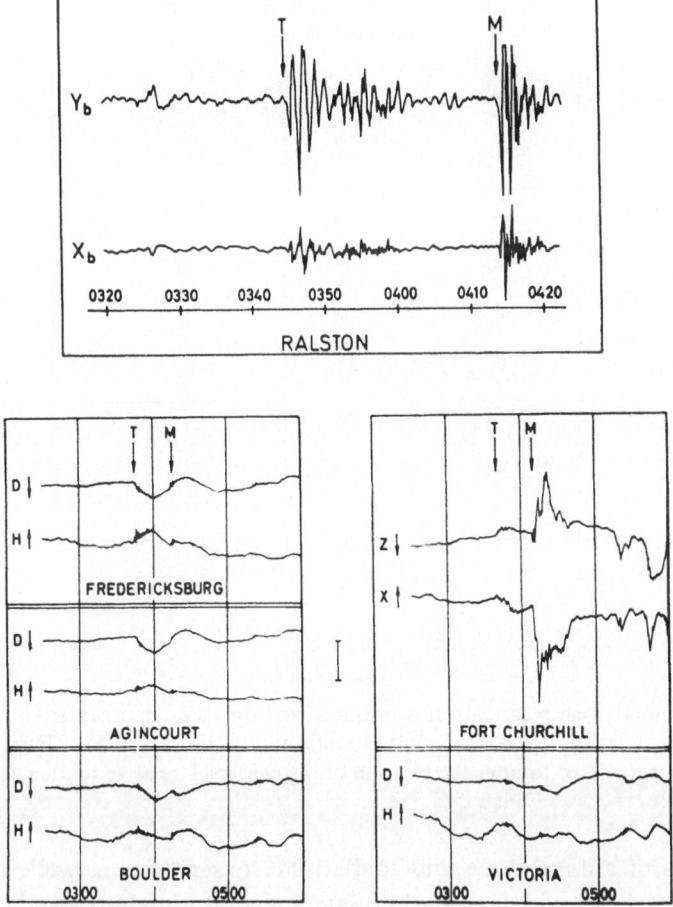

Fig. 7. An example showing the close correlation between the onset of auroral zone negative bays, midlatitude positive bays and Pi 2 bursts (from ROSTOKER, 1968).

relation between auroral zone negative bays, midlatitude positive bays and Pi 2 bursts (see companion paper by Rostoker in this issue) during a double onset substorm. It is now well established that every major substorm onset is accompanied by a mid-latitude Pi 2 burst. As discussed by SAITO *et al.* (1976) this correlation provides a useful tool for timing substorm onsets at midlatitudes overcoming the sensitivity problem mentioned above.

In recent work by PYTTE *et al.* (1976a), it has been established that almost every Pi 2 burst in a multiple onset sequence is accompanied by the usual ground signature of a substorm onset including a westward traveling auroral surge (also see WEINS and ROSTOKER, 1975; SERGEEV, 1974). This fact has led some workers to suggest that each onset is a discrete substorm separated by as little as 10–15 min (KISABETH and ROSTOKER, 1971, 1973, 1974).

Ground magnetic data showing a sequence of four onsets is plotted in Fig. 8 and Fig. 9. Vertical lines show times of Pi 2 bursts but this clearly corresponds to

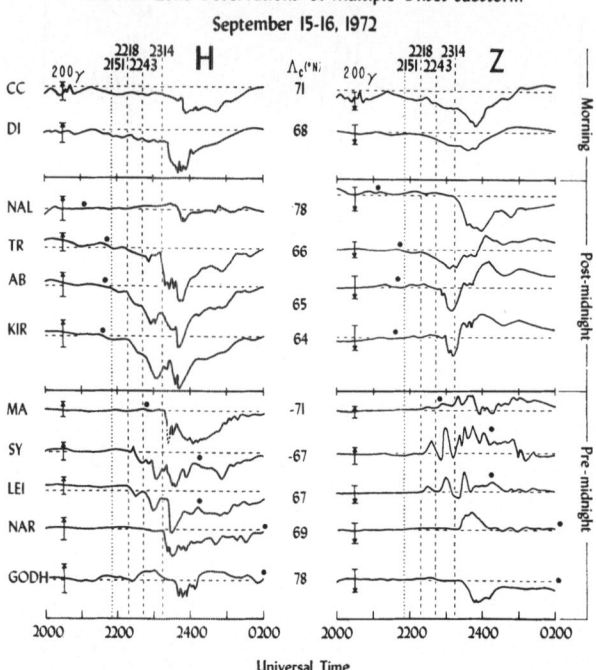

Fig. 8. Auroral zone magnetograms in three local time sectors organized by magnetic latitude. Vertical lines show multiple onesets determined by Pi 2 bursts. The *Z* component is used to infer the location of the electrojet relative to each observatory (from PYTTE *et al.*, 1976a).

the definitions of auroral zone and midlatitude onset times as well. Auroral data for these events show sudden auroral brightening after each burst, and subsequently, westward traveling auroral surges (PYTTE *et al.*, 1976a).

At the present time there is some disagreement about the current systems associated with multiple onset substorms. WEINS and ROSTOKER (1975) suggest that during a sequence of onsets new westward electrojets of localized extent are being formed in a sequence of discrete westward and northward steps. PYTTE *et al.* (1976a) on the other hand suggest that a train of westward propagating surges is generated and that each surge is accompanied by a segment of westward electrojet. KAMIDE *et al.* (1977), however, feel that their "double substorm" has a second poleward expansion formed inside the disturbed region of the first and weaker substorm.

5. A Phenomenological Model for Multiple Onset Substorms

In the substorm model summarized by AKASOFU (1968), the time between expansion onsets is long compared to the duration of the expansion phase. However, the recent work shows that in some substorms the separation of expansion phases is shorter than their expected duration. These observations indicate a need for a new phenomenological model of substorms.

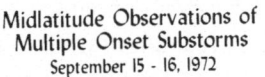

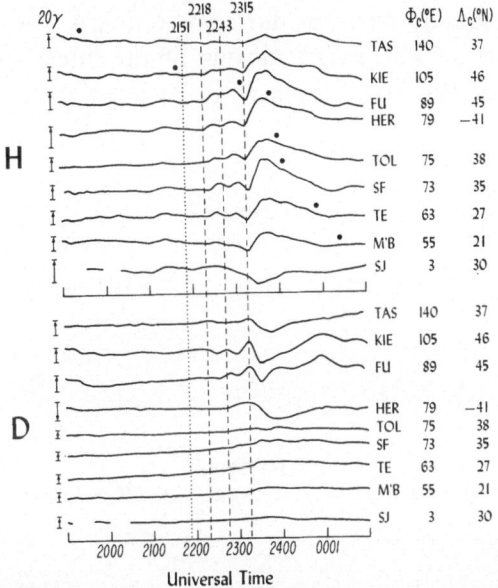

Fig. 9. Midlatitude magnetograms for the same multiple onset substorms shown in Fig. 8. Positive bay onsets are well organized by Pi 2 and correspond to auroral zone onsets of the preceding figure. The *D* component is used to infer the location of the central meridian of the substorm current (cf. Fig. 5) relative to each station (from PYTTE *et al.*, 1976a).

In a phenomenological model outlined by the author and coworkers (McPHERRON *et al.*, 1973b), the substorm model was modified by inclusion of a growth phase (see also IIJIMA and NAGATA, 1972). This phase is one of energy storage preceding substorm onset. Initially the observations of growth phase phenomena were claimed to be expansion phase features (AKASOFU, 1972). Later it was recognized these phenomena do occur and were most likely a consequence of dayside erosion, flux transport, etc., as discussed theoretically by CORONITI and KENNEL (1972a). However, it was argued that these features did not precede every expansion onset so a new name, reconfiguration of the magnetosphere was introduced to describe the growth phase sequence (AKASOFU and KAN, 1973).

As pointed out by McPHERRON *et al.* (1973b) and CAAN *et al.* (1973), the original growth phase model applied only to isolated substorms or the first substorm in a sequence of substorms. Our view is that the full sequence does not repeat between successive onsets but must precede the first onset in a sequence. We suggest in addition that the classic recovery phase does not follow every successive onset but must follow the last onset.

If this view is correct the westward traveling surge and associated expansion is the fundamental entity which replaces the original concept of substorm. Our present view is that the growth phase is a consequence of a southward turning of the interplanetary magnetic field and a reconfiguration of the magnetosphere. Westward traveling surges are generated by a sporadic release of the energy accumulated during the growth phase and, of course, during preceding westward surges. Recovery phase is a consequence of northward turning of the interplanetary magnetic field (IMF) and a return of the magnetosphere to its quiet state. We emphasize that

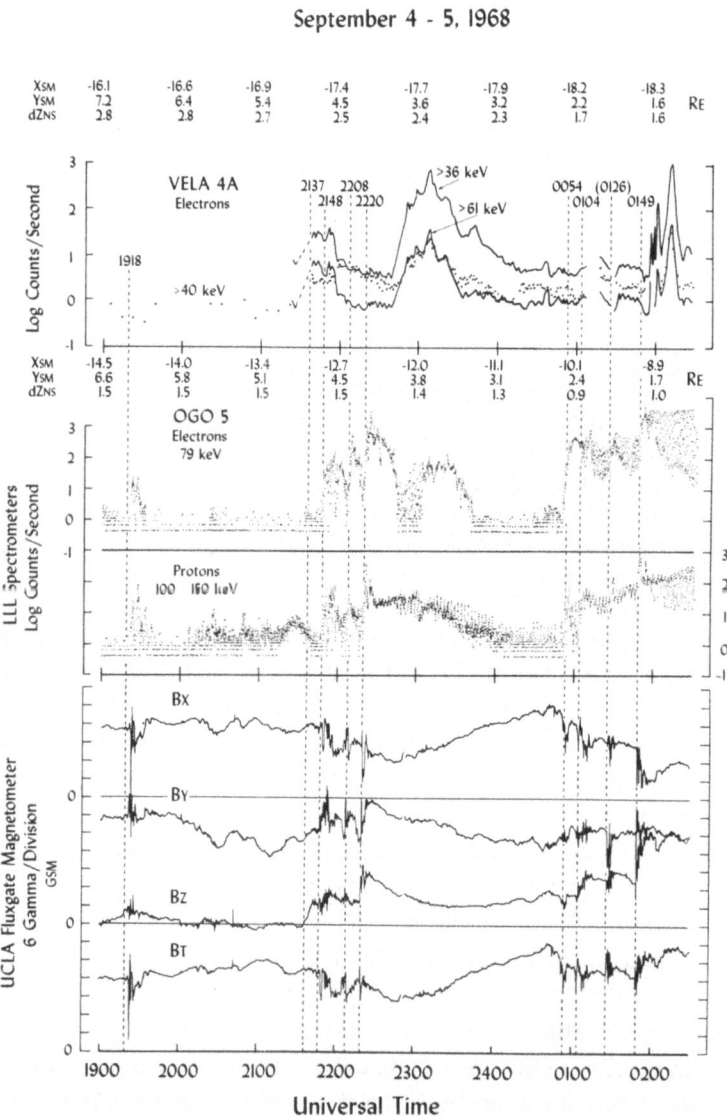

September 4 - 5, 1968

Fig. 10.   Satellite observations contrasting plasma sheet behavior at two radial distances during multiple onset substorms. Drop out of energetic particles (interpreted as plasma sheet thinning) seems to precede each onset in the near earth region (from PYTTE *et al.*, 1976b).

this model distinguishes between the recovery following a westward surge and the recovery caused by the return of the magnetosphere to a quiet state.

The exact sequence of events during a "substorm sequence" depends on several factors. First is the IMF which must remain southward to accumulate energy in the tail. Second is the size and spacing of substorm expansions (westward surges) which release stored energy. Third are possible surge triggering mechanisms for example, sudden impulses (KOKUBUN et al., 1977 and references therein); sudden northward fluctuations of the IMF (CAAN et al., 1976); or possibly onset of instabilities in field-aligned currents (CORONITI and KENNEL, 1972b).

Since the IMF is usually fluctuating it seems likely that there will be many cases in which all the stored energy is released in a single surge. This would constitute a "classic substorm" in the sense that it has a growth, expansion and recovery phase. This is by no means the only sequence which is likely to be observed, as multiple onset sequences seem to be a very frequent occurrence.

6. Discussion

In previous descriptions of the growth phase model it has been suggested that magnetospheric reconfigurations lead to substorm onset. The suggested mechanism is plasma sheet thinning brought about by a demand for closed field lines in the dayside reconnection region. There is some evidence that thinning is important even in multiple onset sequences (PYTTE et al., 1976b; NISHIDA and FUJII, 1976; WALKER et al., 1977). In Fig. 10, data is presented for two satellites on the same meridian near midnight. The closer satellite, OGO-5 at $-12.7\,R_E$ records gradual dropouts in energetic particles prior to each onset. In contrast Vela 4A at $-17.4\,R_E$ records only a single dropout after the first major onset. These observations suggest that each onset is preceded by plasma sheet thinning in a localized region of the near geomagnetic tail.

To determine whether such thinning is of causal importance will require careful studies of future satellite data. Such data must be organized by accurately determined substorm onset times. Ground magnetograms will play an essential role in the determination of these times. The improved magnetometer networks presently under development (LANZEROTTI et al., 1976) will considerably reduce errors in these determinations. However, the most accurate results will be obtained by combining the three major sources of information, auroral, pulsation and ground magnetic data.

Preparation of this manuscript was carried out under the support of the Office of Naval Research No. N00014-75-C-0396 and the support of the Space Environment Laboratory of NOAA in Boulder, Colorado. Original work at UCLA reviewed in this report was performed with the support of ONR N00014-75-C-0396, NASA NGL 05-007-004, and NSF DES 75-10678.

REFERENCES

Akasofu, S.-I., The development of the auroral substorm, *Planet. Space Sci.*, **12**, 273–282, 1964.

Akasofu, S.-I., *Polar and Magnetospheric Substorms*, Springer-Verlag, New York, 1968.

Akasofu, S.-I., Scientific design of a shuttle auroral observatory system, Vol. III, Interim Report on NASA Contract NAS 9-12699, October 1972.

Akasofu, S.-I. and J.R. Kan, Some new thoughts on magnetospheric substorms, *Radio Sci.*, **8** (11), 1049–1057, November 1973.

Akasofu, S.-I. and C.-I. Meng, Low latitude negative bays, *J. Atmos. Terr. Phys.*, **39**, 227–241, 1968.

Akasofu, S.-I. and C.-I. Meng, A study of polar magnetic substorms, *J. Geophys. Res. (Space Phys.)*, **74** (1), 293, 1969.

Akasofu, S.-I. and A.L. Snyder, Comments on the growth phase of magnetospheric substorms, *J. Geophys. Res.*, **77** (31), 6275, 1972.

Akasofu, S.-I., S. Chapman, and C.-I. Meng, The polar electrojet, *J. Atmos. Terr. Phys.*, **27**, 1275–1305, 1965.

Akasofu, S.-I., C.R. Wilson, L. Snyder, and P.D. Perreault, Results from a meridian chain of observatories in the Alaskan sector (1), *Planet. Space Sci.*, **19**, 477–482, 1971.

Bonnevier, B. and R. Bostrom, A three-dimensional model current system for polar magnetic substorms, *J. Geophys. Res. (Space Phys.)*, **75** (1), 107, 1970.

Caan, M.N., R.L. McPherron, and C.T. Russell, Solar wind and substorm-related changes in the lobes of the geomagnetic tail, *J. Geophys. Res.*, **78** (34) 8087, 1973.

Caan, M.N., R.L. McPherron, and C.T. Russell, Characteristics of the association between the interplanetary magnetic field and substorms, *J. Geophys. Res.*, **82** (29), 4837, 1977.

Clauer, C.R. and R.L. McPherron, Mapping the local time-universal time development of magnetospheric substorms using mid-latitude magnetic observations, *J. Geophys. Res.*, **79** (19), 2811, 1974.

Coroniti, F.V. and C.F. Kennel, Changes in magnetospheric configuration during the substorm growth phase, *J. Geophys. Res.*, **77** (19), 3361, 1972a.

Coroniti, F.V. and C.F. Kennel, Polarization of the auroral electrojet, *J. Geophys. Res.*, **77** (16), 2835, 1972b.

Fukushima, N. and Y. Kamide, Partial ring current models for worldwide geomagnetic disturbances, *Rev. Geophys. Space Phys.*, **11** (4), 795–853, 1973.

Iijima, T. and T. Nagata, Signatures for substorm development of the growth phase and expansion phase, *Planet. Space Sci.*, **20**, 1095–1112, 1972.

Kamide, Y. and S. Matsushita, A unified view of substorm sequences, *J. Geophys. Res.*, **83** (A5), 2103, 1978.

Kamide, Y., S.-I. Akasofu, and E.P. Rieger, Coexistence of two substorms in the midnight sector, *J. Geophys. Res.*, **82** (10), 1620, 1977.

Kisabeth, J.L. and G. Rostoker, Development of the polar electrojet during polar magnetic substorms, *J. Geophys. Res.*, **76** (28), 6815, 1971.

Kisabeth, J.L. and G. Rostoker, Current flow in auroral loops and surges inferred from ground-based magnetic observations, *J. Geophys. Res.*, **78** (25), 5573, 1973.

Kisabeth, J.L. and G. Rostoker, The expansive phase of magnetospheric substorms, 1. Development of the auroral electrojets and auroral arc configuration during a substorm, *J. Geophys. Res.*, **79** (7), 972, 1974.

Kokubun, S., R.L. McPherron, and C.T. Russell, Triggering of substorms by solar wind discontinuities, *J. Geophys. Res.*, **82** (1), 74, 1977.

Lanzerotti, L.J., R.D. Regan, M. Suguira, and D.J. Williams, Magnetometer networks during the International Magnetospheric Study, *EOS, Trans. Am. Geophys. Union*, **57** (6), 442–449, 1976.

McPherron, R.L., Growth phase of magnetospheric substorms, *J. Geophys. Res. (Space Phys.)*, **75** (28), 5592, 1970.

McPherron, R.L., C.T. Russell, M.G. Kivelson, and P.J. Coleman, Jr., Substorms in space: The correlation between ground and satellite observations of the magnetic field, *Radio Sci.*, **8** (11), 1059–1076, 1973a.

McPHERRON, R.L., C.T. RUSSELL, and M.P. AURBRY, Satellite studies of magnetospheric substorms on August 15, 1968. 9. Phenomenological model for substorms, *J. Geophys. Res.*, **78** (16), 3131, 1973b.

MENG, C.-I. and S.-I. AKASOFU, A study of polar magnetic substorms. 2. Three-dimensional current system, *J. Geophys. Res. (Space Phys.)*, **74** (16), 4035, 1969.

NISHIDA, A. and K. FUJI, Thinning of the near-earth (10-15 Re) plasma sheet preceding the substorm expansion phase, *Planet. Space Sci.*, **24**, 849-853, 1976.

PYTTE, T., R.L. McPHERRON, and S. KOKUBUN, The ground signatures of the expansion phase during multiple onset substorms, *Planet. Space Sci.*, **24**, 1115-1132, 1976a.

PYTTE, T., R.L. McPHERRON, M.G. KIVELSON, H.I. WEST, Jr., and E.W. HONES, Jr., Multiple-satellite studies of magnetospheric substorms: Radial dynamics of the plasma sheet, *J. Geophys. Res.*, **81** (34), 5921, 1976b.

ROSTOKER, G., Macrostructure of geomagnetic bays, *J. Geophys. Res. (Space Phys.)*, **73** (13), 4217, 1968.

ROSTOKER, G., Polar magnetic substorms, *Rev. Geophys. Space Phys.*, **10** (1), 157-211, 1972a.

ROSTOKER, G., Interpretation of magnetic field variations during substorms, in *Earth's Magnetospheric Processes*, edited by B.M. McCormac, pp. 379-380, D. Reidel Publ. Co., Dordrecht-Holland, 1972b.

ROSTOKER, G., Ground based magnetic signatures of the phases of magnetospheric substorms, in *Magnetospheric Physics*, edited by B.M. McCormac, pp. 325-333, D. Reidel Publ. Co., Dordrecht-Holland, 1974.

SAITO, T., K. YUMOTO, and Y. KOYAMA, Magnetic pulsations Pi 2 as a sensitive indicator of magnetospheric substorm, *Planet. Space Sci.*, **24** (11), 1025-1029, 1976.

SERGEEV, V.A., On the longitudinal localization of the substorm active region and its changes during the substorm, *Planet. Space Sci.*, **22**, 1341-1343, 1974.

WALKER, R.J., K.N. ERICKSON, R.L. SWANSON, and J.R. WINCKLER, Energetic particle flux changes at synchronous orbit and the temporal morphology of substorms, *J. Geophys. Res.*, 1978 (in press).

WEINS, R.G. and G. ROSTOKER, Characteristics of the development of the westward electrojet during the expansive phase of magnetospheric substorms, *J. Geophys. Res.*, **80** (16), 2109, 1975.

Substorm Onset in the Magnetotail

A. Nishida

Institute of Space and Aeronautical Science,
University of Tokyo, Tokyo, Japan

(Received October 17, 1977)

Known signatures of the expansion phase onset in the magnetotail are summarized and discussed. A model is suggested for the nature of the multiple onset events. The importance is emphasized of observing the low-latitude magnetotail at $|X| \sim 10$ to $15\,R_E$ at and shortly before the onset of the expansion phase.

1. Introduction

It has become widely accepted that the magnetotail is the region of vital importance in the substorm physics. It is in the magnetotail that the energy obtained from the solar wind is stored, and it is also in the magnetotail that the stored magnetic field energy is converted explosively to the kinetic energy of charged particles. From this view point the substorm onset in the magnetotail should be a clearly definable concept; the time when the magnetic flux content of the magnetotail suddenly begins to drop and the acceleration process starts to operate should be the moment when a substorm starts, or stated more explicitly, when a substorm enters the expansion phase.

The idea is simple and clear, and there wouldn't be any difficulty in determining the onset time in this way if the magnetotail were monitored continuously by a sufficiently dense network of space probes. Such, however, is not the case, and we usually have to deduce the overall structural change of the magnetotail from observations made at a single point, or, even in most favourable circumstances, two or three points at a time. This limitation of a practical nature has made the definition of the substorm onset in the magnetotail less clear-cut than it could in principle be.

The purpose of this paper is to review our past efforts to overcome this limitation and yield a coherent physical idea of the substorm onset in the magnetotail. Finding relatively rare occasions when more than one satellites were simultaneously orbiting in the magnetotail, or combining statistically the observations made at different occasions, we have been able to construct a reasonably consistent picture of what goes on in the magnetotail at the onset of the expansion phase and during a brief interval that precedes that onset.

These efforts notwithstanding, the limitation of the available information casts shadow on our current understanding. Frst, there still is a missing link in our phenomenology of the magnetotail substorm, because a certain region of the magnetotail which seems to have an utmost importance to the substorm physics has not been explored. Second, while the analyses have been fairly well advanced for relatively large substorms and isolated events, much less is known about the physical nature of smaller substorms and continuous activity. We shall try to point out principal problem areas which remain to be pursued.

Since excellent review articles on magnetotail substorms were written only a few years ago (RUSSELL and McPHERRON, 1973; RUSSELL, 1974), we shall avoid giving a full account of the subject and concentrate on major advances and controversies that have developed in more recent years.

2.   In the Tail Lobe: Reduction in the Field Strength

Rapid decrease in the field magnitude in the tail lobe is one of the basic signatures of the magnetotail substorm (AUBRY and McPHERRON, 1971). The start of this decrease is coincident, within a range of several minutes, with the appearance of an expansion-phase signature on the ground, namely with the earliest onset of a low-latitude positive bay (NISHIDA and NAGAYAMA, 1973; CAAN et al., 1973). At the corresponding moment the tail diameter starts to decrease (MAEZAWA, 1975), and hence the flux content of the magnetotail is indeed reduced by the occurrence of a substorm.

The reduction in the tail flux content seems to occur over a wide longitudinal sector that probably encompasses the entire tail width, as the above correlation has been observed independently of the local time of the observing site in the tail lobe. Dual satellite observations like Fig. 1 supports this contention. The top panel of the figure shows the solar magnetospheric X-component records of two satellites which were separated longitudinally by $>20\,R_E$. One of the satellites (AIMP D, indicated in the figure as D) had a solar magnetospheric Y coordinate of $\sim -18\,R_E$ and was very close to the tail magnetopause on the morning side. The middle panel shows magnetograms from low-latitude stations in the night sector, and the occurrence of at least two cases of low-latitude positive bays can be recognized. Vertical lines are drawn through the starting times of these positive bays. These were also the times when negative bays sharply began to intensify in the night sector of the auroral zone as shown in the bottom panel. Within several minutes of these times, decreases in the X-component of the lobe field started at both satellite positions.

Sometimes the decrease in the field strength in the tail lobe occurs repeatedly in close succession. The decreases indicated by second and third dashed lines in Fig. 2 represent such a case. It is seen in the top panel that the magnitude of the X-component recorded by both satellites had two closely spaced peaks, each of which corresponded to the onset of a low-latitude positive bay as seen in the middle panel. Thus a multiple onset of the expansion phase observed on the ground is associated with a stepwise release of energy from the magnetotail.

In the foregoing example of the double onset, $|X|$-component of the lobe field resumed an increasing trend only several minutes after the first of the double onsets.

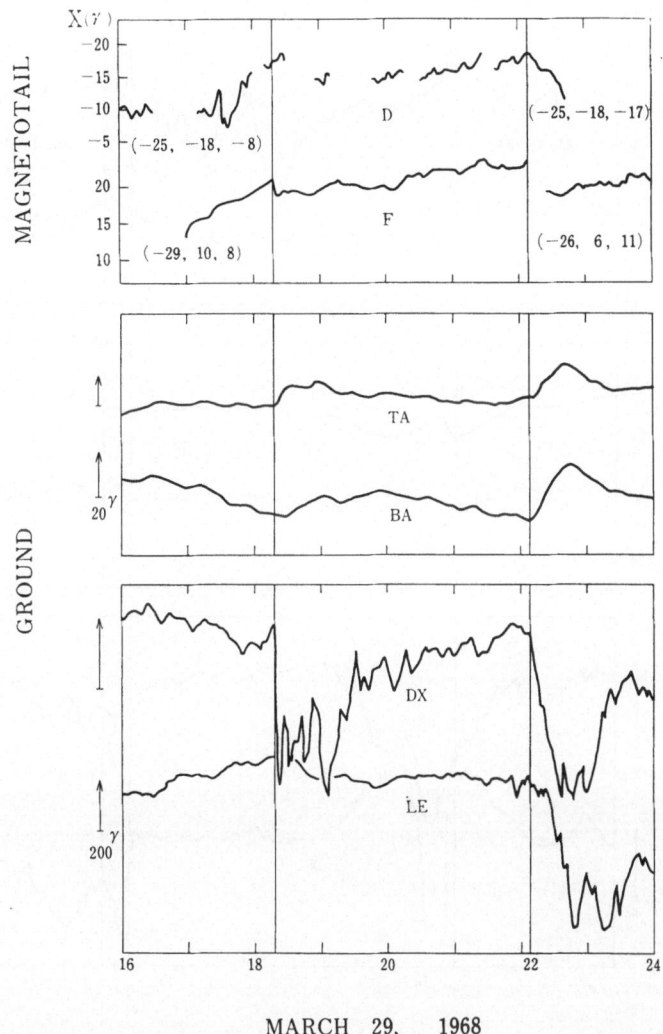

MARCH 29, 1968

Fig. 1. Substorm observations in the tail lobe (top) and on the ground (middle and
bottom). Satellite positions at the beginning and the end of the interval are given
by solar magnetospheric coordinates.

This indicates that the process of the energy release which was initiated at the first
onset was not efficient enough to dispose of the continuing addition of the flux to
the tail. A full-fledged conversion of the stored energy took place after the second
of the double onsets.

That the decrease in the lobe field strength is observed over a wide longitudinal
range at the onset of the expansion phase does not necessarily mean that the energy
conversion process operates in the plasma sheet with an equally wide longitudinal
extent. Even when the collapse of the tail field lines occurs in some limited longitudinal sector, the decrease in the magnetic pressure initiated in that sector is com-
municated to other sectors of the tail by the rarefaction wave, and its effect is

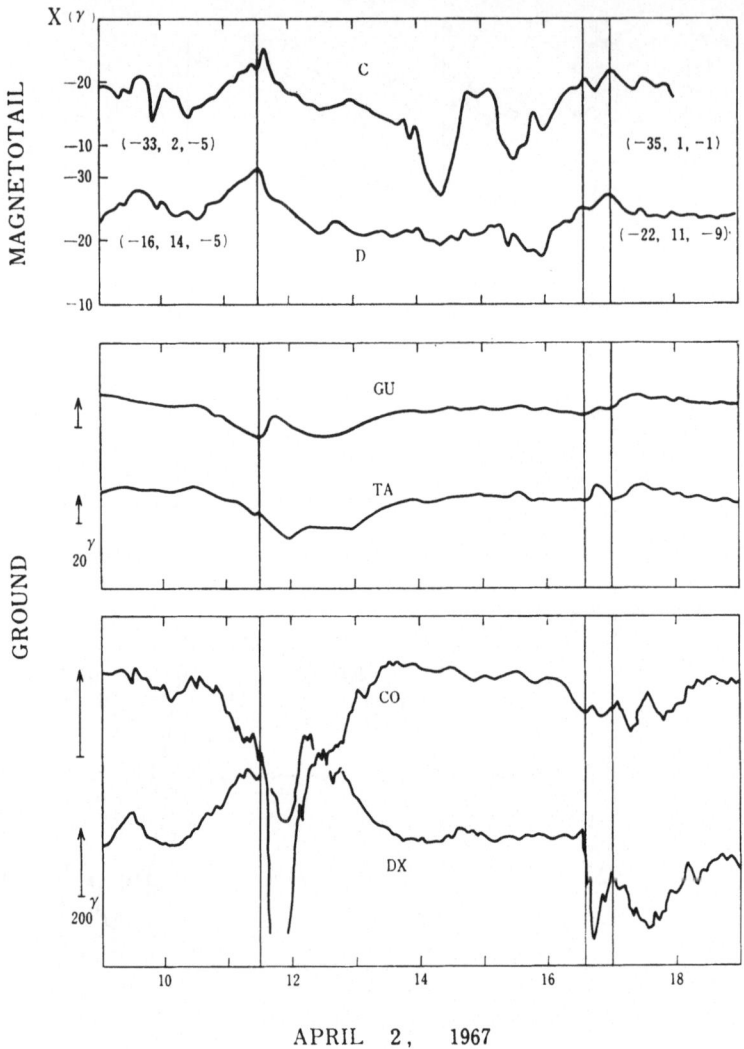

APRIL   2,   1967
Fig. 2.   Same as Fig. 1 (NISHIDA and NAGAYAMA, 1975).

expected to be observable everywhere in the tail lobe with a delay corresponding to the propagation time of the hydromagnetic wave of the fast mode.

3.   In the Near-Earth Plasma Sheet: Resumption of Dipole-Like Field Configuration and Enhancement in Particle Flux

In the near-earth region of the magnetotail at the solar magnetospheric X coordinate of around $-10\,R_E$, the rotation of the magnetic field vector is the characteristic feature observed during substorms.   The rapid rotation toward the more dipolar direction occurs at the expansion-phase onset which is signified, among other things, by the onset of a low-latitude bay, and it is preceded by an interval of 0.5–1 hr in which the field lines become more extended (MCPHERRON et al., 1973).

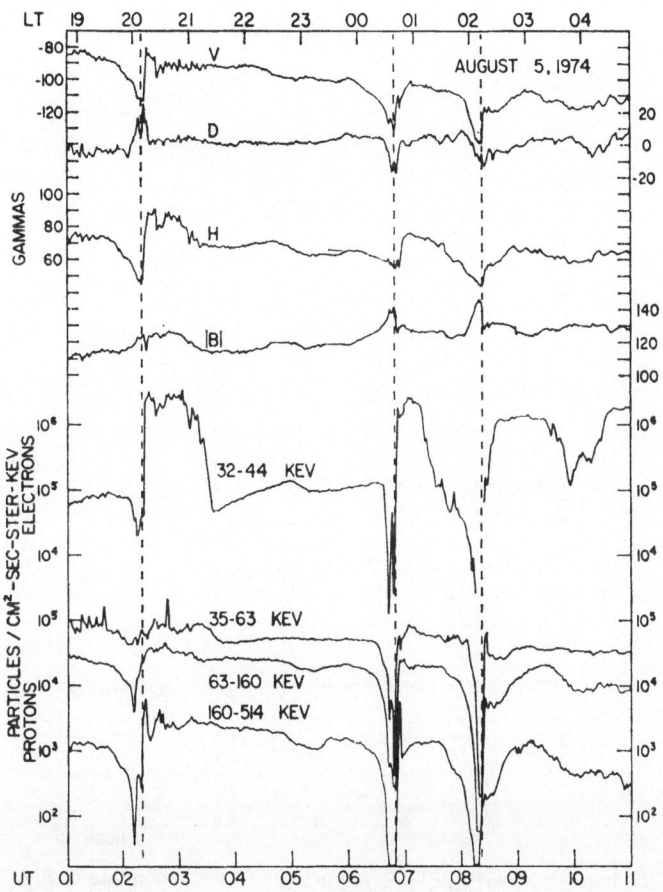

Fig. 3.   Substorm observations at the geosynchronous altitude by ATS-6. Upper four
traces are magnetic field data, and lower four traces are proton data (WALKER
et al., 1976).

These features are frequently observed at the geosynchronous altitude, as shown
by three examples in Fig. 3. The satellite was in the night sector, and three dashed
lines indicate the beginning of abrupt increases in H (which is antiparallel to the
earth's dipole axis) and decreases in $-V$ (which is directed approximately toward
the earth) that correspond to the rotation to more dipolar configuration. These
abrupt changes were preceded by gradual variations having an opposite sense and
starting roughly 1 hr earlier (WALKER *et al.*, 1976). Figure 4 shows low-latitude
ground magnetograms in the corresponding interval, and each dashed line can be
seen to correspond to an onset of a low-latitude positive bay (S. Kokubun, private
communication, 1977).

The records shown in the lower half of Fig. 3 are those of energetic protons.
Each of the field rotations was accompanied by an enhancement of the proton flux;
particularly in second and third events the flux increase was recorded in all energy
channels and was preceded by a flux decrease. Since the energy of the observed
protons is high enough to fall in the energy range of the trapped particles, the

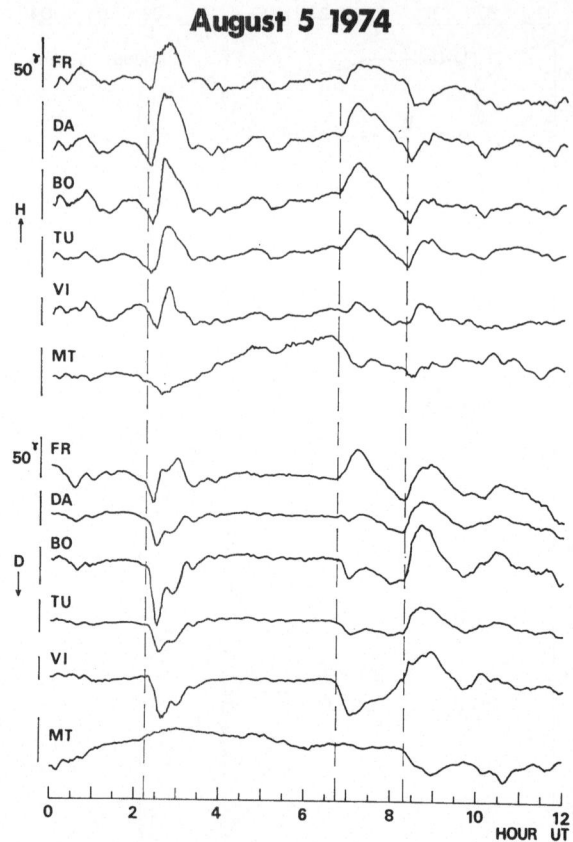

Fig. 4. Low-latitude ground magnetograms corresponding to Fig. 3
(S. Kokubun, private communication, 1977).

decrease-increase sequence of the proton flux was attributed primarily to the earth-ward and then tailward shift of the trapping boundary that results from the change in the magnetic field configuration (BOGOTT and MOZER, 1973), but the anisotropy measurements of these protons indicate that the rapid flux increases are frequently associated with earthward motion, presumably from the plasma sheet (WALKER et al., 1976).

The earthward injection of particles is more evident at lower energies where the trapped population is absent. Namely, at energies ≤ 10 keV, sharp and con-spicuous enhancements in proton and electron fluxes are observed in the near-midnight sector of the geosynchronous orbit within 10 min of the appearance of the expansion phase signatures on the ground (e.g., KAMIDE and McILWAIN, 1974). An example is shown in Fig. 5. In this figure the left-hand column displays electron fluxes at 0.7–1.9 (EA), 1.8–5.4 (EB), 5.9–17.8 (EC) and 17–53 keV (ED), and the central column shows proton fluxes at >5 (PA), >15 (PB) and >38 keV (PC). Expansion-phase onset times defined by the low-latitude positive bay on the ground (right-hand column, middle panel) are indicated by dashed lines in each column. The en-hancement of the flux was observed in every channel at each onset of the

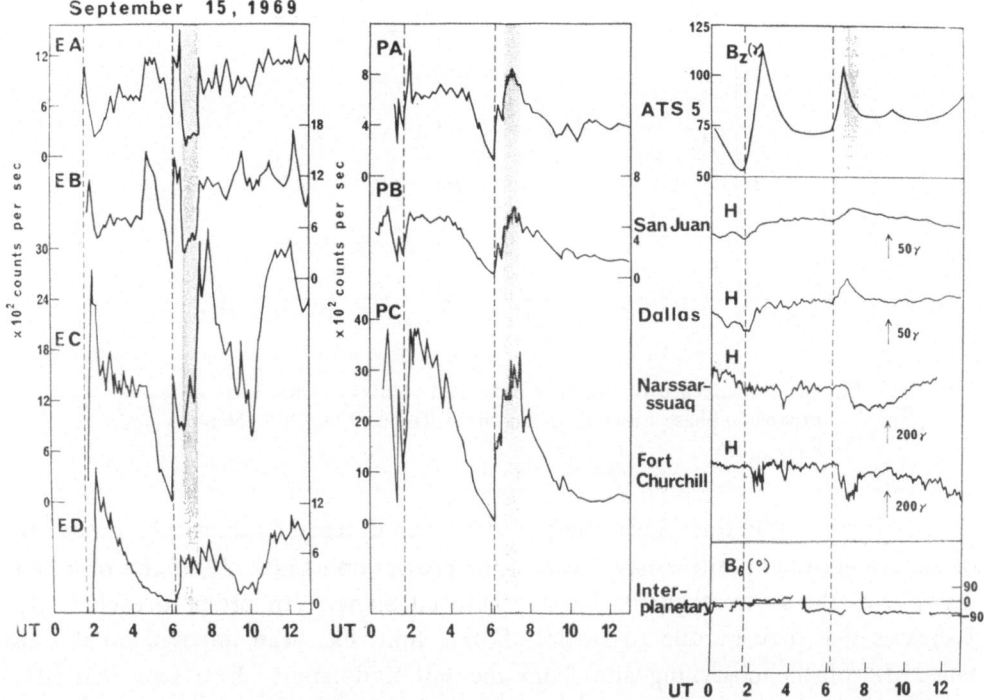

Fig. 5. Substorm observations at the geosynchronous altitude by ATS-5. Left and middle columns show particle data, and right-hand column shows, from the top, magnetic field observed by ATS-5, ground magnetic field data obtained at a pair of low-latitude and a pair of auroral-zone stations, and the inclination θ of the interplanetary magnetic field with respect to the solar magnetospheric equatorial plane (FUJII *et al.*, 1975). Shadings indicate the interval when ATS-5 was in earth's shadow.

expansion phase. This indicates that the inner boundary of the plasma sheet rapidly moved inward.

Sometimes the flux enhancement of these low energy particles is also preceded by an interval of a depressed flux. In Fig. 5 it is seen that in low-energy channels like EA, EB, EC and PA particle fluxes started to drop sharply about 1–2 hr before the flux enhancement at the second expansion-phase onset in the figure. A possible reason for such a depression is the thinning of the plasma sheet; namely, the plasma sheet which had penetrated to the geosynchronous altitude in consequence of the preceding substorm activities briefly reduced its thickness before the onset of the subsequent expansion phase (FUJII *et al.*, 1975; ROSTOKER *et al.*, 1975).

Indications of the plasma sheet thinning before the expansion-phase onset have been obtained also by particle measurements made at slightly larger distance by OGO 5 (KIVELSON *et al.*, 1973). Although energies of the observed particles are rather high ($\sim 80\,\text{keV}$ for electrons and $\sim 100\,\text{keV}$ for protons), at distances of $\geq 10\,\text{R}_E$ from the earth these particles may be considered to comprise high energy tail of the plasma sheet population. The left panel of Fig. 6 shows the time difference

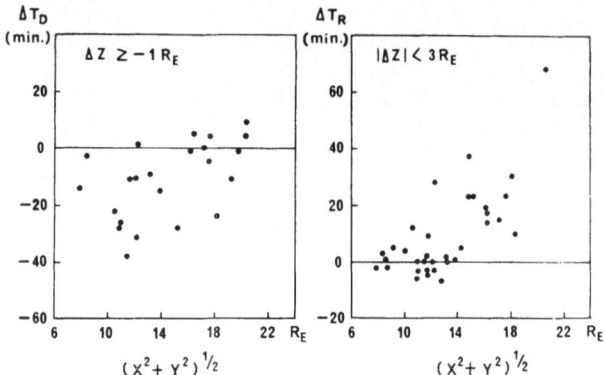

Fig. 6. Dependence of thinning onset and recovery times (relative to the expansion-phase onset time) on the distance $(X^2+Y^2)^{1/2}$ (NISHIDA and FUJII, 1976).

ΔT_D, defined as "the time when the flux started to decrease" minus "the onset time of the low-latitude positive bay," versus the projection $(X^2+Y^2)^{1/2}$ of the observing distance of the solar magnetospheric equatorial plane. (In order to exclude flux decreases that may be due to spatial effect a limit has been imposed on the distance ΔZ of the observing site from the tail midplane.) It is seen that $|\Delta T_D|$ differs from zero at $(X^2+Y^2)^{1/2} \lesssim 15\,R_E$ but at greater distances $|\Delta T_D|$ is nearly zero in most cases. In contrast, the right panel of Fig. 6 shows that ΔT_R, namely "the time of flux increase" minus "the onset time of the low-latitude positive bay," is clustered around zero at $(X^2+Y^2)^{1/2} \lesssim 15\,R_E$ but is tens of minutes long at greater distances. Thus there seems to be a critical distance which separates the plasma sheet into two regions according to its behaviour during substorms; earthward of that distance the plasma sheet thins before, and recovers rapidly at, the expansion-phase onset, while away from that distance the plasma sheet starts to thin at the expansion-phase onset and recovers some tens of minutes after the onset (NISHIDA and FUJII, 1976; HONES et al., 1973). The contrasting behaviours of the near-earth and the distant portions of the plasma sheet during substorms have been demonstrated also by a comparison of simultaneous observations of particle fluxes by OGO 5 and Vela 4A (PYTTE et al., 1976).

4.   In the Distant Plasma Sheet: Southward Turning of the Field and Decrease in the Sheet Thickness

The thinning of the plasma sheet at $(X^2+Y^2)^{1/2} \sim 17\,R_E$ is one of the most thoroughly documented signatures of the expansion-phase onset. At that distance the decrease in particle flux usually begins with an onset of the expansion phase of the auroral substorm (AKASOFU et al., 1971a), although in some occasions the thinning has been found to precede the expansion-phase onset (HONES et al., 1971). Inside the thinned plasma sheet, plasma flows tailward with a velocity of several hundred km/sec

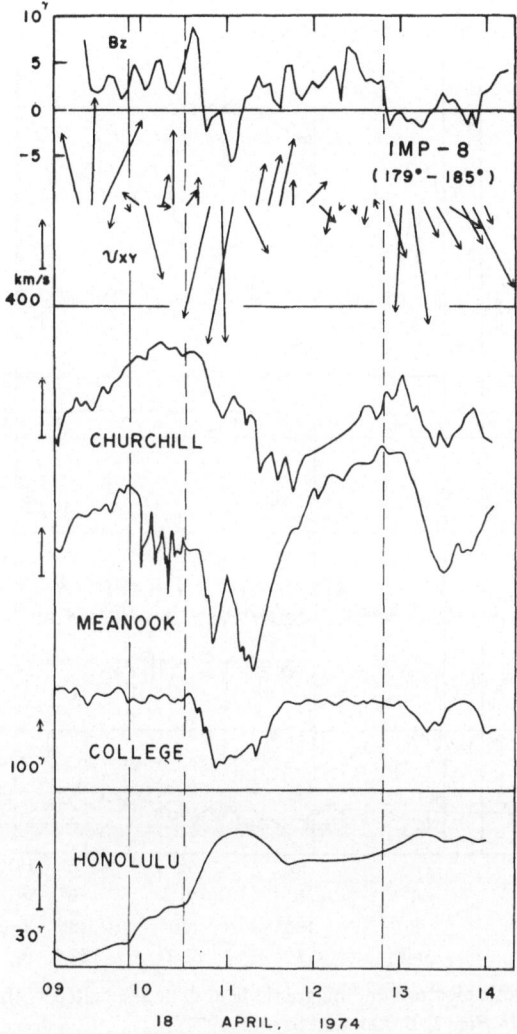

Fig. 7.   Substorm observations in the distant plasma sheet (first two
panels) and simultaneous ground magnetic records.

(HONES et al., 1974).   In addition, an impulsive burst of hard electrons sometimes
occurs at the onset of the expansion phase and lasts for a few tens of minutes
(AKASOFU et al., 1971b).   Unfortunately the Vela series of satellites were not instru-
mented with magnetometers, but other satellites orbiting the distant tail at $|X| \gg 15\,R_E$
have observed a reversal in the north-south component B_z of the low-latitude tail
field; a few minutes after the onset of the expansion phase on the ground, B_z turns
from northward to southward (NISHIDA and NAGAYAMA, 1973).   Recall that in the
corresponding interval the same component of the field is directed northward and
increases sharply at $|X| \sim 7\,R_E$ (Figs. 3 and 5).

A relatively well documented example showing the foregoing features is given
in Figs. 7 and 8.   The top panel of Fig. 7 shows the solar magnetospheric northward

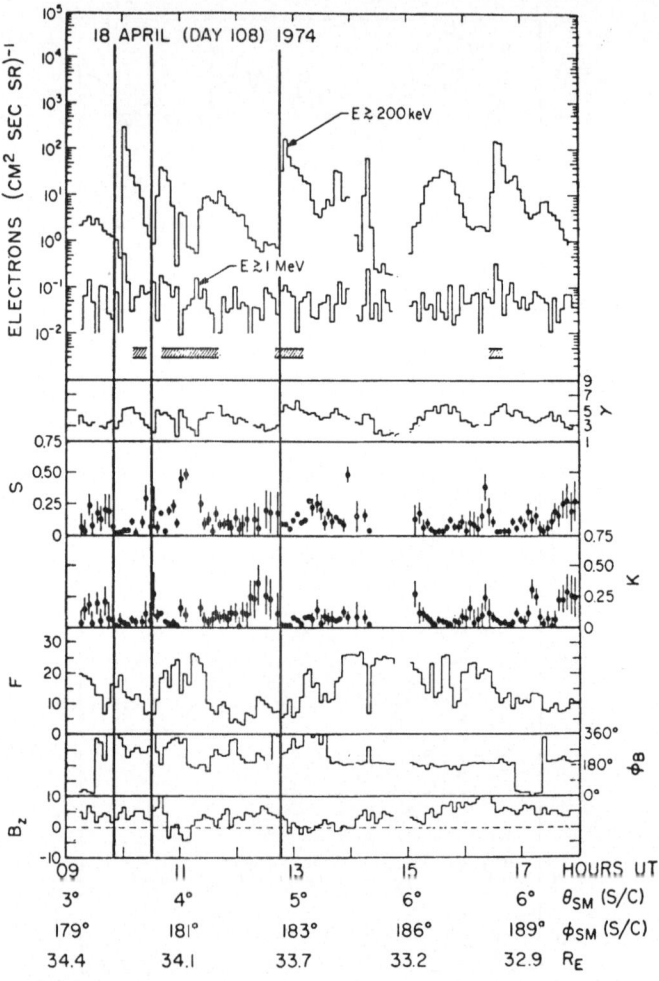

Fig. 8.   Energetic electron and magnetic field data associated with observations
presented in Fig. 7 (BAKER and STONE, 1977).

component of the magnetic field and the plasma flow velocity observed at the radial
distance of $\sim 35\,R_E$ in low latitude (at θ_{SM} of several degrees) near the midnight
meridian.   The bottom panels display ground magnetograms, and expansion-phase
onset times determined by low-latitude positive bay and/or Pi 2 burst are indicated
by vertical lines.   Figure 8 is the energetic electron data in the corresponding in-
terval; the top panel shows fluxes of $\geqslant 200\,\mathrm{keV}$ and $\geqslant 1\,\mathrm{MeV}$ electrons, the second
panel the spectral exponent γ, the third panel the amplitude S of the first harmonic
of the azimuthal variation, and the fourth panel the amplitude K of the second
harmonic.   The remaining three panels are magnetic field data (FRANK et al., 1976;
BAKER and STONE, 1977; HONES, 1978).

Let us look at the third onset first.   There occurred at this time, or shortly
($\lesssim 10$ min) after this, southward turning of B_z, anti-earthward flow of plasma, and

enhancement of $\gtrsim 200$ keV electron flux. These energetic electrons also started to stream tailward shortly after the enhancement.

The earlier two onsets seem to constitute a double onset event. The first of these caused little perturbation at satellite's position except for the occurrence of a very energetic ($E \gtrsim 1$ MeV) electron burst about 10 min later. The second of these, however, caused all the characteristic signatures with a delay of about 10 min. The enhancement of the anisotropy amplitude S is particularly noteworthy in this case.

Although not included in the present example, intense bursts of energetic protons ($E \sim 50$–200 keV) have also been observed in the low-latitude magnetotail at $|X| \sim 35$ R_E. The occurrence of these proton bursts is limited to the dusk sector of the tail, while energetic electron bursts do not show dawn-dusk asymmetry in their occurrences. The onset of the bursts is coincident (within a possible range of several minutes) with the earliest onset of the low-latitude positive bay on the ground (TERASAWA and NISHIDA, 1976; ROELOF et al., 1976).

5. The Neutral Line Model

A phenomenological model which is essentially consistent with foregoing observations is shown in Fig. 9. In this model the formation of an X-type neutral line is assumed at the distance of $|X| \sim 15$ R_E (third panel). Tailward of the neutral line B_z is directed southward and plasma moves tailward. Energetic particles produced at the neutral line fill the magnetic loop shown dotted in the figure. On the earthward side field lines resume more dipolar configuration, plasma sheet expands and its inner boundary moves earthward. The model has been developed by several groups by putting various pieces of evidences together (e.g., RUSSELL and McPHERRON, 1973; HOFFMAN and BURCH, 1973; HONES et al., 1973; TERASAWA and NISHIDA, 1976; HONES, 1976).

On the anti-earthward side of the neutral line the field line structure that is deduced from magnetic field observations is convex toward the neutral line as illustrated in Fig. 10. This structure implies that the plasma flow in the boundary region of the plasma sheet is directed nearly parallel (or, antiparallel) to the magnetic field. This can be seen as follows. The uniformity of the electric field requires that $V_N^{(E)} B_N = V^{(E)} B$, where $V^{(E)}$ denotes the velocity component perpendicular to the magnetic field; the right-hand side is evaluated at an arbitrary point P on the field line passing the tail midplane at point N. The observed field structure implies $B_N < B$, and for the convex field structure to be maintained for a sufficiently long distance from the neutral line the component V of the velocity parallel to the midplane should satisfy $V \geq V_N = V_N^{(E)}$. This means that V is due largely to the field-aligned component of the velocity rather than to $V^{(E)}$, since $V^{(E)} = (B_N/B) V_N^{(E)} < V_N^{(E)}$. The observed flow velocities indeed seem to have a substantial component parallel to the field orientation (HONES et al., 1974, 1976).

The field-aligned, anti-earthward flow of plasma can be an intrinsic part of the magnetotail deformation that drives reconnection in the near-earth region of the

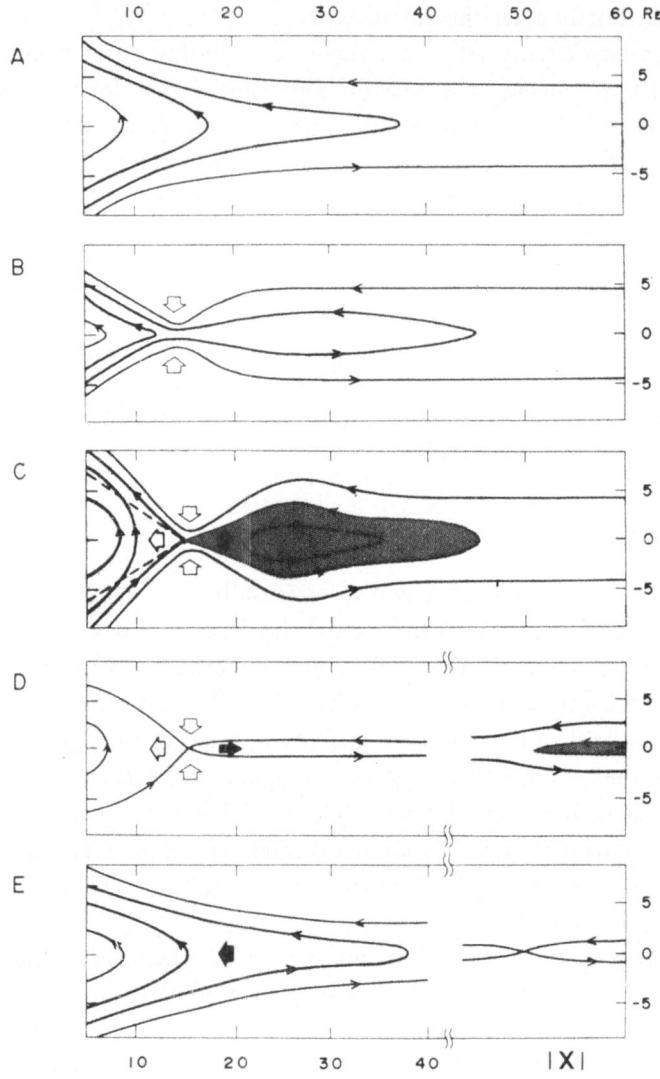

Fig. 9.   Sequence of magnetotail transfiguration during a substorm.   Formation
of the neutral line (between panels B and C) corresponds to the onset of the
expansion phase.

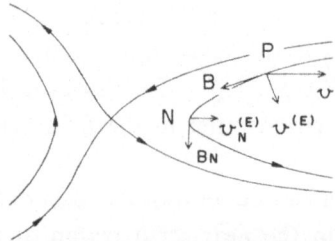

Fig. 10.   Field configuration near the neutral line.

magnetotail. As pointed out by CORONITI and KENNEL (1972), erosion of the dayside magnetosphere and flux pile-up in the magnetotail produces steep inclination in the magnetotail attitude (with respect to the incidence direction of the solar wind) in the transition region between the dayside magnetosphere and the magnetotail. The enhanced pressure of the solar wind applied to this region of the magnetopause induces motion inside the magnetosphere. Reconnection is caused as field lines in northern and southern halves of the tail are pushed toward each other by this pressure. In addition, the pressure produces field-aligned, anti-earthward flow as it acts to push plasma tailward.

The field-aligned flow of similar nature is involved in analytic models of the reconnection produced by Sonnerup or Yeh and Axford (see review by VASYLIUNAS (1975)). The boundary condition which specifies models of this type is that of the cornered flow, and the colliding streams directed toward the interface are deflected at the slow expansion fan where the field-aligned, issuing component is produced in the flow velocity. The presence of this component in the boundary region of the plasma sheet has been predicted also by a numerical model of the reconnection where the compressibility is taken into account (HAYASHI and SATO, 1977). In the compressible model the expansion of the heated plasma seems to intensify the field-aligned issuing flow.

6. Discussion

Before the neutral line model can be established as a proper explanation of the substorm expansion phase, however, it would be essential that the presence of the neutral line be confirmed by in-situ observations. This has not been possible so far, because few satellites have traversed the tail midplane at the distance of $\sim 15\,R_E$ where the neutral line is likely to be formed. The task remains as one of the most urgent missions of the magnetospheric study today.

Nevertheless it can be said that a number of case studies carried out in the past is consistent with the idea that the neutral line is formed in the near-earth portion of the tail at the onset of the substorm expansion phase. Typical examples are second and third expansion phases of Figs. 7 and 8. The first event, however, is an exception, and an explanation is needed why neither the southward B_z nor the anti-earthward plasma flow were observed in that case. The simplest possible way is to assume that the length of the neutral line was quite limited longitudinally at that time and did not reach the near-midnight meridian where the satellite was located (Fig. 11, top). The weakening of the cross-tail current might have occurred in that meridian too but probably it was not severe enough to produce the neutral line. Alternatively it may be considered that the neutral line extended to that meridian but the reconnection stopped to operate after a short while and consequently the field line loop produced thereby was too small to produce a southward B_z at the satellite position (at $|X| \sim 35\,R_E$) (Fig. 11, middle). Strong earthward flow of plasma

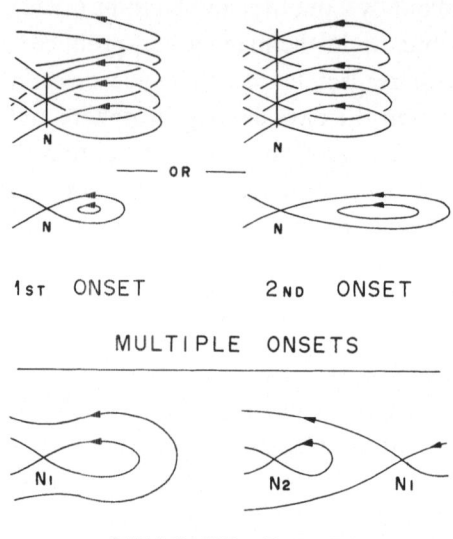

Fig. 11. Top: Two conceivable models of multiple expansion phase onset. The first possibility is the stepwise extention of the neutral line length (above), and the second possibility is the stepwise enlargement of the field line loop (below). Bottom: Model of successive but separate expansion phase onsets. N means neutral line.

from distant magnetotail could be suggested as a possible cause of the suppression of the near-tail reconnection.

The second event which closely followed the first event is likely to have occurred when the weakening of the near-earth portion of the tail current developed further. The multiple onset of the expansion phase probably corresponds to a successive, fuller development of the current weakening and/or the neutral line length (Fig. 11, right).

Conversely, the statement made in the preceding paragraph could be taken as a definition of the "multiple onset" of an expansion phase. Namely, if the region of the weakened cross-tail current increases its longitudinal and/or radial extent in discrete steps and hence the length of the neutral line and/or the dimension of the field line loop are increased in discrete steps, each step can be considered to define multiple onsets of a single expansion phase. When on the other hand neutral lines are formed one after another and are carried successively away to greater distances from the earth (Fig. 11, bottom), it may be more appropriate to regard each neutral-line formation as signifying the onset of a separate expansion phase. In broader terms, one could call a set of expansion-phase onset signatures as multiple onsets if the same physical process continued and merely increased its intensity and/or scale at each step, while the signatures may better be regarded to represent a series of separate events if a new process is initiated at each step following the decay or departure of the earlier process. This definition appears logical, but practically it would be difficult to identify individual multiple onset events conforming to this definition unless field and plasma characteristics in the near-earth portion of the plasma sheet are monitored. (In fact we have used the term multiple onset so far rather loosely without referring to the above definition.)

It has been known that the onset of the low-latitude positive bay on the ground is not globally simultaneous and varies slightly with local time. It is tempting to consider that this feature is the reflection of the successive, longitudinal extension of the neutral line. But comparisons between low-latitude ground observations and low-latitude tail observations at $|X| \gtrsim 17\,R_E$ have not yet confirmed such a simple relationship. Rather, comparisons of simultaneous ground and satellite data have shown that the neutral line signature (namely the southward turning of B_z) observed at any local time tends to be most closely associated with the ground signature observed in the premidnight sector (where the bays tend to show earliest appearance) (NISHIDA and HONES, 1974). More extensive survey is needed to disclose the magnetotail counterpart of the local-time dependent occurrence pattern of low-latitude positive bays. The region of importance would again be the low-latitude magnetotail at the distance of $\sim 15\,R_E$.

If the neutral line extends longitudinally in discrete steps, the earthward injection of plasma is expected to occur from broader local-time sector with the development of the neutral line. Such an indication, however, has not yet been reported. The reason may be that most of available data on the injection was obtained at the geosynchronous altitude which is probably too far from the neutral line to detect the detailed structural change of the neutral line.

In the foregoing model the neutral line was supposed to be formed at $|X| \sim 15\,R_E$. Suggestions have been made that the location of the neutral line is sometimes much further than this. In one of such instances, a prolonged presence of a neutral line beyond $|X| \sim 17\,R_E$ was inferred from a rapid earthward flow of plasma that lasted for almost 10 hr. This interval of earthward flow roughly coincided with an equally prolonged interval of moderately strong geomagnetic activity, but no detailed correlation between individual signatures was said to be recognizable (HONES et al., 1976). Since in the open model of the magnetosphere a neutral line is thought to be permanently present in the distant tail, the above might better be described as an enhancement of the reconnection rate at that distant neutral line. It is not clear, however, why a moderately strong geomagnetic activity, which is presumably associated with a rather expanded auroral oval, is associated with a neutral line located at a large distance. Actually it does not seem to have been established by ground observations that a prolonged activity can be interpreted in terms of a succession of the expansion-phase to recovery-phase sequence of a typical substorm. Hence it is not obvious that the magnetotail process in a prolonged active interval is basically the same as that associated with substorms. Although it would not be correct to say that prolonged activities and substorms are driven by entirely different processes, possibility exists that the principal driving mechanisms for these two types of disturbances are not fully identical. Possible alternative processes include coupling of the DP2 (i.e., IMF driven) electric field with enhanced ionospheric conductivity.

Emphasizing this possibility, Pytte *et al.* (1978) have designated the prolonged geomagnetic activity as "convection bay."

Association with a distant neutral line has been suggested also for small substorms that tend to occur when the auroral oval is contracted. Topologically this association seems to be more plausible than that for prolonged activities, but few detailed case studies seem to have been conducted so far to establish this idea.

As an alternative interpretation of the available information, Lui *et al.* (1977) have proposed that the southward B_z represents a passage of the thinning wave propagating tailward. In their picture the thinning wave has a crest of $\sim 10\,R_E$ width where field lines are more sharply inclined, and this spatially confined feature is suggested to be the nature of the southward B_z observed during the expansion phase. Observationally, however, the southward B_z has been detected from $|X| \sim 25\,R_E$ to beyond $\sim 60\,R_E$ with a duration of as long as 0.5–1 hr, and apparently it does not represent a spatially confined structure as they suggest.

In addition, it was supposed by Lui *et al.* (1977) that the plasma sheet thinning begins everywhere at or after the onset of the expansion phase. This is contrary to the observation that the thinning starts earlier than that onset in the near-earth region of the plasma sheet. The precursory thinning of the plasma sheet has an important bearing on the initiation of the expansion phase, because the ion tearing mode instability is expected to arise and disrupt the cross-tail current when the thickness is reduced to some threshold value (Schindler, 1974). It is important to observationally fix the threshold level when more data from the near-earth region of the plasma sheet become available.

REFERENCES

Akasofu, S.-I., E.W. Hones, Jr., M.D. Montgomery, S.J. Bame, and S. Singer, Association of magnetotail phenomena with visible auroral features, *J. Geophys. Res.*, **76**, 5985–6003, 1971a.

Akasofu, S.-I., E.W. Hones, Jr., and S. Singer, Impulsive energetic electron fluxes in the distant magnetotail associated with the onset of magnetospheric substorms, *J. Geophys. Res.*, **76**, 6976–6979, 1971b.

Aubry, M.P. and R.L. McPherron, Magnetotail changes in relation to the solar wind magnetic field and magnetospheric substorms, *J. Geophys. Res.*, **76**, 4381–4401, 1971.

Baker, D.N. and E.C. Stone, Observations of energetic electrons ($E \geq 200$ keV) in the earth's magnetotail: Plasma sheet and fireball observations, *J. Geophys. Res.*, **82**, 1532–1546, 1977.

Bogott, F.H. and F.S. Mozer, ATS-5 Observations of energetic proton injection, *J. Geophys. Res.*, **78**, 8113–8127, 1973.

Caan, M.N., R.L. McPherron, and C.T. Russell, Solar wind and substorm-related changes in the lobes of the geomagnetic tail, *J. Geophys. Res.*, **78**, 8087–8096, 1973.

Coroniti, F.V. and C.F. Kennel, Changes in magnetospheric configuration during the substorm growth phase, *J. Geophys. Res.*, **77**, 3361–3370, 1972.

Frank, L.A., K.L. Ackerson, and R.P. Lepping, On hot tenuous plasmas, fireballs, and boundary layers in the earth's magnetotail, *J. Geophys. Res.*, **81**, 5859–5881, 1976.

Fujii, K., A. Nishida, R.D. Sharp, and E.G. Shelley, ATS-5 observations of plasma sheet particles before the expansion-phase onset, preprint, 1975.

Hayashi, T. and T. Sato, Magnetic reconnection-acceleration, heating and shock formation, University of Tokyo GRL preprint, 1977.

HOFFMAN, R.A. and J.L. BURCH, Electron precipitation patterns and substorm morphology, *J. Geophys. Res.*, **78**, 2867–2884, 1973.

HONES, E.W., Jr., The magnetotail: Its generation and dissipation, in *Physics of Solar Planetary Environments*, edited by D.J. Williams, p. 559, AGU, Washington, D.C., 1976.

HONES, E.W., Jr., Substorm processes in the magnetotail: Comments on 'On hot tenuous plasmas, fireballs, and boundary layers in the earth's magnetotail,' by L.A. Frank, K.L. Ackerson and R.P. Lepping, *J. Geophys. Res.*, **83**, 1183–1186, 1978.

HONES, E.W., Jr., J.R. ASBRIDGE, and S. J. BAME, Time variations of the magnetotail plasma sheet at 18 R_E determined from concurrent observations by a pair of Vela satellites, *J. Geophys. Res.*, **76**, 4402–4419, 1971.

HONES, E.W., Jr., J.R. ASBRIDGE, S.J. BAME, and S. SINGER, Substorm variations in the magnetotail plasma sheet from $X_{SM} \simeq -6\,R_E$ to $X_{SM} \simeq -60\,R_E$, *J. Geophys. Res.*, **78**, 109–132, 1973.

HONES, E.W., Jr., A.T.Y. LUI, S.J. BAME, and S. SINGER, Prolonged tailward flow of plasma in the thinned plasma sheet observed at $r \approx 18\,R_E$ during substorms, *J. Geophys. Res.*, **79**, 1385–1392, 1974.

HONES, E.W., Jr., S.J. BAME, and J.R. ASBRIDGE, Proton flow measurements in the magnetotail plasma sheet made with Imp 6, *J. Geophys. Res.*, **81**, 227–234, 1976.

KAMIDE, Y. and C.E. McILWAIN, The onset of magnetospheric substorms determined from ground and synchronous satellite records, *J. Geophys. Res.*, **79**, 4787–4790, 1974.

KIVELSON, M.G., T.A. FARLEY, and M.P. AUBRY, Energetic electrons, spatial boundaries, and wave-particle interactions at Ogo 5, *J. Geophys. Res.*, **78**, 3079–3092, 1973.

LUI, A.T.Y., C.-I. MENG, and S.-I. AKASOFU, Search for the magnetic neutral line in the near-earth plasma sheet. 2. Systematic study of IMP 6 magnetic field observations, *J. Geophys. Res.*, **82**, 1547–1565, 1977.

MAEZAWA, K., Magnetotail boundary motion associated with geomagnetic substorms, *J. Geophys. Res.*, **80**, 3543–3548, 1975.

McPHERRON, R.L., M.P. AUBRY, C.T. RUSSELL, and P.J. COLEMAN, Jr., Ogo 5 magnetic field observations, *J. Geophys. Res.*, **78**, 3068–3078, 1973.

NISHIDA, A. and N. NAGAYAMA, Synoptic survey for the neutral line in the magnetotail during the substorm expansion phase, *J. Geophys. Res.*, **78**, 3782–3798, 1973.

NISHIDA, A. and E.W. HONES, Jr., Association of plasma sheet thinning with neutral line formation in the magnetotail, *J. Geophys. Res.*, **79**, 535–547, 1974.

NISHIDA, A. and N. NAGAYAMA, Magnetic field observations in the low-latitude magnetotail during substorms, *Planet. Space Sci.*, **23**, 1119–1125, 1975.

NISHIDA, A. and K. FUJII, Thinning of the near-earth (10–15 R_E) plasma sheet preceding the substorm expansion phase, *Planet. Space Sci.*, **24**, 849–853, 1976.

PYTTE, T., R.L. McPHERRON, M.G. KIVELSON, H.I. WEST, Jr., and E.W. HONES, Jr., Multiple-satellite studies of magnetospheric substorms: Radial dynamics of the plasma sheet, *J. Geophys. Res.*, **81**, 5921–5933, 1976.

PYTTE, T., R.L. McPHERRON, E.W. HONES, Jr., and H.I. WEST, Jr., Multiple-satellite studies of magnetospheric substorms: distinction between polar magnetic substorm and convection-driven negative bays, *J. Geophys. Res.*, **83**, 663–679, 1978.

ROELOF, E.C., E.P. KEATH, and T. IIJIMA, Fluxes of >50 keV protons and >30 keV electrons at ~35 R_E. 4. Association of intense bursts of energetic particles in the magnetotail with the expansion phase of geomagnetic substorms, preprint, JHU/APL 76-05, 1976.

ROSTOKER, G., J.L. KISABETH, R.D. SHARP, and E.G. SHELLEY, The expansive phase of magnetospheric substorms. 2. The response at synchronous altitude of particles of different energy ranges, *J. Geophys. Res.*, **80**, 3557–3570, 1975.

RUSSELL, C.T., The solar wind and magnetospheric dynamics, in *Correlated Interplanetary and Magnetospheric Observations*, edited by D.E. Page, D. Reidel Publ. Co., 1974.

RUSSELL, C.T. and R.L. McPHERRON, The magnetotail and substorms, *Space Sci. Rev.*, **15**, 205–266, 1973.

SCHINDLER, K., A theory of the substorm mechanism, *J. Geophys. Res.*, **79**, 2803–2810, 1974.

TERASAWA, T. and A. NISHIDA, Simultaneous observations of relativistic electron bursts and neutral-line signatures in the magnetotail, *Planet. Space Sci.*, **24**, 855–866, 1976.

VASYLIUNAS, V.M., Theoretical models of magnetic field line merging (1), *Rev. Geophys. Space Phys.*, **13**, 303–336, 1975.

WALKER, R.J., K.N. ERICKSON, R.L. SWANSON, and J.R. WINCKLER, Substorm-associated particle boundary motion at synchronous orbit, *J. Geophys. Res.*, **81**, 5541–5555, 1976.

ELECTROMAGNETIC AND ELECTROSTATIC INSTABILITIES ON AURORAL FIELD LINES

A Review of Electrostatic Wave Measurements on Auroral Magnetic Field Lines

Michael C. KELLEY

*School of Electrical Engineering, Cornell University,
Ithaca, New York, U.S.A.*

(Received October 17, 1977)

This review emphasizes experimental evidence for electrostatic waves on auroral zone magnetic field lines. Data were obtained from radar, balloon, rocket, and satellite observations. The paper is organized around three topics: (1) the turbulent density and velocity fields of the magnetospheric flow, (2) instabilities associated with the convection electric field and its interactions with the neutral atmosphere, and (3) waves and discontinuities in the dynamic region at about one Earth radius above the auroral zone. Measurements of both the electric field and density fluctuation component of such waves will be included. Although this is not a theoretical study, an attempt is made to organize the measurements about the existing theoretical framework.

1. Introduction

The purpose of this review is to gather experimental data concerning electrostatic waves observed on the magnetic field lines which thread the auroral zone. An attempt will be made to organize the observations around the existing theoretical framework. Data presented here were obtained by balloon, rocket, satellite, and radar techniques. Reference will be made to measurements of both the electric field component of such waves and the plasma density fluctuation component. These two quantities, $\delta\bar{E}$ and $\delta n/n$, play an analogous role to $\delta\bar{E}$ and $\delta\bar{B}$ in an electromagnetic wave. The observations span more than eight orders of magnitude in wavelength and ten orders of magnitude in frequency.

The paper is organized as much as possible about physical processes. In order are discussed (1) the turbulent velocity field as well as the irregular density variations imbedded in the magnetospheric flow, (2) instabilities associated with the perpendicular electric field and the resultant interaction between moving plasma and the neutral atmosphere, (3) waves and discontinuities in the extremely dynamic and important region at about one Earth radius altitude above the auroral oval.

2. Turbulent and Irregular Magnetospheric Flow

The magnetic Reynolds number for the magnetospheric flow is sufficiently high that it is not surprising that it is turbulent. Thus, although there is a mean flow which forms a two-cell convection pattern with anti-sunward motion in the polar cap and sunward motion at auroral and sub-auroral latitudes, the motion at a given time and place can deviate drastically from the mean, much as do the eddies and back flows in a mountain stream. An example of the rapidly varying velocity of two barium clouds is presented in Fig. 1. The measurements were made near local magnetic noon in northern hemisphere winter. The poleward cloud broke into numerous patches which drifted chaotically along the ionospheric projection of the polar cusp (KELLEY *et al.*, 1977).

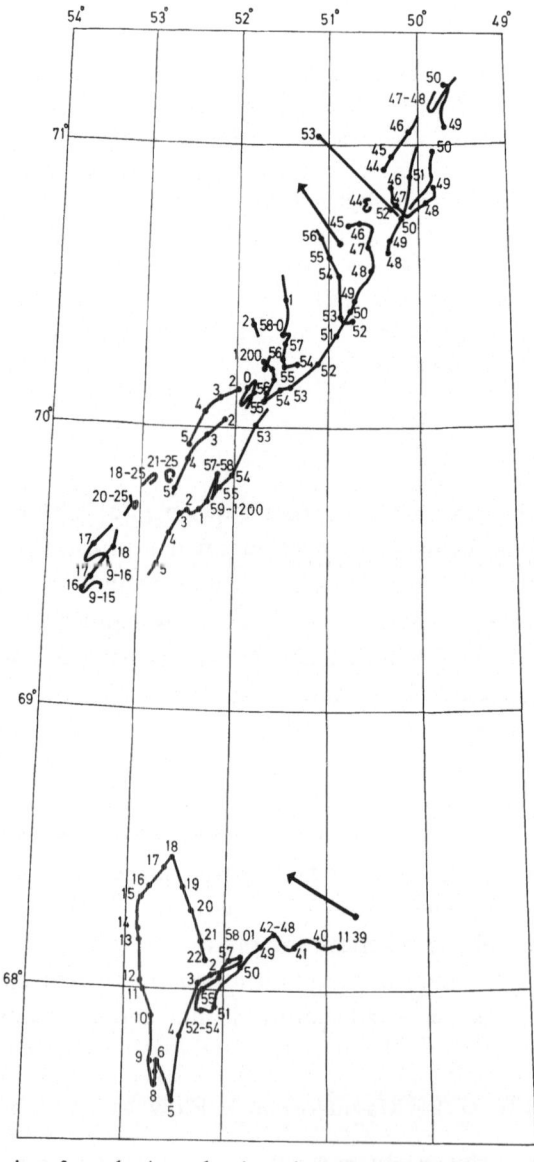

Fig. 1.   Trajectories of two barium clouds and patches deployed at 230 km latitude just prior to magnetic noon and near the polar cusp. The numbers are minute markers past 1100 or 1200 UT. (Reprinted by permission of *J. Atmos. Terr. Phys.*)

Another way to describe such a spatially varying velocity field is through its spatial Fourier transform, $V^2(k)$. A prime goal of theory and experiment is to determine the spectral form which this function takes. Dimensional analysis has proven to be very successful in predicting the wave number spectrum of two and three dimensional turbulence in neutral fluids. The physics is somewhat obscured but the result from energy and angular momentum conservation is that when a fluid is excited at a particular wave number k_0, and if over a certain range near k_0 there is no dissipation, then the nearby wave numbers will be excited by mode-mode coupling and a less ordered, turbulent state results.

It has been argued (see review by MONTGOMERY, 1976) that low β magnetized plasmas should behave very much like two-dimensional fluids. Turbulence in such fluids displays a power law dependence for $V^2(k)$ with index $-5/3$ at small k changing to -3 for large k. Balloon (MOZER, 1971) and satellite (KINTNER, 1976) electric field measurements have been combined by KELLEY and KINTNER (1978) to describe the wave number spectrum of magnetospheric velocity turbulence $(V^2(k)=E^2(k)/B_0^2)$ and the results plotted in Fig. 2. The predictions of two-dimensional turbulence theory fit the data very well. The break between spectral indices theoretically occurs at the wave number where turbulence is injected. Kelley and Kintner argued that the wave numbers indicated in Fig. 2 are typical of the inverse scale size of auroral folds and curls. Since auroral electric fields are approximately perpendicular

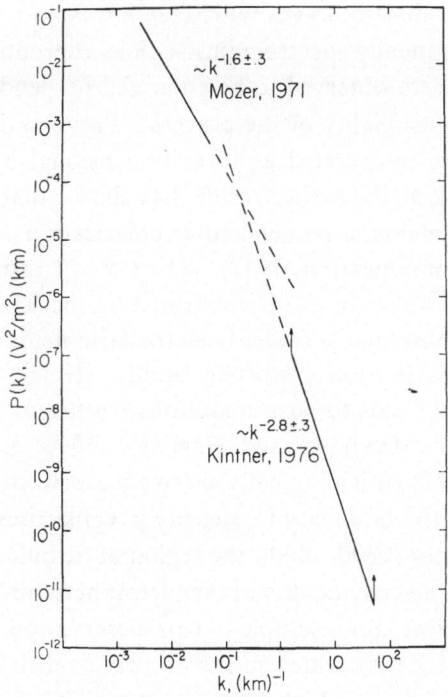

Fig. 2. Spatial power spectrum of the magnetospheric electric field deduced from balloon and satellite observations.

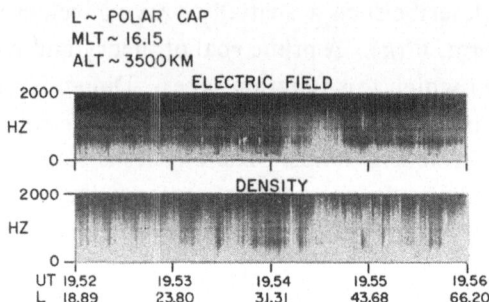

Fig. 3. Frequency time spectrogram of electric field and plasma
density fluctuations in the polar cap. The lightest areas correspond to the signal and the dark fingerprint-like whorls are
the interference pattern.

to auroral forms, such visual structures may indeed map the velocity turbulence.
SWIFT (1977) has also recently appealed to this type of process in an attempt to
quantitatively model the complex region discussed below in Section 4.

Crucial to this description is the concept that the electric fields are spatially
varying with zero or very low frequency in the frame moving with the mean plasma
flow velocity. Recent S3-3 satellite measurements of turbulent electric fields have
convincingly shown this to be the case (TEMERIN, 1978). The detectors used had a
separation distance d between spherical electrodes of 37 m. As the spacecraft spun
an interference pattern due to waves with $\lambda = d/n$ was seen in the data as spin dependent nulls in the frequency spectrograms such as reproduced in Fig. 3. The irregular electric fields were observed to be polarized perpendicular to $\vec{B}$ which also
argues for the two-dimensionality of the process. Previous high-latitude measurements of this type were interpreted as three-dimensional by KELLEY and MOZER
(1972) but a second look at the earlier results has shown that, although not as clear
cut as the S3-3 measurements, a perpendicular polarization was indicated (Temerin
and Kelley, private communication, 1977). The OVI-17 results were interpreted as
very low phase velocity waves, a result confirmed by the S3-3 results.

The S3-3 results show that turbulent electrostatic fields are present at least up
to 8,000 km altitude. Earlier lower altitude satellite electric field detectors results
also showed fluctuating E fields to be a ubiquitous feature of the high-latitude ionosphere (HEPPNER, 1969; MAYNARD and HEPPNER, 1970; LAASPERE et al., 1971;
KELLEY and MOZER, 1972) with a typically sharp equatorward boundary (see Fig. 4)
statistically identical to the boundary for density irregularities discussed below. Except for the morning sector (0800–1400), the region of turbulence corresponds to the
entire region where the electric field is of magnetospheric origin, and is not limited
to within the auroral oval, for example. This observation is consistent with the
concept that the turbulence is created in the auroral oval, is imbedded in the flow
and moves with it throughout the high-latitude region. An interesting, and thus far
unresolved question is related to the short wavelength cutoff of the turbulence. Since
the turbulence is two-dimensional, likely candidates are the ion or electron gyro

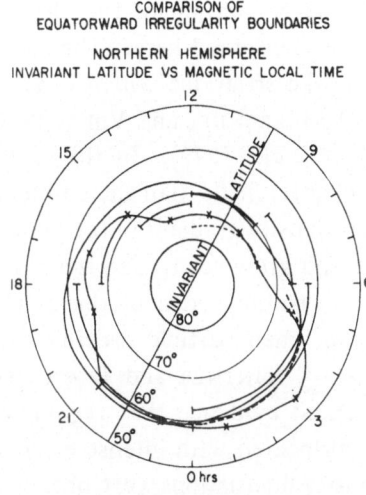

COMPARISON OF
EQUATORWARD IRREGULARITY BOUNDARIES

NORTHERN HEMISPHERE
INVARIANT LATITUDE VS MAGNETIC LOCAL TIME

├──┤ IRREGULARITY BOUNDARY (ISIS-I)
───── SCINTILLATION BOUNDARY
---- IRREGULARITY BOUNDARY (ALOUETTE 2)
─*─ ELECTROSTATIC TURBULENCE BOUNDARY

Fig. 4. Comparison of various equatorward boundary for high-latitude irregularities in density and electric field.

radii. Experimentally determining the dissipation scale length is important since computer models are limited by mesh constraints to relatively long wavelengths.

Also imbedded in the magnetospheric flow are irregular plasma density patterns which have been detected by numerous low-altitude satellite probe experiments (DYSON, 1969; DYSON et al., 1974; SAGALYN et al., 1974). They are observed throughout the polar cap, auroral, and sub-auroral zones often with extremely sharp equatorward boundaries at all local times. The boundary shape as indicated by several in situ techniques is plotted in Fig. 4 along with the scintillation boundary determined from ground observations (AARONS, 1973). It is very near the auroral oval between 0800–1400 but at all other local times is considerably equatorward of it. It seems unlikely that a single plasma instability is operating throughout this vast region. Most likely the irregular density patterns are created in the auroral oval and convect out of those regions to fill the observed volume. It should be noted that turbulent velocity fields alone cannot create density structure since the flow is incompressible. Velocity turbulence near horizontal density gradients, created for example by auroral zone particle precipitation, could be quite efficient in producing density irregularities. Equatorial spread F irregularities seem to be generated in this way by electrostatic turbulence acting on the steep vertical zero order gradient on the bottomside of the F peak. Particle precipitation itself very likely accounts for some of the irregular density patterns by direct spatially varying production. The power spectrum of high-latitude density irregularities is typically reported to vary as k^{-2} (DYSON et al., 1974; SAGALYN et al., 1974). One must be careful in interpreting this, however, since the spatial Fourier transform of a single irregularity with a sharp edge varies as k^{-2}. The k^{-2} power spectrum of equatorial spread F irregularities, for example, have been shown to be due to such nonlinear steepened structures (COSTA and KELLEY, 1977).

Various microprocesses have been suggested for the generation of the electrostatic waves and irregularities, but only a few will be mentioned here. The cross-
field instability has been very successful in explaining the striations which develop
in barium cloud releases (LINSON and WORKMAN, 1970; HAERENDEL and VOLK, 1971;
PERKINS et al., 1973; ZABINSKY et al., 1973; GOLDMAN et al., 1976). In this instability enhanced density regions have slightly smaller electric fields than the ambient,
and hence drift more slowly. If there is a horizontal density gradient in the plasma
parallel to $\bar{E} \times \bar{B}$ the denser regions will lag behind and grow relative to the background density. Fingers of ionization develop in the trailing edge of the cloud.
This process should operate in the auroral F region when particle precipitation
creates localized regions of high plasma density. Also, KINTNER and D'ANGELO
(1976) and KELLEY and CARLSON (1977) have presented evidence that regions of
zero order velocity shear in the auroral plasma are co-located with intense electrostatic fluctuations. Both electric field and plasma density fluctuations were observed
in the latter case. As discussed in Section 4 below some of the zero order velocity
turbulence is probably created at about $1 R_E$ altitude where very intense quasi D.C.
electric fields are observed.

3.   Instabilities in the Lower Ionosphere Driven by Perpendicular Electric Fields

Both the linear and nonlinear theories for electrostatic waves in the equatorial
electrojet are much further developed than are theories for the auroral electrojet.
This is primarily due to the excellent data set provided the theorist by radar backscatter measurements at the Jicamarca Radar Observatory (FEJER et al., 1975). In
recent years auroral VHF radars with the capability of determining the Doppler
shift of the returned echo have been operated in Alaska by groups at NOAA and at
the Stanford Research Institute, over eastern Canada by Cornell University, and in
Scandinavia by a group at the Max Planck Institut für Aeronomie at Lindau. Many
previous auroral radar measurements have been made without Doppler information
(e.g., UNWIN, 1966) but are less useful for wave studies. Balloon, rocket, and HF
scatter measurements have also been used in the study of the polar cap E region
instabilities.

Since the equatorial case is better understood it will be briefly reviewed. The
linear dispersion relation has two driving terms, the two stream and gradient-drift
terms. The former operates most efficiently at relatively short wavelengths and leads
to instabilities when the perpendicular electron drift relative to the ions exceeds the
sound speed. Radar observations show that the waves propagate at the sound speed
even though the electrons apparently attain even higher velocities, a result not predicted by the linear theory. Another nonlinear feature is that the Doppler shift is
nearly independent of angle to the current. That is, even at right angles to the drift,
waves have been observed at the sound speed. The gradient-drift term has no clear
threshold but requires $\bar{E}$ parallel to $\bar{\nabla}n$ to operate and has the highest growth rates
at long wavelengths (several hundred meters). At the equator the relevant gradients

and E fields are vertical. Dimensional arguments on the k dependence of the power spectrum of density irregularities due to the gradient-drift instability by OTT and FARLEY (1974) are in good agreement with numerical simulations by MACDONALD et al. (1975) and by SUDAN and KESKINEN (1977), and with experimental results by PRAKASH et al. (1971). Thus the concept of mode-mode coupling and cascading of energy from the fastest linearly growing mode seems to work in this two-dimensional turbulent situation caused by the gradient-drift process. Unlike the two-stream case, the gradient-drift driven irregularities seem to have phase velocities proportional to the electron drift speed.

Both the equatorial and auroral waves which backscatter radar signals have wave vectors nearly perpendicular to $\bar{B}$. Thus the geometry for receiving echoes is constrained and in the auroral case requires that the radar be well south of the echoing region. Since the strong electrojet currents are east-west, the radars seldom look 'head-on' into the current where the most intense waves are expected to occur. Thus most of the results deal with 'diffuse radar aurora' a signature which occurs when the radar is directed at considerable angles (even perpendicular) to the current. The primary or linearly unstable modes are usually not detected by VHF radar and one must be careful interpreting the results. Thus rocket data and HF backscatter in the polar cap seem most relevant to the primary process and will be discussed first.

The consensus seems to be that the primary wave process operating at high latitudes is the two-stream instability rather than the gradient drift. The zero order vertical density gradient at high latitudes is nearly parallel to $\bar{B}$ and hence the geometry for obtaining $\bar{E}\|\bar{V}n$ is less advantageous. In the polar cap the dawn-dusk directed electric field has a vertical component at dawn and dusk due to the slight inclination of $\bar{B}$, but is perpendicular to the vertical at noon and midnight. Thus one would predict an asymmetry if the gradient-drift process were operating. However, according to TSUNODA et al. (1976), there is no evidence for enhanced echoes at dawn and dusk. Also rocket electric field and zero-order density measurements by OLESEN et al. (1976) in the polar cap showed strong waves where $\bar{E}$ was antiparallel to $\bar{V}n$ (bottomside) and very weak ones when $\bar{E}$ was parallel to $\bar{V}n$ on the topside of the E region density profile, exactly the opposite of the linear theory prediction for the gradient-drift process. Horizontal gradients may play an important role, particularly near auroral arcs, but it is very difficult to do definitive experiments. Perhaps simultaneous measurements with the EISCAT incoherent scatter and STARE auroral radar will be able to address this issue.

On the other hand, a clear electric field threshold for observation of HF slant range echoes has been shown by OLESEN et al. (1975) and IVERSEN et al. (1975) using simultaneous balloon electric field and HF backscatter measurements. One of the figures in the latter paper is reproduced in Fig. 5 which displays $|\bar{E}|$ along with the time of occurrence of anomalous HF reflections. In these and other examples the enhanced echoes occur during times of large E fields. The authors quote a threshold of $\sim 25 \, mV/m$ is in good agreement with the ion acoustic threshold. Since these were polar cap measurements the ionosphere was well behaved and the possibility

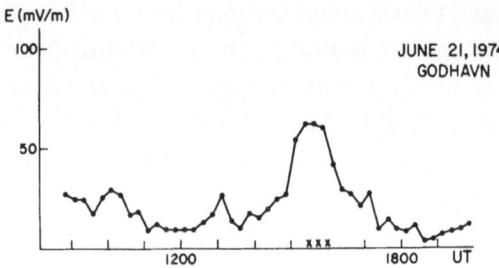

Fig. 5.   One example of a threshold effect for the slant E condition
and the ambient D.C. electric field as measured by balloon borne
detectors. (Reprinted by permission of *J. Geophys. Res.*)

of great changes in the acoustic speed due to electron heating less likely. Tsunoda
and Presnell (1976) have shown a 30 mV/m threshold for backscatter at 398 MHz
by comparing their auroral zone diffuse echoes to simultaneous Chatanika E field
measurements. Thus it seems that the Farley-Buneman two-stream instability is the
primary mechanism at the high latitudes.

The strong HF radar echoes which are observed when the radar is directed approximately parallel or anti-parallel to the polar cap current systems are probably
due to primary two-stream waves in the wavelength range of 10–20 m (Olesen *et al.*,
1975; Tsunoda *et al.*, 1976). Simultaneous rocket D.C. and A.C. electric field measurements as well as A.C. plasma density measurements by Kelley and Mozer (1973)
in the auroral zone showed that the primary waves are indeed electrostatic and also
showed that the peak intensity is in waves with wave vector parallel to the current.
Some of their data is reproduced in Fig. 6. The raw electric field signal in a narrow
frequency band centered at 290 Hz (6 m) is plotted in the upper inset and displays
peaks twice per rotation. The peaks occur at zero crossings of the D.C. electric
field (also plotted) measured by the same antenna. The lower plot shows the direction of the electrojet current deduced from the measured D.C. E field and a neutral
model along with the direction of the A.C. electric field ($\pm 180°$) which in an electrostatic wave is parallel to $\bar{k}$. The most straightforward explanation is that the most
intense waves propagated parallel to the current. Care must be taken however, since
measurements such as those in Fig. 6 are at fixed frequency, not wavelength. For
waves propagating perpendicular to the current, the phase velocity in the rocket
frame will in general be different from the phase velocity of parallel propagating
waves. Thus as the rocket spins it may be sampling different portions of the wave
number spectrum. The E field signal intensity parallel to the current peaked between
100–200 Hz, which corresponds to the range of 10–20 m, in agreement with the HF
backscatter results for the primary wavelength regime. The rocket results of Olesen
et al. (1976) were very similar, although their signals peaked at wavelengths of several meters.

The envelope of the raw data in Fig. 6 is nearly sinusoidal at twice the spin
frequency but does not go to zero between peaks. Since the diffuse radar aurora
echoes are detected at large angles to the current, the non-zero response perpendicular

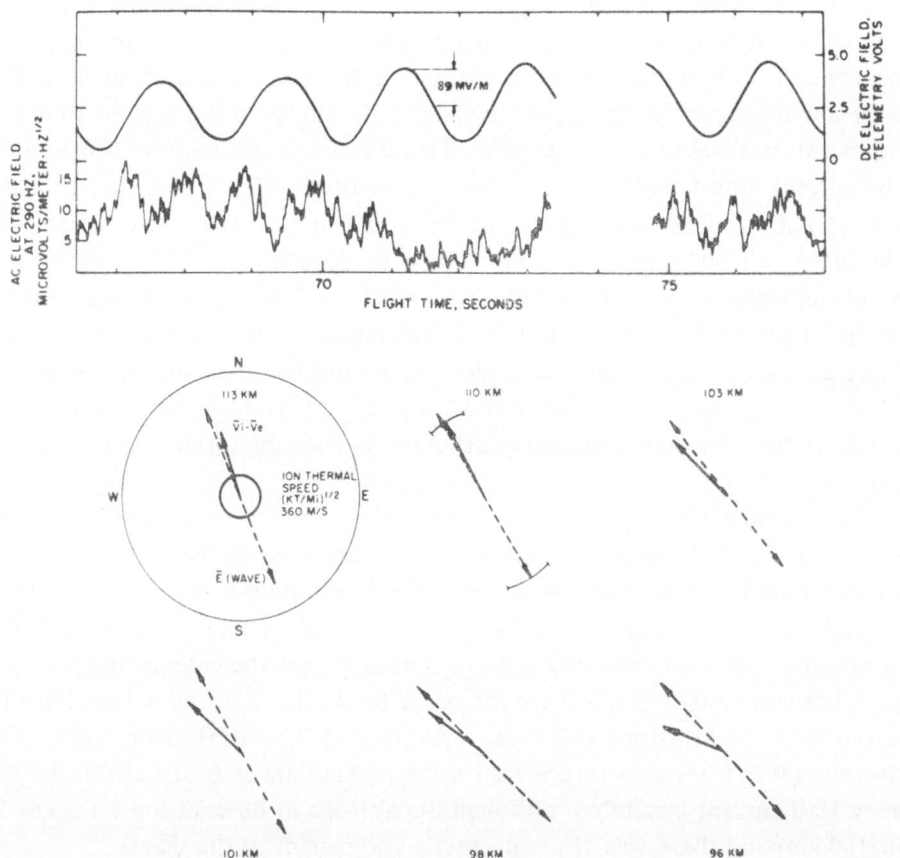

Fig. 6.   D.C. and A.C. electric field measurements in the auroral electrojet during a magnetospheric substorm.   (Reprinted by permission of *J. Geophys. Res.*)

to *j* may indicate the waves responsible for diffuse backscatter.   GREENWALD (1974)
has argued that the primary waves may be so strong that secondary gradient drift
instabilities operate to produce waves at the perpendicular angles.   He estimated
that density fluctuation intensities $\geq 5\%$ are required.   The density perturbation
levels reported by Kelley and Mozer were in the range 5–15% and the broadband
electric field fluctuations were 1–3 mV/m.

Doppler shift data as a function of azimuth presented by TSUNODA (1975) and
by MOORCROFT and TSUNODA (1978) indicate that the wave phase velocity often
saturates at the ion acoustic speed over a wide range of angles much as it does at
the equator.   At 50 MHz (3 m) this does not seem to occur in the aurora zone, since
the phase velocity has been shown to be proportional to the projection of the electron
drift velocity (deduced from Chatanika D.C. *E*-field measurements) on the radar look
direction (ECKLUND *et al.*, 1977) and since Doppler shifts up to 2,000 m/sec have
been observed at 50 MHz (GREENWALD and ECKLUND, 1975).   One explanation is
that secondary gradient-drift waves dominate at 3 m as suggested by Greenwald but

that secondary two-stream waves operate at 0.37 m. An explanation for the observation of two-stream waves at right angles to the equatorial electrojet suggested by SUDAN et al. (1973) is that a primary gradient-drift wave electric field exceeds the two-stream threshold for first order drifts at right angles to the current thus generating two-stream waves at right angles. This process would require 25 mV/m fields in the auroral zone which are an order of magnitude higher than the broadband 'primary' A.C. E field reported by Kelley and Mozer. The measurements were made during a 1,000 γ magnetic substorm with 50–80 mV/m D.C. electric fields, certainly an extreme case. Hopefully more theoretical and experimental work will soon shed light on the fully nonlinear development of the two-stream instability both at equatorial and auroral latitudes. It should be noted that the most recent auroral radar results of MOORCROFT and TSUNODA (1978) show that the auroral electrojet is as fully two-dimensionally turbulent as the equatorial case (FEJER et al., 1975).

The auroral electrojet wave data presented by KELLEY and MOZER (1973) extended into the VLF range and showed that electric fields were generated at frequencies up to 7.5 kHz, which was about the lower hybrid frequency. This was predicted for the two-stream instability by LEE et al. (1971) when the differential drift velocity is several times the acoustic speed, as was the case during the rocket flight. The wavelengths implied are the order to 20 cm. Auroral echoes have been received from the Millstone Hill radar operating at 23 cm (HAGFORS et al., 1971). UNGSTRUP (1975) has also interpreted some rocket data in terms of this 'high frequency Hall current instability' although the altitude of observation was very high (140–180 km) and there was also a magnetic component to the waves.

'Discrete' radar auroral echoes may occur when the electrojet turns more nearly into the radar beam and primary waves are observed directly. They may also be due to other plasma processes occurring at the edge of an auroral arc, however. CHATURVEDI (1976) has suggested that intense field-aligned currents at the edge of auroral arcs could generate oblique ion cyclotron waves in the E and F regions which could thus satisfy the radar backscatter geometry over a considerable altitude range. KELLEY et al. (1975) have reported intense ion cyclotron waves at 380 km near the edge of an arc and KELLEY and CARLSON (1977) reported intense electrostatic waves with broader spectra associated with the velocity shear at the same arc boundary. Thus there may be some F region contamination in the radar data and care should be taken in interpreting such data.

OTT and FARLEY (1975) have pointed out another instability process which can occur when large electric fields drive plasma through a neutral gas. If the differential drift exceeds 1.8 times the neutral acoustic speed (requiring an electric field ≥ 50 mV/m in the auroral zone) collisions with neutrals create an anisotropic ion velocity distribution which is unstable to generation of waves near the lower hybrid frequency via the Post-Rosenbluth instability. These should occur in the 150–250 km altitude range but have not yet been conclusively detected.

4.   Electrostatic Waves and Discontinuities at 1 R_E above the Auroral Ionosphere

It is somewhat early to write a review concerning the myriad of processes occurring at about 1 R_E above the Earth on auroral zone magnetic field lines. Nonetheless the region is so important that it seems worthwhile to at least present some of the observations. In this context I expand the definition of electrostatic waves to also include discontinuities in the potential structure. Whether called electrostatic shocks, double layers, or whatever, they constitute a major electrostatic perturbation in the macroscopic plasma properties which evolve somehow from a microscopic electrostatic process.

First indications of intense perpendicular electric fields which do not map to ionospheric altitudes came from auroral television studies (HALLINAN and DAVIS, 1970) which showed auroral forms moving very rapidly along the edges of auroral arcs. Direct evidence has since accrued from barium shaped charge releases (WESCOTT et al., 1975, 1976; JEFFRIES et al., 1975) and in situ D.C. electric field probes (MOZER et al., 1977) which have shown the existence of perpendicular D.C. electric fields exceeding 500 mV/m at altitudes of 4,000–8,000 km. If these fields mapped along magnetic field lines as equipotentials, ionospheric fields of many volts per meter would occur, contrary to measurements. Therefore, there must be significant potential drop along the field lines. This drop has been observed by both shaped charge and probe techniques. An example of the trajectory of a barium jet

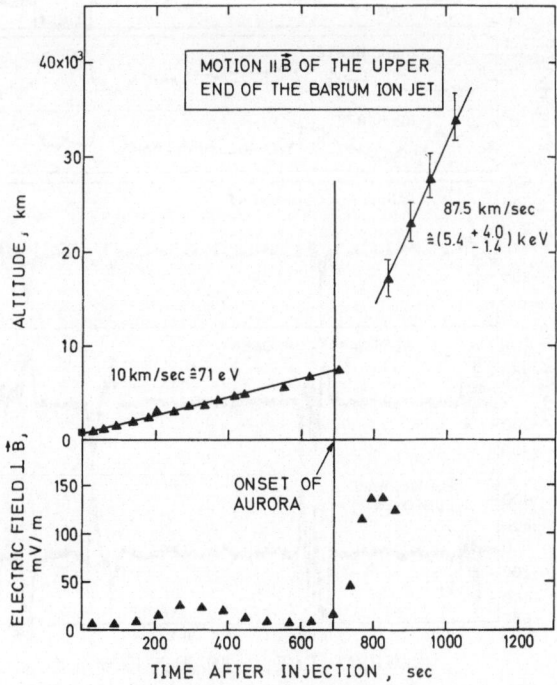

Fig. 7.   Velocity of a barium jet parallel to the Earth's magnetic field
as a function of time and altitude after injection from a sounding
rocket near the polar cusp.

launched above the polar cusp by the Max Planck (Garching) group is presented in
Fig. 7.   Near 7,000 km altitude the barium ions passed through a 5 keV potential
difference which accelerated the ions upward (HAERENDEL *et al.*, 1976).   Simultaneously auroral electron precipitation was observed and the D.C. electric field near
the foot of the flux tube increased to 100 mV/m.   An example of the probe electric
field results is shown in Fig. 8.   Perpendicular fields in excess of 500 mV/m are indicated with parallel components of several tens of mV/m.   Thus a severe departure
from frozen-in magnetic field occurred several thousand kilometers above the auroral oval.

Most theoretical efforts to date involve an instability when the electron drift
velocity exceeds some threshold.   Since the electron density decreases with altitude
faster than magnetic field lines diverge, the drift velocity, associated with field-
aligned currents, $V=J/ne$, must increase.   The group at the Royal Institute in
Stockholm along with Hannes Alfven (now at the University of California at San

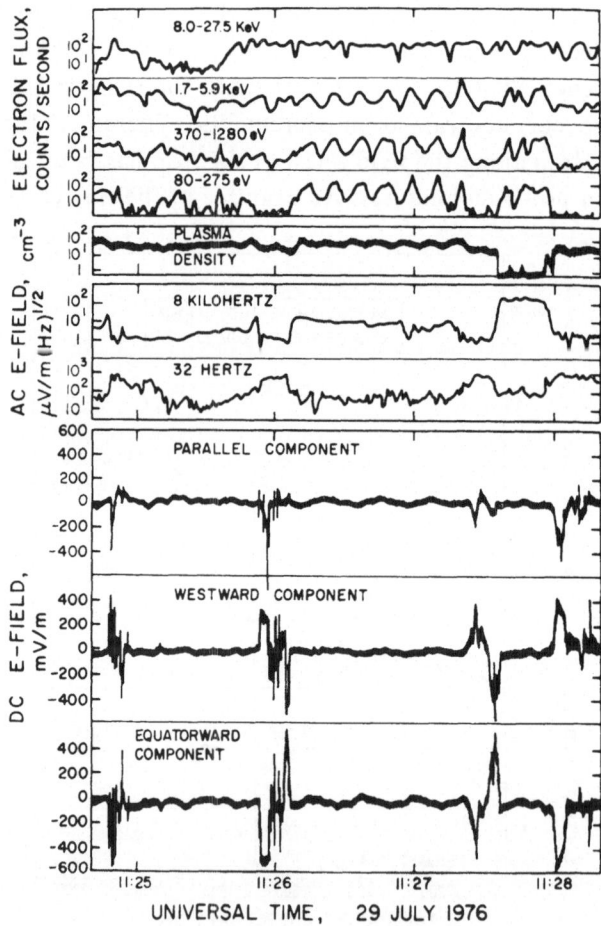

Fig. 8.   A.C. and D.C. electric fields along with electron fluxes measured
during an example of shock like structures about 1 R_E above the auroral oval. (Reprinted by permission of *Phys. Rev. Lett.*)

Diego) have long advocated a current interruption via 'double layer' formation, a process observed in laboratory experiments (CARLQUIST and BOSTROM, 1970; BLOCK, 1972). Such layers were originally thought to be quite thin parallel to $\bar{B}$ and hence hard to detect in a horizontally moving spacecraft. Common detection by S3-3 (MOZER *et al.*, 1977) indicates considerable extent along $\bar{B}$, however. Other models invoking larger oblique shock structures (KAN, 1975; SWIFT *et al.*, 1976; SWIFT, 1975) have also been recently proposed and it seems likely that considerably more theoretical effort will be expended in the near future.

Even earlier, KINDEL and KENNEL (1971) predicted that plasma in the altitude range of one Earth radius should first go unstable to electrostatic ion cyclotron waves. The electric field and plasma density fluctuation component of hydrogen ion cyclotron waves have indeed been detected using the S3-3 broad-band data transmission link (KINTNER *et al.*, 1977) in this altitude range. The observations fit the ion cyclotron dispersion relation remarkably well and can even be used to determine the electron temperature. An event shown in Fig. 9 was co-located with 3 of the electrostatic discontinuities discussed above and at least three harmonics of the hydrogen cyclotron frequency are clearly visible in one of the cases. This may not be surprising since the threshold for formation of the shock-like structures exceeds that for ion cyclotron emission in most theories. These observations add further evidence that field-aligned currents may be responsible for both processes. Furthermore, HUDSON and MOZER (1977) have calculated the anomalous resistivity due to such intense ion cyclotron waves and shown that they are capable of supporting 1 mV/m parallel E fields at 6,000 km. Thus these waves may play an important role

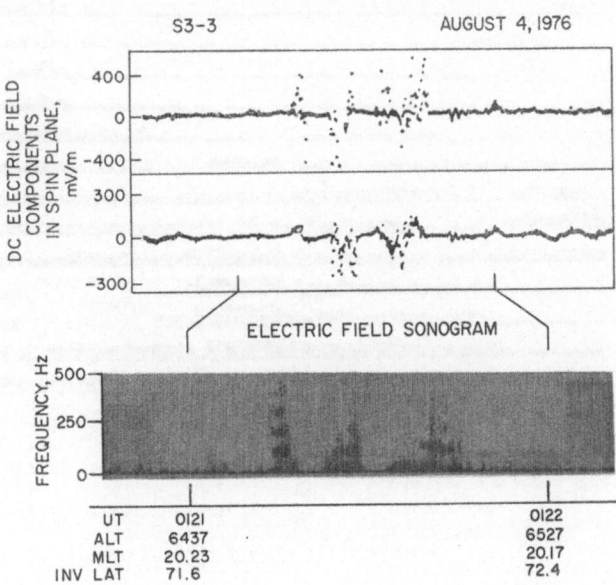

Fig. 9. Intense hydrogen cyclotron waves co-located with electrostatic shock-like structures in the D.C. field.

in the microscopic physics which determines the transport coefficients of the plasma. SWIFT (1977) has suggested a cascade process which transports energy from small to large wavelengths in two-dimensional systems. In his view energy initially injected

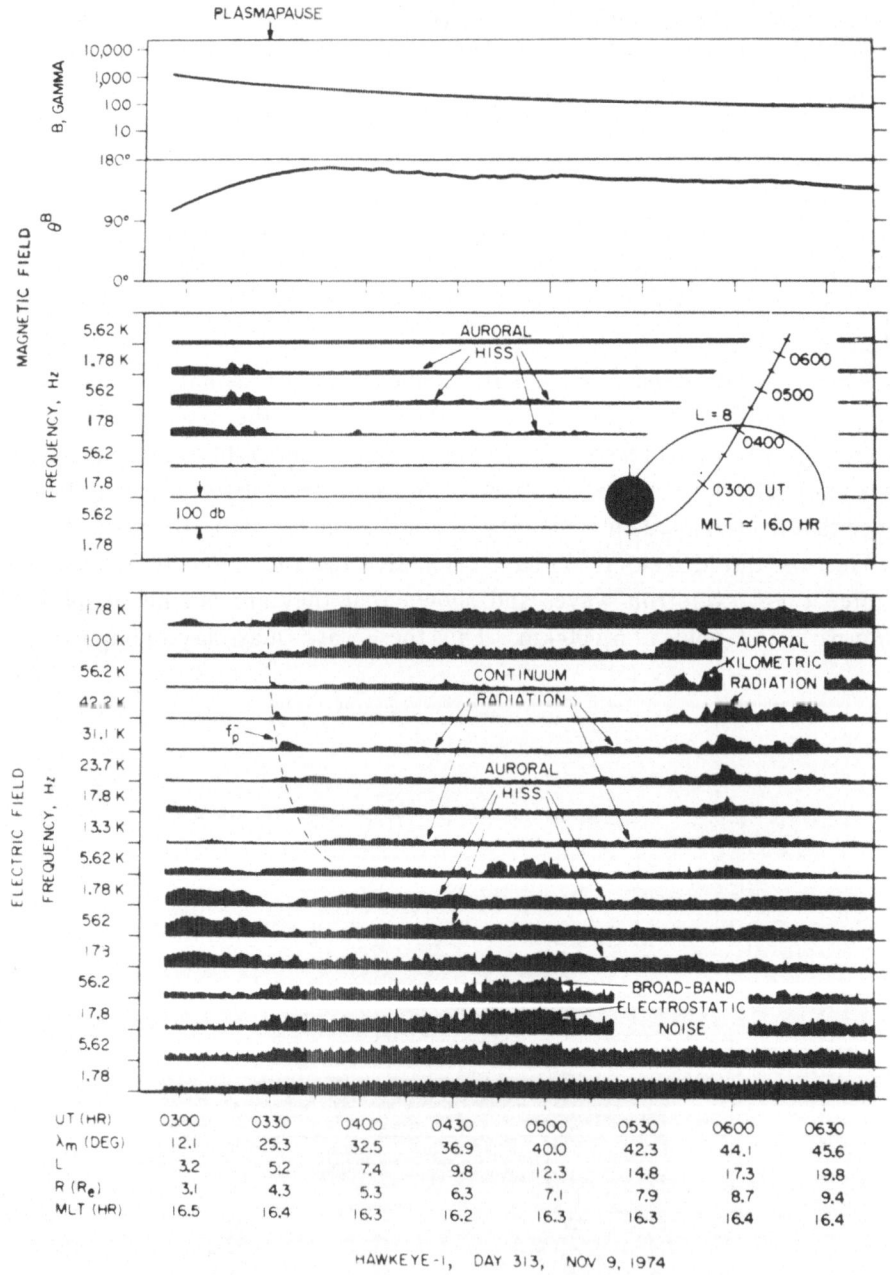

Fig. 10.   A.C. electric and magnetic field measurements made on Hawkeye on November 9, 1974. (Reprinted by permission of *J. Geophys. Res.*)

at wavelengths near the ion gyro radius is the ultimate source of the macroscopic D.C. field pattern. The observations of KINTNER et al. (1977) show that the cyclotron waves indeed have wavelengths near ρ_i.

Often the discontinuities are imbedded in a region of strong (≥ 50 mV/m) electric field turbulence which extends for a degree or more in invariant latitude (TORBERT and MOZER, 1978). The origin of this turbulence is not yet understood but may correspond to some 'spin-off' from the major structures. CAUFFMAN and GURNETT (1971) also noticed intense rapidly fluctuating D.C. electric fields which were most intense at the highest altitudes reached by Injun V. Just as in the case of the most intense localized perpendicular fields, discussed above, these potential contours do not map unattenuated to ionospheric altitudes, since low-altitude measurements do not show the corresponding large fields. Rapidly moving patches of auroral emission with apparent drift much greater than the local $\bar{E} \times \bar{B}$ (PERREAULT and AKASOFU, 1977) may be responding to drifts of the source at the 6,000 km altitude.

A very similar (identical?) feature has recently been reported by GURNETT and FRANK (1977) at slightly higher altitudes and dubbed broadband electrostatic noise. An example is reproduced in Fig. 10. Peak signals are observed in the tens of Hertz range with broadband amplitudes often exceeding 10 mV/m. (Note that each division indicates a 100 db range in signal). The waves were observed at all local times and correlated well with the characteristic V-shaped frequency cutoff in auroral hiss, an ELF electromagnetic hiss band and field-aligned currents. Co-location of the auroral hiss band suggests that the region is identical with inverted V electron precipitation events. This agrees well with the S3-3 low frequency electrostatic wave observations which are seen on both the high- and low-latitude sides of the electric potential discontinuities. GURNETT and FRANK (1977) pointed out that more intense waves were probably located at the slightly lower altitude where S3-3 operated. Although the wave mode is not yet clear, the intensities are, according to FREDERICKS et al. (1973), sufficiently high to create anomalous resistivity and potential drops the order of 1–2 keV. Important open questions involve the polarization of the waves and their detailed relationship to the proton gyro and plasma frequencies. This study must include Doppler shift effects upon low phase velocity electrostatic waves.

Both the S3-3 and Hawkeye results indicate a relation between these low-frequency waves, or at least their region of occurrence, and auroral hiss. It is possible that both the electromagnetic hiss emission and the electrostatic waves are parametric decay products of a higher-frequency emission. An electron Langmuir wave, for example, can decay to another electron Langmuir wave and an ion acoustic wave. The Arecibo heating experiments show that a broad range of acoustic waves are generated in a cascade-like process which starts with an electromagnetic pump decaying to Langmuir waves which in turn decay to a Langmuir plus and acoustic wave (PERKINS, 1974). Beam-generated Langmuir waves might undergo such a process.

Auroral kilometric radiation has been shown to originate in the altitude range of 1–2 Earth radii but the generation mechanism is not yet clear. Again electron Langmuir waves have been invoked as an intermediary. However, a search for such waves using the Hawkeye and IMP6 instruments has not shown the intense waves expected (Kurth, private communication, 1977).

Laboratory experiments (BOSWELL, 1977; FREDERICKS, 1977) indicate that electron Bernstein waves and the related $(n+1/2)$ electron gyro harmonic emissions may be a very important source for electromagnetic waves by a parametric process. Electrostatic emissions at $(n+1/2)\Omega_e$ with $n=1$, 2, and 3 have been observed in the high-altitude magnetospheric on OGO-5 (KENNEL et al., 1970; SCARF et al., 1972).

5. Conclusion

The important role which electrostatic waves and discontinuities play in determining both the macroscopic and microscopic structure of plasma near the Earth is rapidly becoming clear. Anomalous resistivity, cross magnetic field diffusion, particle acceleration by shock-like as well as stochastic processes, and formation of plasma irregularities are just a few of the processes related to electrostatic waves. In all the areas discussed here, further development of nonlinear theories are required. Experimental data are becoming of sufficient quality that details in competing theories can be used to test their validity. Both the theory and experiment are further advanced for phenomena at lower altitudes. This is partly due to the larger experimental data set from relatively inexpensive rocket and radar studies. Both linear and nonlinear theory is also relatively advanced in the field of electrojet instabilities.

The single-most important area for further study is the region near one Earth radius altitude. Although in an awkward altitude range, it is nonetheless close enough to the Earth for detailed investigation. Such a study is justified not only by its importance for magnetospheric physics but for possible solar flare and astrophysical applications as well.

This research was supported by the Atmospheric Research Section at the National Science Foundation under Grant number ATM77-04518 and NASA Grant NGR33-010-161. The author thanks those who gave permission to use their data in this review.

REFERENCES

AARONS, J., A descriptive model of F layer high-latitude irregularities as shown by scintillation observations, *J. Geophys. Res.*, **78**, 7441, 1973.

BLOCK, L.P., Potential double layers in the ionosphere, *Cosmic Electrodyn.*, **3**, 349, 1972.

BOSWELL, R.W., Laboratory astro plasma physics and the measurement of turbulence, *Trans. Am. Geophys. Union*, **473**, 1977.

CARLQVIST, P. and R. BOSTROM, Space charge regions above the aurora, *J. Geophys. Res.*, **75**, 7140, 1970.

CAUFFMAN, D. and D. GURNETT, Double probe measurements of convection electric fields, *J. Geophys, Res.*, **76**, 6014, 1971.

CHATURVEDI, P.K., Collisional ion cyclotron waves in the auroral ionosphere, *J. Geophys. Res.*, **81**, 6169, 1976.

COSTA, E. and M.C. KELLEY, Nonlinear steepening in equatorial spread *F*, Reprint 77-1, School of Electrical Engineering, Cornell Univ., Ithaca, New York, 1977.

DYSON, P.L., Direct measurements of the size and amplitude of irregularities in the topside iono-sphere, *J. Geophys. Res.*, **74**, 6291, 1969.

DYSON, P.L., J.P. McCLURE, and W.B. HANSON, In situ measurements of the spectral characteristics of *F* region ionospheric irregularities, *J. Geophys. Res.*, **79**, 1497, 1974.

ECKLUND, W.L., B.B. BALSLEY, and R.A. GREENWALD, Crossed beam measurements of the diffuse radar aurora, *J. Geophys. Res.*, **80**, 1805, 1975.

ECKLUND, W.L., B.B. BALSLEY, and D.A. CARTER, A preliminary comparison of *F* region plasma drifts and *F* region irregularity drifts in the auroral zone, *J. Geophys. Res.*, **82**, 195, 1977.

FEJER, B.G., D.T. FARLEY, B.B. BALSLEY, and R.F. WOODMAN, Radar observations of two-dimen-sional turbulence in the equatorial electrojet, *J. Geophys. Res.*, **81**, 130, 1975.

FREDERICKS, R.W., F.L. SCARF, and C.T. RUSSELL, Field aligned currents, plasma waves, and anoma-lous resistivity in the disturbed polar cusp, *J. Geophys. Res.*, **78**, 2133, 1973.

FREDERICKS, R.W., Laboratory experiments of interest to space plasma physics, Post Deadline Ab-stract, Spring AGU Meeting, 1977.

GOLDMAN, S.R., L. BAKER, S.L. OSSAKOW, and A.J. SCANNAPIECO, Striation formation associated with barium clouds in an inhomogeneous medium, *J. Geophys. Res.*, **81**, 5097, 1976.

GREENWALD, R.A., Radar aurora and the gradient drift instability, *J. Geophys. Res.*, **74**, 4807, 1974.

GREENWALD, R.A. and W.L. ECKLUND, A new look at radar auroral motion, *J. Geophys. Res.*, **80**, 3642, 1975.

GREENWALD, R.A., W.L. ECKLUND, and B.B. BALSLEY, Diffuse radar aurora: Spectral observations of non-two-stream irregularities, *J. Geophys. Res.*, **80**, 131, 1975.

GURNETT, D.A. and L.A. FRANK, A region of intense plasma wave turbulence on auroral field lines, *J. Geophys. Res.*, **82**, 1031, 1977.

HAGFORS, T., R.G. JOHNSON, and R.A. POUIER, Simultaneous observation of proton precipitation and auroral radar echoes, *J. Geophys. Res.*, **76**, 6093, 1971.

HAERENDEL, G. and H. VOLK, Striations in ion clouds, *J. Geophys. Res.*, **76**, 4591, 1971.

HAERENDEL, G., E. RIEGER, A. VALENZUELA, H. FOPPL, H.C. STENBAEK-NEILSEN, and E.W. WESCOTT, First observation of electrostatic acceleration of barium ions into the magnetosphere, Proceedings of a Symposium on European Programmes on Sounding Rocket and Balloon Re-search in the Auroral Zone, ESA SP 115, Schlob, Elmau, Germany, 3-7 May, 1976.

HALLINAN, T.J. and T.N. DAVIS, Small-scale auroral arc distortion, *Planet. Space Sci.*, **18**, 1735, 1970.

HEPPNER, J.P., Magnetospheric convection patterns inferred from high latitude activity, in *Atmos-pheric Emissions*, edited by B.M. McCormac and A. Omholt, Van Nostrand Reinhold, New York, 1969.

HUDSON, M.K. and F.S. MOZER, Magnetic field aligned potential differences due to electrostatic ion cyclotron turbulence, *Trans. Am. Geophys. Union*, **473**, 1977.

IVERSEN, I.B., N.D. ANGELO, and J.K. OLESEN, Further evidence for the Farley-Buneman instability in the polar cap ionosphere, *J. Geophys. Res.*, **80**, 3713, 1975.

JEFFRIES, R.A., W.H. ROACH, E.W. HONES, Jr., E.M. WESCOTT, H.C. STENBAEK-NIELSEN, T.N. DAVIS, and J.D. WINNINGHAM, Two plasma injections into the northern magnetospheric cleft, *Geophys. Res. Lett.*, **3**, 285, 1975.

KAN, J.R., Energization of auroral electrons by electrostatic shock waves, *J. Geophys. Res.*, **80**, 2089, 1975.

KELLEY, M.C. and F.S. MOZER, A satellite survey of vector electric fields in the ionosphere at fre-quencies of 10 to 500 Hz. 1. Isotropic, high-latitude electrostatic emissions, *J. Geophys. Res.*, **77**, 4158, 1972.

KELLEY, M.C. and F.S. MOZER, Electric field and plasma density oscillations due to the high-fre-quency Hall current instability in the auroral regions, *J. Geophys. Res.*, **78**, 2214, 1973.

KELLEY, M.C., E.A. BERING, and F.S. MOZER, Evidence that the electrostatic ion cyclotron instability is saturated by ion heating, *Phys. Fluids*, **18**, 1950, 1975.

KELLEY, M.C. and C.W. CARLSON, Observations of intense velocity shear and associated electrostatic waves near an auroral arc, *J. Geophys. Res.*, **82**, 2343, 1977.

KELLEY, M.C., T.S. JORGENSEN, and I.S. MIKKELSEN, Thermospheric wind measurements in the polar region, *JATP*, **39**, 211, 1977.

KELLEY, M.C. and P. KINTNER, Two-dimensional turbulence in a low β cosmic scale plasma, *Astrophys. J.*, **220**, 339, 1978.

KENNEL, C.F., F.L. SCARF, R.W. FREDERICKS, J.H. McGEHER, and F.V. CORONITI, VLF electric field observations in the magnetosphere, *J. Geophys. Res.*, **75**, 6136, 1970.

KINDEL, J.M. and C.F. KENNEL, Topside current instabilities, *J. Geophys. Res.*, **76**, 3055, 1971.

KINTNER, P.M., Observations of velocity shear driven plasma turbulence, *J. Geophys. Res.*, **81**, 5114, 1976.

KINTNER, P.M. and N. D'ANGELO, Transverse Kelvin-Helmholtz instability in a magnetized plasma, *J. Geophys. Res.*, **82**, 1628, 1977.

KINTNER, P.M., M.C. KELLEY, and F.S. MOZER, Electrostatic hydrogen cyclotron waves near one Earth radius altitude in the polar magnetosphere, *Geophys. Res. Lett.*, **5**, 139, 1978.

LAASPERE, T., W.C. JOHNSON, and L.C. SEMPREBON, Observations of auroral hiss, LHR noise, and other phenomena in the frequency range 20 Hz–540 kHz on OGO-6, *J. Geophys. Res.*, **76**, 4477, 1971.

LEE, K., C.F. KENNEL, and J.M. KINDEL, High-frequency Hall current instability, *Radio Sci.*, **6**, 209, 1971.

LINSON, L.M. and J.B. WORKMAN, Formation of striations in ionospheric barium clouds, *J. Geophys. Res.*, **75**, 3211, 1970.

MAYNARD, N.C. and J.P. HEPPNER, Variations in electric fields from polar orbiting satellites, in *Particles and Fields in the Magnetosphere*, edited by B.M. McCormac, p. 247, D. Reidel, Dordrecht, Netherlands, 1970.

McDONALD, B.E., T.P. COFFEY, S.L. OSSAKOW, and R.N. SUDAN, Numerical studies of type 2 equatorial electrojet irregularity development, *Radio Sci.*, **10**, 247, 1975.

MOORCROFT, D.R. and R.T. TSUNODA, Rapid scan Doppler velocity maps of the UHF diffuse radar aurora, *J. Geophys. Res.*, **83**, 1482, 1978.

MONTGOMERY, D., Plasma kinetic processes in a strong magnetic field, *Physica*, **82**, B&C, 125, 1976.

MOZER, F.S., Power spectra of the magnetospheric electric field, *J. Geophys. Res.*, **76**, 3651, 1971.

MOZER, F.S., C.W. CARLSON, M.K. HUDSON, R.B. TORBERT, B. PARADY, J. YATTEAU, and M.C. KELLEY, Observations of paired electrostatic shocks in the polar magnetosphere, *Phys. Rev. Lett.*, **38**, 292, 1977.

OLESEN, J.K., F. PRIMDAHL, F. SPANGSLEV, and N. D'ANGELO, On the Farley instability in the polar cape region, *J. Geophys. Res.*, **80**, 696, 1975.

OLESEN, J.K., F. PRIMDAHL, F. SPANGSLEV, E. UNGSTRUP, A. BAHNSEN, O. FAHLESON, C.G. FALTHAMMAR, and A. PEDERSEN, Rocket-borne wave, field, and plasma observations in unstable polar cap h region, *Geophys. Res. Lett.*, **3**, 399, 1976.

OTT, E. and D.T. FARLEY, Microinstabilities and the production of short wavelength irregularities in the auroral F region, *J. Geophys. Res.*, **80**, 4599, 1975.

PERKINS, F.W., A theoretical model for short scale field aligned plasma density striations, *Radio Sci.*, **9**, 1065, 1974.

PERKINS, F.W., N.J. ZABINSKY, and J.H. DOLES, III, Deformation and striation of plasma clouds in the ionosphere (1), *J. Geophys. Res.*, **78**, 697, 1973.

PERREAULT, P.D. and S.I. AKASOFU, On the relationship between electric fields measured by Chatanika radar and simultaneous auroral substorm features, *Trans. Am. Geophys. Union*, **477**, 1977.

PRAKASH, S., S.P. GUPTA, and B.H. SUBBARAYA, Experimental evidence for cross-field instability in the equatorial ionosphere, *Space Res.*, **11**, 1139, 1971.

SAGALYN, R.C., M. SMIDDY, and M. AHMED, High latitude irregularities in the topside ionosphere based on ISIS 1 thermal probe data, *J. Geophys. Res.*, **79**, 4252, 1974.

SCARF, F.L., R.W. FREDRICKS, I.M. GREEN, and C.T. RUSSELL, Plasma waves in the dayside polar cusp. 1. Magnetospheric observations, *J. Geophys. Res.*, **77**, 2274, 1972.

SUDAN, R.M., J. AKINRIMISI, and D.T. FARLEY, Generation of small scale irregularities in the equatorial electrojet, *J. Geophys. Res.*, **78**, 240, 1973.

SUDAN, R.N. and M. KESKINEN, Theory of strongly turbulent equatorial electrojet irregularities, *Trans. Am. Geophys. Union*, **58**, 449, 1977.

SWIFT, D.W., On the formation of auroral arcs and acceleration of auroral electrons, *J. Geophys. Res.*, **80**, 2096, 1975.

SWIFT, D.W., H.C. STENBACK-NIELSEN, and T.J. HALLINAN, An equipotential model for auroral arcs, *J. Geophys. Res.*, **81**, 3931, 1976.

SWIFT, D.W., Turbulent generation of electrostatic fields in the magnetosphere, *J. Geophys. Res.*, **82**, 5143, 1977.

TEMERIN, M., The polarization, frequency and wavelengths of high latitude turbulence, *Geophys. Res. Lett.*, **83**, 2609, 1978.

TORBERT, R. and F.S. MOZER, Electrostatic shocks as the source of discrete auroral arcs, *Geophys. Res. Lett.*, **5**, 139, 1978.

TSUNODA, R.T., Electric field measurements above a radar scattering volume producing 'diffuse' auroral echoes, *J. Geophys. Res.*, **80**, 4297, 1975.

TSUNODA, R.T. and R.I. PRESNELL, On a threshold electric field associated with the 398 MHz diffuse radar aurora, *J. Geophys. Res.*, **81**, 88, 1976.

TSUNODA, R.T., P. PERREAULT, and J.C. HODGES, Asimuthal distribution of HF slant echoes and its relationship to the polar cap electric field, *J. Geophys. Res.*, **81**, 3834, 1976.

UNGSTRUP, E., Narrow band VLF electromagnetic signals generated in the auroral ionosphere by the high frequency two-stream instability, *J. Geophys. Res.*, **80**, 4272, 1975.

UNWIN, R.S., The morphology of the VHF radio aurora at sunspot maximum. 2. The behavior of different echo types, *J. Atmos. Terr. Phys.*, **28**, 1183, 1966.

WESCOTT, E.M., H.C. STENBAEK-NIELSEN, T.N. DAVIS, W.B. MURCRAY, H.M. PEEK, and P.J. BOTTOMS, The Oosik barium plasma injection experiment and magnetic storm of March 7, 1972, *J. Geophys. Res.*, **80**, 951, 1975.

WESCOTT, E.M., H.C. STENBAEK-NIELSEN, T.J. HALLINAN, T.N. DAVIS, and H.M. PEEK, The Skylab injection experiments. 2. Evidence for a double layer, *J. Geophys. Res.*, **81**, 4495, 1976.

ZABINSKY, N.J., J.H. DOLES, III, and F.W. PERKINS, Deformation and striation of ionospheric clouds in the ionosphere. 2. Numerical simulation of a nonlinear two-dimensional model, *J. Geophys. Res.*, **78**, 711, 1973.

Diffuse Auroral Precipitation

M. Ashour-Abdalla[*],[***] and C.F. Kennel[*],[**],[***]

*Department of Physics, University of California at Los Angeles,
Los Angeles, California, U.S.A
**Center for Plasma Physics and Fusion Technology,
University of California at Los Angeles,
Los Angeles, California, U.S.A.
***Institute of Geophysics and Planetary Physics,
University of California at Los Angeles,
Los Angeles, California, U.S.A.

(Received October 17, 1977)

In this review, we summarize one at least plausibly coherent view of the self-consistent coupling of convection in the plasma sheet to particle precipitation and the ionosphere. We focus upon our understanding of the plasma instabilities responsible for diffuse auroral precipitation. At present, the electrostatic ion and electron cyclotron harmonic loss-cone instabilities seem to be the best candidates. However, they depend sensitively upon the cold electron density and temperature deep in space on auroral field lines, parameters about which we have little or no experimental or theoretical information.

1. Introduction

Scientific interest in the *aurora borealis* extends back to the ancient Greeks. Naturally, the first observations of aurora were made by the naked eye. The eye's spectral response, and particularly its sensitivity to contrast, thus shaped our conceptions of the aurora. Since its limitations are similar to those of the eye, the all-sky camera, which replaced visual observations for systematic auroral studies in the 1950's, served to systematize the morphology primarily of the discrete aurora so prominent to the naked eye. Only with the advent of satellite-borne energetic particle detectors and more recently scanning photometers did our awareness of a broad region of diffuse particle precipitation and light emission, corresponding roughly to the auroral oval, emerge. Because of its great extent, this diffuse auroral precipitation can be the dominant form of auroral energy dissipation (ANGER and LUI, 1973; LUI and ANGER, 1973; LUI et al., 1973; HULTQVIST, 1975).

The diffuse aurora seems to be caused primarily by the precipitation of 1–10 keV electrons. Moreover, recent studies have established that ~1–100 keV proton precipitation is an integral part of the diffuse aurora (LUI and ANGER, 1973; LUI et al., 1973; EATHER et al., 1976). The zones of soft electron and proton precipitation generally coincide (HULTQVIST, 1975) and define an auroral oval similar to that deduced from discrete arc observations. While protons typically contribute no more than 30% of the diffuse auroral light intensity (Mende and Eather, 1976; Hultqvist, 1975), the zone of diffuse proton precipitation can extend equatorward of the diffuse electron precipitation in the pre-midnight sector so that proton induced light emissions can predominate over a small range of invariant latitude in this sector (Fukunishi, 1975). Figure 1 shows a simultaneous ESRO 1A observation of precipitating soft electrons and protons which illustrates the above conclusions (Hultqvist, 1975).

We can ask two questions about the discrete aurora. First, where do the auroral particles come from and why do they precipitate where they do? Second, what causes the precipitation of electrons and protons? Because neutral and Coulomb collisions cannot be responsible for the observed precipitation, the second question can be refined to, what plasma instability and ensuing plasma wave turbulence causes the diffuse auroral precipitation? In this regard, it will be important to note that the auroral electron and protons generally have isotropic pitch angle distributions when they are observed in or near the auroral ionosphere (Hultqvist, 1975; Bernstein *et al.*, 1974; Hultqvist *et al.*, 1974).

In this paper, we selectively review the development of our still partial answers to these two questions. Since we will try to illuminate one at least plausible chain of arguments linking observation with theory, our review is by no means exhaustive. We apologize in advance for the sins of commission and especially omission inherent in our methodology.

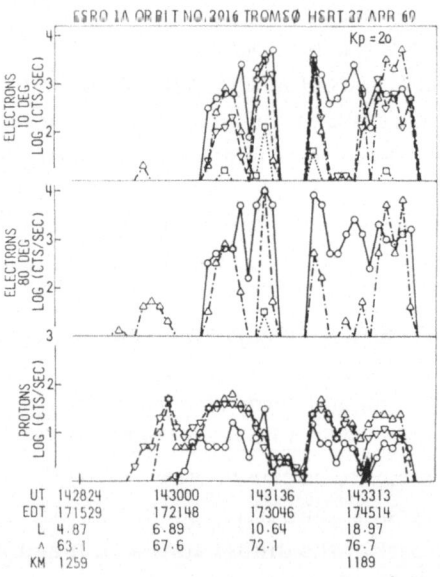

Fig. 1. Proton and electron precipitation profiles. This figure illustrates the general similarity of the electron and proton diffuse auroral precipitation profiles (taken from Hultqvist, 1975).

2. Coupling of Convection and Precipitation

The general morphology and energy spectra of the diffuse auroral precipitation generally suggest a plasma sheet origin for the night side diffuse auroral particles (EATHER and MENDE, 1972a, b; EATHER et al., 1976). While there have been no simultaneous measurements of the energetic particle fluxes in the ionosphere and deep in space on identifiably the same auroral field line, the magnitudes of the precipitating particle fluxes are generally comparable to those found near the inner edge of the plasma sheet, a fact consistent with the measured pitch angle isotropy of the precipitating particle fluxes. Finally, the observed position of the sharp inner edge to the electron plasma sheet (VASYLIUNAS, 1968) is consistent with its mapping to the equatorward edge of the diffuse auroral electron precipitation.

The question of the origin and location of the diffuse auroral precipitation was first attacked by PETSCHEK and KENNEL (1966) and subsequently independently by KENNEL (1969) and VASYLIUNAS (1969). The physical content of their essentially identical models involves two fundamental statements. First, earthward convection must replenish the supply of electrons and protons lost by precipitation to the auroral ionosphere. Second, because the particle pitch angle distributions are nearly isotropic, the precipitation lifetime approaches the strong diffusion limit (KENNEL and PETSCHEK, 1966; KENNEL, 1969), independent of the mode of plasma turbulence responsible for the precipitation. Combining these two statements led to the physical consequences discussed below.

A flux tube in the plasma sheet would initially lose a small fraction of its particles as it convects towards the earth, even though isotropic particle fluxes are precipitating from it, because the strong diffusion minimum lifetime, which varies at L^{-4}, is much longer than a characteristic flow time. Since at first the convection is virtually loss-free, the electrons and protons are heated by adiabatic compression. As the flow penetrates further into an increasing magnetic field, it reaches a point where the minimum lifetime becomes comparable to the flowtime. Since the electron minimum lifetime is much smaller than the proton lifetime this happens first for

STRONG DIFFUSION MODEL OF DIFFUSE AURORAL PRECIPITATION
KENNEL - VASYLIUNAS 1969

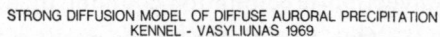

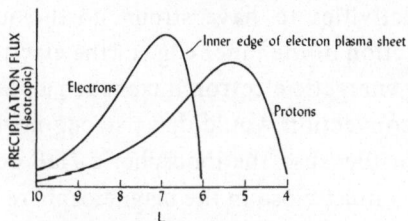

• Inner edge occurs when $T_{min} \leq T_{flow}$
  (hot electron exchange with cold electrons)
• Plasma wave turbulence not specified, but strong diffusion
• No corotational electric field
• No line-tying of flow to ionosphere
• No effects of ring current on flow

Fig. 2. Coupling of strong diffusion particle losses with purely sunward convection near local midnight. Shown here is the qualitative precipitation profile expected in the least elaborate coupling model. The precipitation of energetic electrons has a sharp equatorward edge (corresponding to the inner edge of the electron plasma sheet) when the minimum lifetime becomes short enough to precipitate particles before they penetrate the earth's field further. The prediction that protons would form a sharp inner edge equatorward of the electrons has proven false.

electrons. At this point, electrons are rapidly removed from the convecting flux tube by precipitation. Since, for equal temperatures, the isotropic precipitation flux of electrons exceeds that of protons by a factor 43, the difference between electron and proton precipitation fluxes must be made up by a return flux of colder electrons streaming out of the ionosphere. Since a hot electron is exchanged for a colder one, the mean energy of the electrons in the convecting flux tubes decreases. Thus, this model predicts a sharp decrease of electron temperature at the 'inner edge of the electron plasma sheet' whose computed location generally agrees with obser- vation (VASYLIUNAS, 1968) and with the general location of the equatorward edge of the diffuse auroral electron precipitation (KENNEL, 1969). The precipitation profiles expected in this model are sketched in Fig. 2.

Since the proton minimum lifetime exceeds that of electrons, the protons should penetrate further than the electrons and form an inner edge deeper within the mag- netosphere. A sketch of the latitude variation of electron and proton precipitation predicted by this simplest of all models of convection-precipitation coupling is sketched in Fig. 2; if the electron inner edge is at $L \approx 6$, the proton inner edge would be near $L \approx 4$. This last prediction of the theory has proven false: the electron and proton precipitation zones more or less coincide. However, refinements of the convection model may be able to ameliorate this discrepancy. By including the coro- tation electric field, coupling to the finite conducting ionospheric plasma and the effects of finite proton pressure, JAGGI and WOLF (1973) and subsequently others have shown that protons, originally convecting radially inwards can be deflected azimuthally around the earth as soon as they penetrate to L-shells where the high ionospheric conductivity created by electron precipitation has diminished. Thus, on the night side, one would not expect protons to penetrate significantly beyond the electron inner edge (D.J. Southwood, private communication, 1977). Since energetic protons can be precipitated only from flux tubes containing energetic protons, the zones of electron and proton precipitation should more or less coincide, with protons extending slightly equatorward of the electrons, particularly in the pre-midnight region towards which protons are carried by their magnetic field gradient drifts.

One further consequence of the above convection-precipitation coupling model was discussed by CORONITI and KENNEL (1972). One would expect the night-side auroral ionosphere Pedersen and Hall conductivities to have strong north-south gradients with a peak just poleward of the projection of the inner edge of the electron plasma sheet, because the ionizing precipitating energetic electron fluxes such a peak. The idealized east-west electric field driving convection would drive strong north- south Hall currents in the auroral ionosphere. Because the ionospheric Hall con- ductivity is inhomogeneous, these Hall currents must close in the magnetosphere by field-aligned currents into the ionosphere at the equatorward edge and out of the poleward edge of the diffuse auroral precipitation ragion. The field-aligned current density maximizes on those field lines which connect to the maximum Hall conductivity gradients. They then speculated that strong convection could then increase the field-aligned current density above its stability limits leading to regions of resistive dissipation and parallel electric field at high altitudes on the strong current field lines. This would then polarize the lower ionosphere in turn to produce a broad westward electrojet near the equatorward edge of the diffuse electron precipitation. Another physical consequence is implicit in this model: if ions are accelerated downwards by the parallel electric field, one would expect structured non-isotropic ion precipitation at the equatorward edge of the diffuse electron precipitation, whose intensity possibly correlates with the intensity of the diffuse electrojet.

Clearly the above model was oversimplified. It cannot describe the entire pattern of field-aligned currents threading the auroral oval, because it leaves out those currents that connect to the magnetopause and drive magnetospheric convection. Its geometrical assumptions are best suited to the local midnight sector. Even there, since the reaction of the induced ionospheric conductivity back onto the convective flow pattern was not taken into account—as well as finite proton pressure—its usefulness is primarily illustrative. It illustrates that field-aligned currents, parallel electric fields (and therefore particle acceleration) emerge naturally from a model of diffuse auroral precipitation. It also suggests a role for parallel electric fields in electrojet formation.

In summary, our first question—of the origin and location of the night-side diffuse aurora—seems now to have been answered in a general way. Soft auroral electrons and protons are precipitated from flux tubes which are convecting from the plasma sheet towards the earth's night side. Strong diffusion precipitation naturally forms a sharp inner edge to the electron plasma sheet. Because of their long minimum lifetimes, the protons do not form a sharp inner edge; their precipitation patterns should generally mirror proton convection patterns in deep space.

We also derive several hints from the model of diffuse auroral precipitation above which will guide our discussion of the plasma instabilities responsible for it. First, the general equality and isotropy of plasma sheet and diffuse auroral electron fluxes, the fact that the inner edge of the electron plasma sheet can be observed at a variety of magnetic latitudes (VASYLIUNAS, 1968), and the magnitude of the observed decrease in electron mean energy at the inner edge, all argue that nearly all electrons of plasma sheet origin which penetrate sufficiently closely to the earth are eventually precipitated. Furthermore, this strongly suggests that the instability responsible for diffuse auroral electron precipitation must occur near the geomagnetic equator (less than say 45° geomagnetic latitude). Waves near the equator can interact with all electrons on the flux tube; conversely, it is difficult to see how waves occurring only at high latitudes can reduce the 1–10 keV electron fluxes needed to account for the observations of the inner edge. The case is less clear for protons, but we will pursue the consequences of assuming an equatorial location for ion plasma turbulence shortly. Second, since the electron precipitation number flux must be neutralized by a return flux of colder electrons, plasma sheet convection and precipitation helps determine not only the hot plasma, but also the cold electron environment in which the instabilities responsible for the auroral precipitation operate. This environment should permit two instabilities, one for electrons and one for protons, to grow. Finally, since the diffuse aurora is present at quiet as well as disturbed times, these instabilities must not require unusual circumstances for their growth. A source of free energy must nearly always be present to drive them.

3.   Free Energy Sources

A self-consistent model of the coupling of convection to particle precipitation must demonstrate that convection naturally generates a source of free energy for the instabilities responsible for particle precipitation. In this section, we discuss these free energy sources, and in the next, some of the instabilities created by these sources.

One can divide up the free energy sources for plasma instabilities into a few general types. Among these are:
1)   Electrical currents
2)   Fast ion beams
3)   Fast electron beams
4)   Density and temperature gradients
5)   Pitch angle anisotropy
6)   Loss cone distribution

Given the great activity of the field lines connecting to the auroral oval, it is not surprising that all six are present there, at least at times. However, an extension of the chain of reasoning outlined in the preceding section indicates that all six are plausible theoretical consequences of the convection and precipitation model discussed in Section 2.

Even when no net electrical current need flow, a return current of cold electrons must stream out of the ionosphere to balance the difference between electron and ion precipitation fluxes. KINDEL and KENNEL (1971) suggested that this return current could be unstable to electrostatic ion cyclotron waves in the far topside ionosphere at altitudes above 1,000 km. When a net electrical current must also flow, the case for instability is even stronger. It is currently thought that ion cyclotron waves are involved in the formation of regions of strong parallel electric fields in the auroral topside ionosphere. Therefore sources 2) and 3)—fast ion and electron beams— may in practice be associated with regions of strong field-aligned current on auroral lines of force.

Several arguments suggest that field-aligned currents and fast electron or ion beams may not be the free energy source responsible for the diffuse aurora. Field-aligned currents tend to be spatially structured, and the precipitation we wish to understand is diffuse. Even if the field-aligned currents—or the precipitation return currents are diffuse, our present understanding suggests that they will create instabilities far from the geomagnetic equator that can interact with only a small fraction of the particles on a given flux tube. If fast electron and ion beams are created by parallel electric fields associated with strong field-aligned currents—they too would be a highly structured free energy source which is probably strongest at disturbed times. Such beams have been observed primarily near the ionosphere and would probably produce turbulence at high latitudes. Finally, the first effect of beam-generated turbulence is to scatter particles out of the beam—from the loss cone onto trapped orbits. While such instabilities—which have not been discussed at all for ions in the auroral context—may be important for populating the auroral field lines with ions of ionospheric origin—it seems unlikely that we can ask them to drive the reverse process which precipitates ions back into the ionosphere.

The coupling of precipitation and convection naturally produces a sharp spatial gradient in electron temperature which is the inner edge of the electron plasma sheet. CORONITI and KENNEL (1970b) and CHANCE et al. (1973) argued that this temperature gradient could be unstable to the growth of drift Alfven waves, thereby producing micropulsations with 5–15 sec periods. However, unless the spatial gradient scale length is as small as an ion Larmor radius, drift waves must have frequencies well below the ion cyclotron frequency, so low that the particles first adiabatic invariant must be conserved. The isotropy of the observed precipitation fluxes suggest that the first adiabatic invariant is violated. The best such low frequency instabilities seem capable of is modulating the growth of the instability responsible for the particle precipitation. CORONITI and KENNEL (1970a) suggested that drift and hydromagnetic wave modulation of whistler growth might account for the modulation of 40 keV electron and X-ray fluxes occasionally observed.

This leaves us with the last two, related, free energy sources—thermal anisotropy and a loss cone property. An initially thermally isotropic plasma will become anisotropic as it flows towards the earth on more or less dipolar field lines. Eventually, the perpendicular temperature $T_\perp$ will exceed the parallel temperature $T_{//}$. Sufficiently large thermal anisotropies are known to be unstable to the growth of both electromagnetic and electrostatic waves.

The most familiar example of a loss cone distribution—whence it derives its name—occurs in laboratory mirror machines (HARRIS, 1959). There since the mirror ratio is small, of order two, the loss cone is large, and end losses ensure that the particle distribution contains virtually no particles with small perpendicular velocities $v_\perp$. In other words, $F(v_\perp = 0) = 0$. At the edge of the loss cone, $\partial F/\partial v_\perp > 0$. This positive gradient in the perpendicular velocity distribution can destabilize waves with perpendicular phase velocities $\omega/k_\perp$ lying in the positive gradient region. In space, a strong loss cone property cannot be created by particle losses. The equatorial loss cone is small, and the precipitating particles are observed to be isotropic. However, ASHOUR-ABDALLA and COWLEY (1974) showed by including the effects of magnetic gradient drifts into a calculation of the distribution functions of a plasma convecting into an increasing magnetic field that a region of positive $\partial F/\partial v_\perp$ could be generated in the main part of the velocity distributions of ions or electrons. Thus, a loss cone property should be created in the convective flow of plasma towards the earth.

Our discussion of free energy sources is summarized pictorially in Fig. 3. A

SOURCES OF FREE ENERGY ON AURORAL FIELD LINES

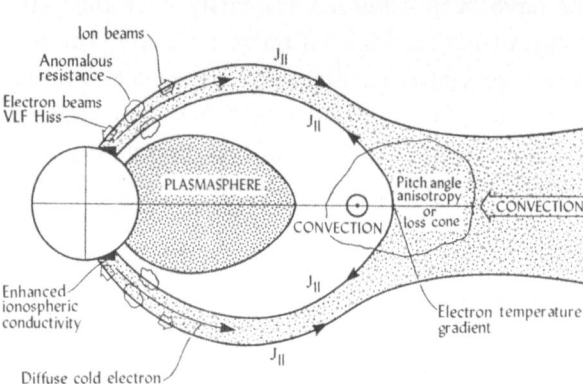

Fig. 3. Sources of free energy on auroral field lines. Sketched here are sources of free energy which emerge conceptually from the convection-precipitation coupling model discussed in the text. In our opinion, pitch angle anisotropy or loss-cone instabilities seem to be the best candidate for explaining diffuse, though not structured, auroral precipitation.

series of heuristic arguments has led us to reject all but thermal anisotropy and a loss cone property as plausible free energy sources for the instabilities causing diffuse auroral precipitation. Both are consequences of flow into an increasing magnetic field, both can be present in the equatorial particle velocity distributions of electrons and ions.

We note that, despite many detailed observations of ion and electron distributions on auroral lines of force, we know of none that have specifically addressed the question of whether the loss cone property exists near the geomagnetic equator. By plotting the pitch angle distribution at constant energy, one can deduce something about thermal anisotropy. However, the quantity that enters instability calculations is the perpendicular velocity distribution at constant parallel velocity; to find the loss cone property, it is necessary that the data be plotted in this way.

4. Instabilities Responsible for the Diffuse Aurora

We have now refined our second question about the diffuse aurora to: "What instabilities capable of violating the particles' first adiabatic invariant use thermal anisotropy or a loss cone property as a free energy source?" To violate the first invariant, the waves should have frequencies near the particles' gyrofrequencies. This leaves only four possible instabilities: those of the electromagnetic and electrostatic ion and electron cyclotron waves. Both electromagnetic and electrostatic cyclotron waves can be unstable to either free energy source.

Electromagnetic ion cyclotron waves (EMIC) and electromagnetic electron cyclotron waves (whistler mode waves) can be unstable when the pitch angle anisotropy of the ions and electrons, respectively, which resonate with them is sufficiently large (KENNEL and PETSCHEK, 1966; CORNWALL, 1966). However, the work of CORNWALL et al. (1970, 1971) made it clear by implication that these waves could not be responsible for the precipitation of 1–10 keV electrons and 1–100 keV protons on auroral lines of force. Beyond the plasmapause, the cold electron and ion density is so low that only higher energy electrons and protons can have cyclotron resonances with electromagnetic cyclotron waves. This is the first appearance of a theme underlying the rest of this paper: the determining influence of cold plasma, not only on which instability will occur, but also on the properties of the instabilities that do occur.

Now we turn to electrostatic cyclotron waves. The electrostatic wave experiment on OGO-5—the first to operate successfully beyond the plasmapause—detected a number of different plasma waves with frequencies exceeding the local electron cyclotron frequency. The dominant class had frequencies between multiples of the cyclotron and has come to be known as 'odd-half harmonic' or, more simply, '3/2' emissions. The facts that electrostatic waves interact with the main part of the particle distribution and not a high energy tail, and that the observed half-harmonic waves were often, or nearly always, present near the equatorial plane on auroral lines of force led the TRW group to suggest that electrostatic electron cyclotron harmonic waves were responsible for the diffuse electron aurora (KENNEL et al., 1970). LYONS (1974) showed by explicit calculation that the observed 1–10 mV/m amplitudes of electrostatic electron cyclotron waves were sufficient to put 1–10 keV electrons on strong diffusion, necessary to account for the observed isotropy of the precipitating electron fluxes.

Much theoretical effort has been devoted to explaining odd half-harmonic emissions. We can only briefly summarize it here. FREDRICKS (1971) first suggested the electron loss cone instability as a candidate explanation of odd half-harmonic emissions. To make his point, he chose a very strong loss cone distribution function with no spread in parallel velocity. YOUNG et al. (1973) were the first to realize the extreme importance of cold electrons to the behavior of this instability. Addition of cold electrons—even with a density, a small fraction of the 1–10 keV hot electron density—enables much gentler loss cone distributions to destabilize the 3/2 emission predominantly observed. The work of KARPMAN and his co-workers (KARPMAN et al., 1973, 1975) continued to emphasize the importance of cold electrons. They found that the cold electron density controls the frequency of the instability. The instability occurs only when the upper hybrid frequency based upon the cold electron density exceeds the electron cyclotron frequency, and then, the unstable frequencies never exceed the cold upper hybrid frequency. More recently ASHOUR-ABDALLA and KENNEL (1975, 1976a, b, 1978) showed that the cold electron temperature controls the spatial growth rate. When the cold electron temperature is sufficiently small, the spatial growth rates are very large, and very gentle loss cone distributions with almost filled loss cones can have appreciable growth rates. Thus, this instability seems capable of operating in strong diffusion conditions. Finally, ASHOUR-ABDALLA and Kennel (1977) argued that non-resonant quasilinear heating of the cold electrons plays a role equally as important as hot electron precipitation in reducing the spatial growth rates to the small values consistent with the maintenance of a steady state turbulence level.

Having given a general outline of our understanding of diffuse auroral electron precipitation, we now turn to the waves which are the best candidate at present for explaining diffuse auroral proton precipitation. In this case, theory preceded observation. Coroniti et al. (1972) suggested that a quasi-electrostatic mode could be excited at frequencies near but below the ion plasma frequency by an ion loss-cone distribution. In retrospect, this work suffered from the same defect as the equally pioneering work of Fredricks (1971) on the electron electrostatic cyclotron waves: it used too strong-loss cone distribution to be entirely realistic for the magnetosphere, especially during quiet times. Armed with previous experience on electron waves, Ashour-Abdalla and Thorne (1977) undertook a refined calculation of spatial and temporal growth rates, using a gentle ion loss cone distribution modeled upon observations and including cold electrons and ions. Just as their preliminary results appeared in print, so also did the Hawkeye observation of Gurnett and Frank (1977).

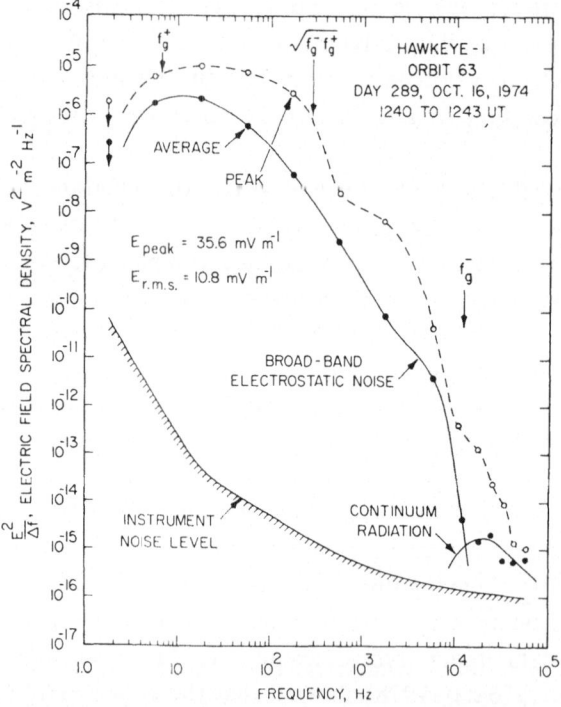

Fig. 4.   Broadband electrostatic noise on auroral lines of force. Reproduced here from Gurnett and Frank (1977) is a spectrum of broadband electrostatic noise, which peaks at several times the proton cyclotron frequency. This noise was typically observed at intermediate latitudes on auroral lines of force.

In Fig. 4, we reproduce a spectrum of what GURNETT and FRANK (1977) call broadband electrostatic noise. We note the following features:

1) The spectrum peaks at two to three times the ion cyclotron frequency.

2) The ion waves are observed on auroral field lines, consistent with the observed location of diffuse proton precipitation.

3) The observed amplitudes are sufficient to put protons on strong diffusion.

The fact that Hawkeye could only observe the broadband electrostatic noise at intermediate latitudes ($\lambda \sim 45°$) makes theoretical arguments concerning their free energy source somewhat ambiguous. We could try to extend primarily ionospheric free energy sources (currents, beams) outwards, or we could extend equatorial sources (loss cone) to intermediate latitudes. Hopefully, this ambiguity will be cleared up by future observations, especially near the equatorial plane. We also note that electron cyclotron harmonic waves were not present at intermediate latitudes, which seems *prima facie* inconsistent with the observed simultaneous precipitation of electrons and protons. However, there is some indication that electrostatic electron cyclotron waves may be confined near the equatorial plane (KENNEL *et al.*, 1970) which needs to be clarified by further observation (FREDRICKS and SCARF, 1973).

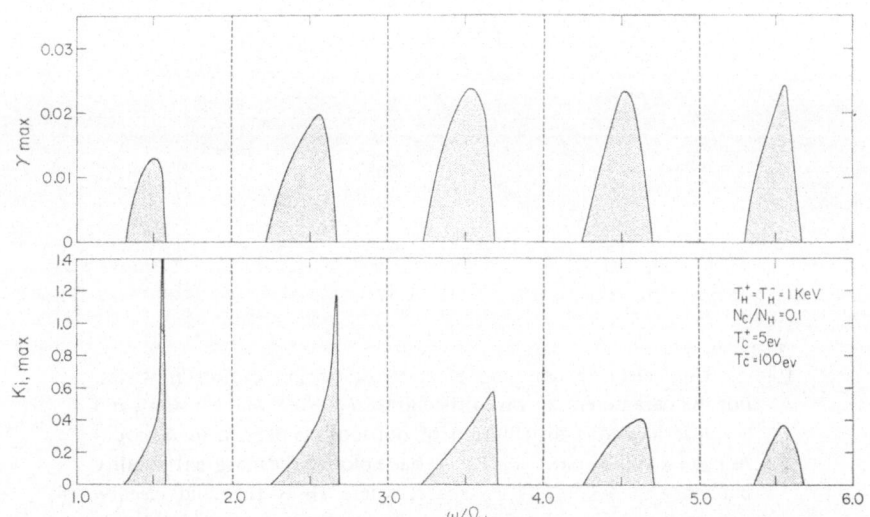

Fig. 5. Temporal and spatial growth rate of the ion loss-cone instability. Reproduced here from ASHOUR-ABDALLA and THORNE (1977) is the temporal growth rate (top panels) and the spatial growth rate (bottom panels) plotted against frequency for the first few ion harmonics. The parameters chosen correspond to $L=10$ at the geomagnetic equator. N_C and N_H are the cold and hot ion and electron densities. T_H^+, T_H^- are the hot ion and electron temperature. To have instability at all, it is necessary to choose $T_C^- \simeq 100$ eV when $N_C/N_H \simeq 0.1$. To have large spatial growth rates given instability, it is necessary that the cold ions remain cold (in this case 5 eV). For these conditions, the spatial growth rate favors the first few ion harmonics, consistent with observation (GURNETT and FRANK, 1977).

On the hypothesis that electrostatic ion cyclotron waves are responsible for diffuse auroral proton precipitation, we choose to consider them from the point of view of an equatorial free energy source. This requires that the broadband electrostatic noise extend from midlatitudes through the equatorial plane, so that it can interact with a large portion of the auroral proton population. With these preliminaries, we now return to the calculations of Ashour-Abdalla and Thorne (1977). They found that the cold ion temperature controls the spatial growth rate of the ion harmonic modes, in complete analogy with the electron modes. Moreover, as is shown in Fig. 5, the spatial growth rate peaks at low ion harmonics, in agreement with the above observations, whereas the temporal growth rate would peak at much higher frequencies, consistent with the original temporal growth rate calculation of Coroniti et al. (1972). Since in a spatially inhomogeneous system in steady state, one must use the spatial growth rate, we conclude that the electrostatic ion cyclotron harmonic loss cone instability can account for the observed spectrum, provided that

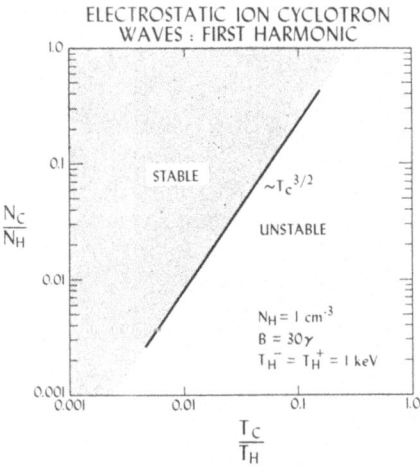

Fig. 6. Cold electron damping of electrostatic ion cyclotron waves. For the parameters shown in the figure, $T_C^+ = 5$ eV and constant, and a gentle loss-cone distribution of hot ions (as defined in Ashour-Abdalla and Thorne (1977)), we have plotted the marginal stability boundary in N_C/N_H, T_C/T_H space, where T_C is the cold *electron* temperature. This plot illustrates the importance of cold electron Landau damping. The $T_C^{3/2}$ dependence may be understood as follows. A critical number of electrons with speeds near the ion wave phase speed, roughly the hot ion thermal speed, can damp the wave. For larger T_C, this critical number comprises a smaller fraction of the cold electron distribution, and so a larger total N_C is required to damp the wave. Changing the strength of the hot ion loss-cone distribution can move this marginal stability boundary up and down, but will not change our general conclusion. Since we believe N_C/N_H cannot be too small on auroral field lines (N_C comes from precipitation neutralization) T_C must be of the oder of 100 eV to permit the ion loss-cone instability to grow.

nonlinear effects do not selectively suppress lower harmonic modes (which seems unlikely).

Whereas ions play no role in the electrostatic electron cyclotron instability, cold electrons determine whether or not the ion cyclotron harmonic waves can be unstable at all. Even a few electrons with speeds near the hot ion thermal speed can suppress the ion harmonic instability by their Landau damping. In Fig. 6 we plot the marginal stability boundary in a $(N_C/N_H, T_C/T_H)$ space where N_C and N_H are the cold electron and ion cold and hot densities, T_C is the cold electron temperature and T_H is the hot ion mean energy. The ions had a loss-cone distribution similar to those used in ASHOUR-ABDALLA and THORNE (1977). For the parameters chosen, a fractional density $N_C/N_H \sim 10^{-3}$ of $T_C = 1$ eV electrons will suppress the instability. Since $0.1 \lesssim N_C/N_H \leq 1$ on auroral field lines, ion loss-cone instability is possible only if the cold electron temperature is of the order of 100 eV. That is why we chose 100 eV in the preceding plot (Fig. 5).

Now let's go through the logic. Broadband electrostatic ion harmonic waves exist. Our hypothesis of an equatorial loss-cone free energy source thus stands or falls on the question of the cold electron temperature deep in space on auroral field lines. We have no experimental information on this crucial parameter. We doubt that the electrons will be as cold as they are in the ionosphere ($\simeq 1$ eV), for several reasons:

1)   Precipitation neutralization could actually be carried out by backscattered secondary electrons with a few hundred eV electrons.

2)   Let us suppose that precipitation neutralization can actually be carried out by 1 eV particles initially. It is difficult to imagine that these electrons arrive at the equator with 1 eV after passing through the highly turbulent high latitude regions of the auroral lines of force.

3)   Should the electrons manage to arrive at the equatorial plane with 1 eV energy, they would induce strong spatial growth rates for the electron cyclotron harmonic waves, which we can infer are present from the simultaneous observation of electron and proton diffuse aurora. A 1 eV electron can be heated to 100 eV in about 100 sec by 1 mV/m electron waves (ASHOUR-ABDALLA and KENNEL, 1978).

5.   Discussion

We have highlighted attempts over the years to construct a self-consistent conceptual model of the interaction of the convecting plasma sheet and the night-side auroral ionosphere which includes the plasma turbulence responsible for particle precipitation. Up to now, this set of ideas has started with the density and temperature of the plasma sheet and a given driving convection electric field and then involves:

1)   Particle precipitation fluxes and charge neutralization return fluxes
2)   Inhomogeneous ionospheric conductivity profiles

3)   Action of finite ionospheric conductivity and plasma pressure on the convective flow pattern

4)   Locating field-aligned currents (exclusive of those in smaller scale structures such as arcs)

5)   Implicitly, through field-aligned currents, current instabilities, parallel electric fields, and the generation and instability of fast ion and electron streams

6)   Providing sources of free energy for the instabilities responsible for particle precipitation

7)   Estimating self-consistently the plasma environment in which instabilities grow and saturate

A sketch of the elements of such a conceptual model is shown in Fig. 7. For every box in Fig. 7 there has been considerable theoretical and experimental effort over the past decade, but a complete linkage has to be made. In this paper, we have concentrated upon the right-hand side of this sketch, where the plasma turbulence fits in.

Clearly, more experimental and theoretical work on all elements of this conceptual model would be valuable. In the case of the electrostatic ion and electron cyclotron harmonic instabilities, which in our opinion are the best candidates for explaining diffuse auroral precipitation, we may list two important experimental objectives:

a)   Determination of the cold electron temperature, and better yet, the distribution function on auroral field lines. We have concluded that further theoretical

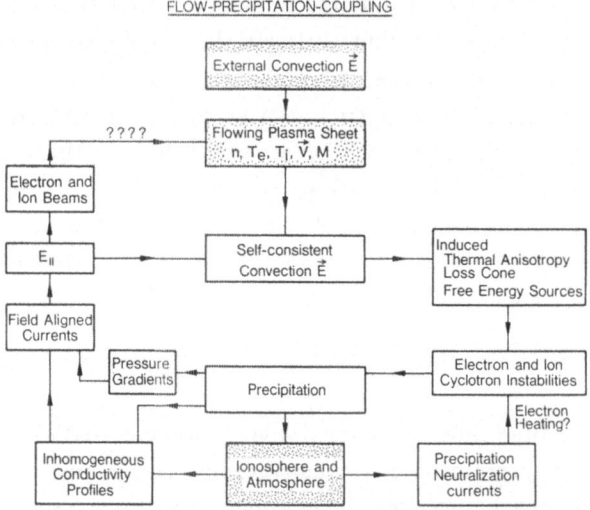

Fig. 7.   Elements of self-consistent models. Coupling convection, precipitation, and the ionosphere. Here is our version of the flow diagram relating the elements of these models. We believe important questions concerning electrons and ions of ionospheric origin flowing into the magnetosphere must be answered experimentally before theory can complete the linkages between elements sketched above.

IONOSPHERIC SOURCES OF AURORAL FIELD LINE PLASMA

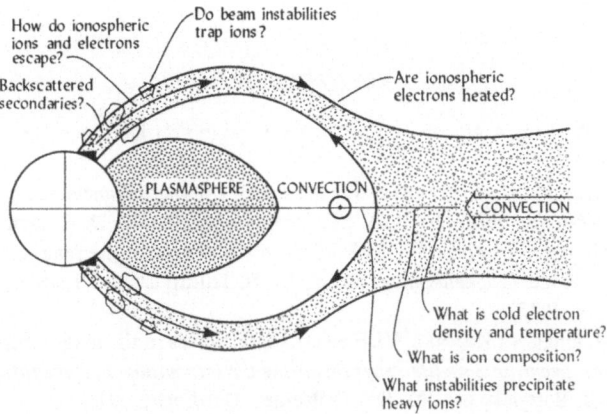

Fig. 8.   Ionospheric sources of auroral field line plasma. Sketched here are some
of the ionospheric sources discussed in the text and some of the questions
associated with them.

progress on cyclotron harmonic instabilities will be limited until we know the
electron temperature.

  b)   A complete phenomenological understanding of ion harmonic modes. Are
they found near the equator, or on the dayside where the cold electron density should
be higher? The situation with electron cyclotron harmonic instabilities is somewhat
better, but still needs improvement.
We may also list two theoretical objectives:
  a)   Modelling the cold electron distribution on auroral field lines.
  b)   Improving our understanding of nonlinear saturation of the instabilities.
In particular, since the wave energy density can exceed the cold plasma energy
density, it is unlikely that an originally locally smooth cold plasma density profile
can remain so.

  One final comment for the future. Up to now, people have generally assumed
that the ionospheric plasma source was a passive partner in the chain of interaction.
We have already seen that this is not so. Cold electrons of ionospheric origin control the properties of electrostatic cyclotron instabilities. Clearly, we have under-
estimated the importance of the ionospheric source. We may not even know the
ion composition on auroral field lines. Heavy ions have been observed streaming
out of the auroral ionosphere and have also been observed in the geomagnetic tail
and in the ring current at disturbed times. How are these ions accelerated? Do
beam instabilities scatter them onto trapped orbits? How are these ions lost? Do
they convect to the magnetopause, are they charge exchanged, do they have instabilities which can reprecipitate them to the ionosphere? These questions are sketched
in Fig. 8. Questions such as these we believe are on the frontier of the problem of
the coupling of the plasma sheet to the auroral ionosphere.

It is a pleasure to acknowledge many fruitful conversations with F.V. Coroniti and R.M. Thorne. This work was supported by National Science Foundation grant ATM 76-13792 and NASA grant NGL 05-007-190.

REFERENCES

Anger, C.D. and A.T.Y. Lui, A global view at the polar region on December 18, 1971, *Planet. Space Sci.*, **21**, 873-878, 1973.

Ashour-Abdalla, M. and S.W.H. Cowley, Wave particle interactions near the geostationary orbit, in *Magnetospheric Physics*, edited by B.M. McCormac, pp. 241-270, D. Reidel, Dordrecht, 1974.

Ashour-Abdalla, M. and C.F. Kennel, VLF electrostatic waves in the magnetosphere, in *Physics of the Hot Plasma in the Magnetosphere*, edited by B. Hultqvist and L. Stenflow, pp. 201-227, Plenum, New York, 1975.

Ashour-Abdalla, M. and C.F. Kennel, VLF electrostatic waves in the magnetospheres of the Earth and Jupiter, in *The Scientific Satellite Program during the International Magnetospheric Study*, edited by K. Knott and B. Bottrick, pp. 303-325, D. Reidel, Dordrecht, 1976a.

Ashour-Abdalla, M. and C.F. Kennel, Convective cold upper hybrid instabilities, in *Magnetospheric Particles and Fields*, edited by B.M. McCormac, pp. 181-196, D. Reidel, Dordrecht, 1976b.

Ashour-Abdalla, M. and C.F. Kennel, Nonconvective and convective electron cyclotron harmonic instabilities, *J. Geophys, Res.*, 1531-1543, 1978.

Ashour-Abdalla, M. and Thorne, R.M., The importance of electrostatic ion-cyclotron instability for quiet-time proton auroral precipitation, *Geophys. Res. Lett.*, **4**, 45-48, 1977.

Bernstein, W., B. Hultqvist, and H. Borg, Some implications of low altitude observations os isotropic precipitation of ring current protons beyond the plasma pause, *Planet. Space Sci.*, **22**, 767-776, 1974.

Chance, M.S., F.V. Coroniti, and C.F. Kennel, Finite-β drift Alfven instability, *J. Geophys. Res.*, **78**, 7521-7530, 1973.

Cornwall, J.M., Micropulsations and the outer radiation zone, *J. Geophys. Res.*, **71**, 2185-2199, 1966.

Cornwall, J.M., F.V. Coroniti, and R.M. Thorne, Turbulent loss of ring current protons, *J. Geophys. Res.*, **75**, 4699-4709, 1970.

Cornwall, J.M., F.V. Coroniti, and R.M. Thorne, Unified theory of SAR arc formation at the plasmapause boundary, *J. Geophys. Res.*, **76**, 4428-4445, 1971.

Coroniti, F.V., R.W. Fredericks, and R. White, Instability of ring current protons beyond the plasmapause during injection events, *J. Geophys. Res.*, **77**, 6243-6248, 1972.

Coroniti, F.V. and C.F. Kennel, Electron precipitation pulsations, *J. Geophys. Res.*, **75**, 1279-1289, 1970a.

Coroniti, F.V. and C.F. Kennel, Auroral micropulsation instability, *J. Geophys. Res.*, **75**, 1863-1878, 1970b.

Coroniti, F.V. and C.F. Kennel, Polarization of the auroral electroject, *J. Geophys. Res.*, **77**, 2835-2850, 1972.

Eather, R.H. and S.B. Mende, Systematic auroral energy spectra, *J. Geophys. Res.*, **77**, 660-673, 1972a.

Eather, R.H. and S.B. Mende, High latitude particle precipitation and source regions in the magnetosphere, in *Magnetosphere-Ionosphere Interactions*, edited by K.V. Folkestad, pp. 139-154, Universitets Forlaget, Oslo, 1972b.

Eather, R.H., S.B. Mende, and R.J.R. Judge, Plasma injection at synchronous orbit and spacial and temporal auroral morphology, *J. Geophys. Res.*, **81**, 2805-2824, 1976.

Fredricks, R.W., Plasma instability at $(n+1/2)f_{ce}$ and its relationship to some satellite observations, *J. Geophys. Res.*, **76**, 5344-5348, 1971.

Fredricks, R.W. and F.L. Scarf, Recent studies of magnetospheric electric field emission above the electron gyrofrequency, *J. Geophys. Res.*, **78**, 310-314, 1973.

Fukunishi, H., Dynamic relationship between proton and electron auroral substorms, *J. Geophys. Res.*, **80**, 553-574, 1975.

GURNETT, D.A. and L.A. FRANK, A region of intense plasma wave turbulence on auroral field lines, *J. Geophys. Res.*, **82**, 1031–1050, 1977.

HARRIS, E.G., Unstable plasma oscillations in a magnetic field, *Phys. Rev. Lett.*, **2**, 34–36, 1959.

HULTQVIST, B., The aurora, in *The Magnetospheres of the Earth and Jupiter*, edited by V. Formisano, pp. 77–111, D. Reidel Publ. Co., Dordrecht, Holland, 1975.

HULTQVIST, B., H. BORG, P. CHRISTOPHERSEN, W. RIEDLER, and W. BERNSTEIN, Energetic protons in the keV energy range and associated keV electrons observed at various local times and disturbance levels in the upper ionosphere, NOAA Technical Report, ERL 305-SEL29, January 1974.

JAGGI, R.K. and R.A. WOLF, Self-consistent calculation of the motion of a sheet of ions in the magnetosphere, *J. Geophys. Res.*, **78**, 2852–2866, 1973.

KARPMAN, V.I., Ju. K. ALEKHIN, N.D. BORISOV, and N.A. RJABOVA, Electrostatic waves with frequencies exceeding the gyrofrequency in the magnetosphere, *Astrophys. Space Sci.*, **22**, 267–278, 1973.

KARPMAN, V.I., Ju. K. ALEKHIN, N.D. BORISOV, and N.A. RJABOVA, Electrostatic electron cyclotron waves in plasma with a loss cone distribution, *Plasma Phys.*, **17**, 361–372, 1975.

KENNEL, C.F. Consequences of a magnetospheric plasma, *Rev. Geophys.*, **7**, 379–419, 1969.

KENNEL, C.F. and H.E. PETSCHEK, Limit on stably trapped particle fluxes, *J. Geophys. Res.*, **71**, 1–28, 1966.

KENNEL, C.F., F.L. SCARF, R.W. FREDRICKS, J.G. McGEHEE, and F.V. CORONITI, VLF electric field observations in the magnetosphere, *J. Geophys. Res.*, **75**, 6136–6152, 1970.

KINDEL, J. and C.F. KENNEL, Topside current instabilities, *J. Geophys. Res.*, **76**, 3055–3078, 1971.

LUI, A.T.Y. and C.D. ANGER, A uniform belt of diffuse auroral emission seen by the Isis-2 scanning photometer, *Planet. Space Sci.*, **21**, 799–809, 1973.

LUI, A.T.Y., P. PERREAULT, S.I. AKASOFU, and C.D. ANGER, The diffuse aurora, *Planet. Space Sci.*, **21**, 857–861, 1973.

LYONS, L.R., Electron diffusion driven by magnetospheric electrostatic waves, *J. Geophys. Res.*, **79**, 575–580, 1974.

MENDE, S.B. and R.H. EATHER, Monochromatic all-sky observations and auroral precipitation patterns, *J. Geophys. Res.*, **81**, 3771–3780, 1976.

PETSCHEK, H.E. and C.F. KENNEL, Tail flow, auroral precipitation, and ring currents, *Trans. Am. Geophys. Union*, **47**, 137–138, 1966.

VASYLIUNAS, V.M., A survey of low energy electrons in the evening sector at the magnetosphere with OGO-1 and OGO-3, *J. Geophys. Res.*, **73**, 2839–2884, 1968.

VASYLIUNAS, V.M., Unpublished work described in KENNEL, 1969.

YOUNG, T.S.T., J.D. CALLEN, and J.E. McCUNE, High frequency electrostatic waves in the magnetosphere, *J. Geophys. Res.*, **78**, 1082–1099, 1973.

Electromagnetic Plasma Wave Emissions from the Auroral Field Lines

Donald A. GURNETT

*Department of Physics and Astronomy, The University of Iowa,
Iowa City, Iowa, U.S.A.*

(Received October 17, 1977)

Several types of electromagnetic waves are known to be emitted by charged particles on the auroral field lines. In this paper we review the most important types of auroral radio emissions, both from a historical perspective as well as considering the latest results. Particular emphasis is placed on four types of electromagnetic emissions which are directly associated with the plasma on the auroral field lines. These emissions are (1) auroral hiss, (2) saucers, (3) ELF noise bands, and (4) auroral kilometric radiation. Ray tracing and radio direction finding measurements indicate that both the auroral hiss and auroral kilometric radiation are generated along the auroral field lines relatively close to the earth, at radial distances from about 2.5 to 5 R_E, probably in direct association with the acceleration of auroral particles by parallel electric fields. The exact mechanism by which these radio emissions are generated has not been firmly established. For the auroral hiss the favored mechanism appears to be amplified Cerenkov radiation. For the auroral kilometric radiation several mechanisms have been proposed, usually involving the intermediate generation of electrostatic waves by the precipitating electrons.

1. Introduction

For many years it has been known that certain types of electromagnetic emissions from the earth's magnetosphere are closely associated with the occurrence of auroras. As early as 1933, BURTON and BOARDMAN (1933) reported observations of bursts of very-low-frequency (VLF) 'static' which were closely correlated with flashes of auroral light. Later investigations using ground based VLF radio receivers firmly established that auroral disturbances at high latitudes are often accompanied by intense bursts of broad-band radio noise at frequencies from a few hundred Hz to over 100 kHz (ELLIS, 1957; DUNCAN and ELLIS, 1959; DOWDEN, 1959; MARTIN *et al.*, 1960; JØRGENSEN and UNGSTRUP, 1962; MOROZUMI, 1963; HARANG and LARSEN, 1964). Because of the close association of these radio emissions with aurora and their broad bandwidth these emissions came to be known as auroral hiss, following the classification scheme of HELLIWELL (1965). The first satellite observations of auroral hiss were reported by GURNETT (1966) who showed that auroral hiss is closely correlated with intense fluxes of precipitating electrons with energies less than 10 keV. Subsequent studies have firmly established that auroral hiss is generated along the auroral field lines by intense fluxes of electrons precipitating into the ionosphere with energies in the range from a few hundred eV to several keV (HARTZ, 1970; GURNETT and FRANK, 1972a; HOFFMAN and LAASPERE, 1972). Since the auroral hiss emissions occur at frequencies below the local electron gyrofrequency these waves must be propagating in the whistler mode. Simple ray tracing considerations show that the auroral hiss appears to be propagating downward from a source at an altitude of about 5,000 to 10,000 km. Poynting flux measurements by MOSIER and GURNETT (1969) showed that another type of emission, called a saucer, also occurs on the auroral field lines, propagating upward from a source at altitudes of approximately 1,400 km. These VLF saucer emissions have been studied in greater detail by JAMES (1976). Other electromagnetic emissions have been observed along the auroral field lines at even lower frequencies, from 100 to 300 Hz, by GURNETT and FRANK (1972b). These emissions, which are called ELF noise bands, have a very narrow bandwidth and are also closely associated with the auroral electron precipitation.

Since the auroral hiss, saucers, and ELF noise bands consist of internally trapped plasma wave modes these emissions cannot escape from the earth's magnetosphere. At higher frequencies, above the local characteristic frequencies of the plasma, radio emissions of auroral origin have been observed escaping from the earth. Since the ionosphere effectively blocks all radiation at frequencies below the local electron plasma frequency from reaching the earth's surface, escaping electromagnetic emissions of this type can only be observed by a satellite above the ionosphere. The first evidence of intense auroral-related radio emissions escaping from the earth's magnetosphere was obtained from the Elektron 2 and 4 satellites by BENEDIKTOV et al. (1965, 1968). These observations showed that bursts of radio noise at 725 kHz and 2.3 MHz were originating from the earth in close association with geomagnetic storms. Later DUNCKEL et al. (1970) reported similar bursts of radio noise, also associated with high-latitude magnetic disturbances, at frequencies below 100 kHz. The first complete determination of the spectrum of these radio emissions was provided by the IMP-6 spacecraft which showed that the maximum intensities are in the frequency range from about 100 to 500 kHz (STONE, 1973; BROWN, 1973) and that the total power radiated from the earth is sometimes as large as 10^9 watts (GURNETT, 1974), comparable to the decametric (3.0 to 30 MHz) radio emission from Jupiter. It was also determined that the intense radio bursts from the earth are directly associated with the occurrence of discrete auroral arcs and that the angular distribution of the escaping radiation is consistent with generation along the auroral field lines on the night side of the earth. Because of the close association of these radio emissions with auroral processes and the kilometer wavelength of the radiation, KURTH et al. (1975) have referred to this radiation as auroral kilometric radiation, which is the terminology that will be used in this paper. Other names for this radiation include 'high pass noise' (DUNCKEL et al., 1970), 'mid-frequency noise' (BROWN, 1973), and 'terrestrial kilometric radiation' (GURNETT, 1974; ALEXANDER and KAISER, 1976).

These four types of electromagnetic emissions (auroral hiss, saucers, ELF noise bands, and auroral kilometric radiation) constitute all of the known electromagnetic plasma wave emissions from the auroral field lines. The purpose of this paper is to review the present state of knowledge of these electromagnetic plasma wave emissions and to discuss the mechanisms by which these waves are thought to be generated.

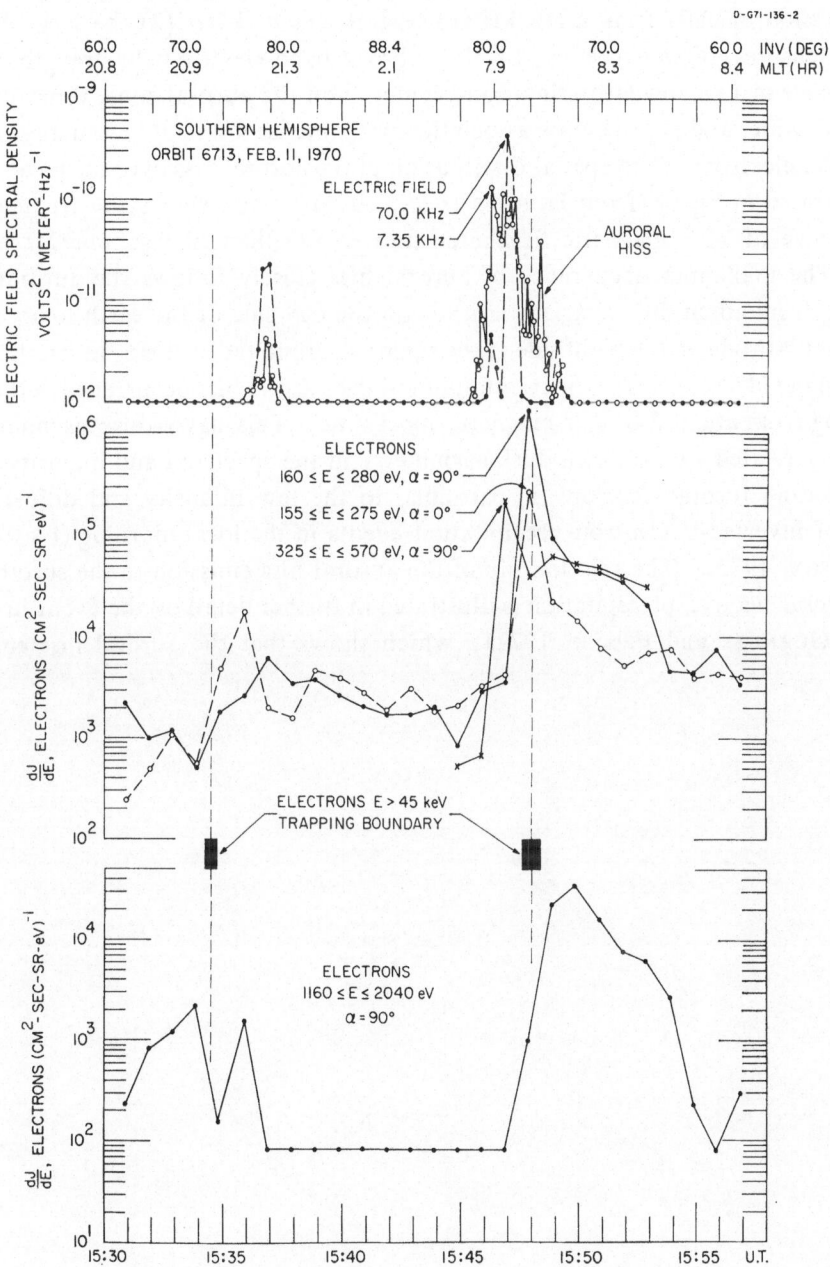

Fig. 1. An example of an intense broad-band auroral hiss emission associated with intense fluxes of low energy, a few hundred eV, auroral electrons. The auroral hiss is indicated by the enhanced 7.35 and 70.0 kHz electric field intensities from about 1546 to 1549 UT, slightly poleward of the $E > 45$ keV electron trapping boundary.

2. Auroral Hiss, Saucers and ELF Noise Bands

A typical example of an auroral hiss event detected by a low-altitude polar-orbiting satellite is shown in Fig. 1. This event illustrates the primary identifying characteristics of auroral hiss, consisting of (1) the very broad frequency range of the emission, usually from a few kHz to several tens of kHz, (2) the large electric field intensities, often exceeding $1\,\mathrm{mV\,m^{-1}}$ broad band electric field strength, (3) the occurrence in a narrow latitudinal band centered on the auroral zone, typically only 5 to 10° wide, and (4) the close association with intense fluxes of low energy, 100 eV to 1 keV, electrons. The spatial distribution of the auroral hiss over the polar region is illustrated in Fig. 2 (from HUGHES *et al.*, 1971), which shows the frequency of occurrence of VLF magnetic field intensities greater than $10^{-12}\,\mathrm{gamma^2\,Hz^{-1}}$ at 9.6 kHz. The maximum occurrence of auroral hiss closely follows the auroral oval, varying from about 80° invariant latitude on the day side of the earth to about 72° invariant latitude on the night side of the earth. A pronounced dawn-dusk asymmetry is clearly evident, with a distinct minimum in the auroral hiss occurrence in the local morning from about 2 to 8 hr magnetic local time. This dawn-dusk asymmetry is probably related to the dawn-dusk asymmetry in the spectrum and intensity of the precipitating auroral electrons, in particular to the low intensity and diffuse character of inverted-V electron precipitation events in the local morning (FRANK and ACKERSON, 1972). The relationship of the auroral hiss emission to the spectrum of the auroral electron precipitation is illustrated in further detail by the event in Fig. 3 (from GURNETT and FRANK, 1972a), which shows that the auroral hiss emission

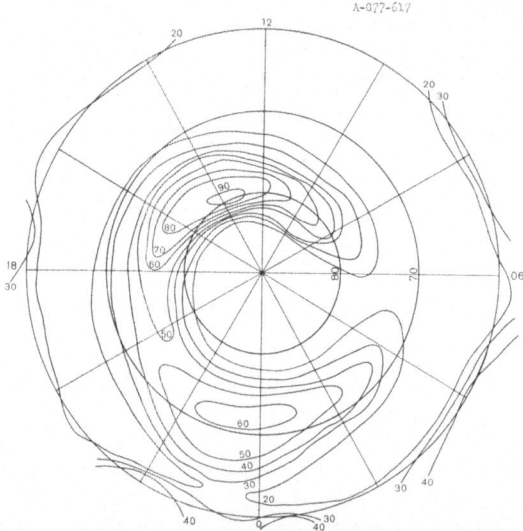

Fig. 2. The frequency of occurrence distribution of auroral hiss at 9.6 kHz as a function of invariant latitude and magnetic local time. Only events with magnetic field intensities greater than $10^{-12}\,\mathrm{gamma^2\,Hz^{-1}}$ are counted (from HUGHES *et al.*, 1971).

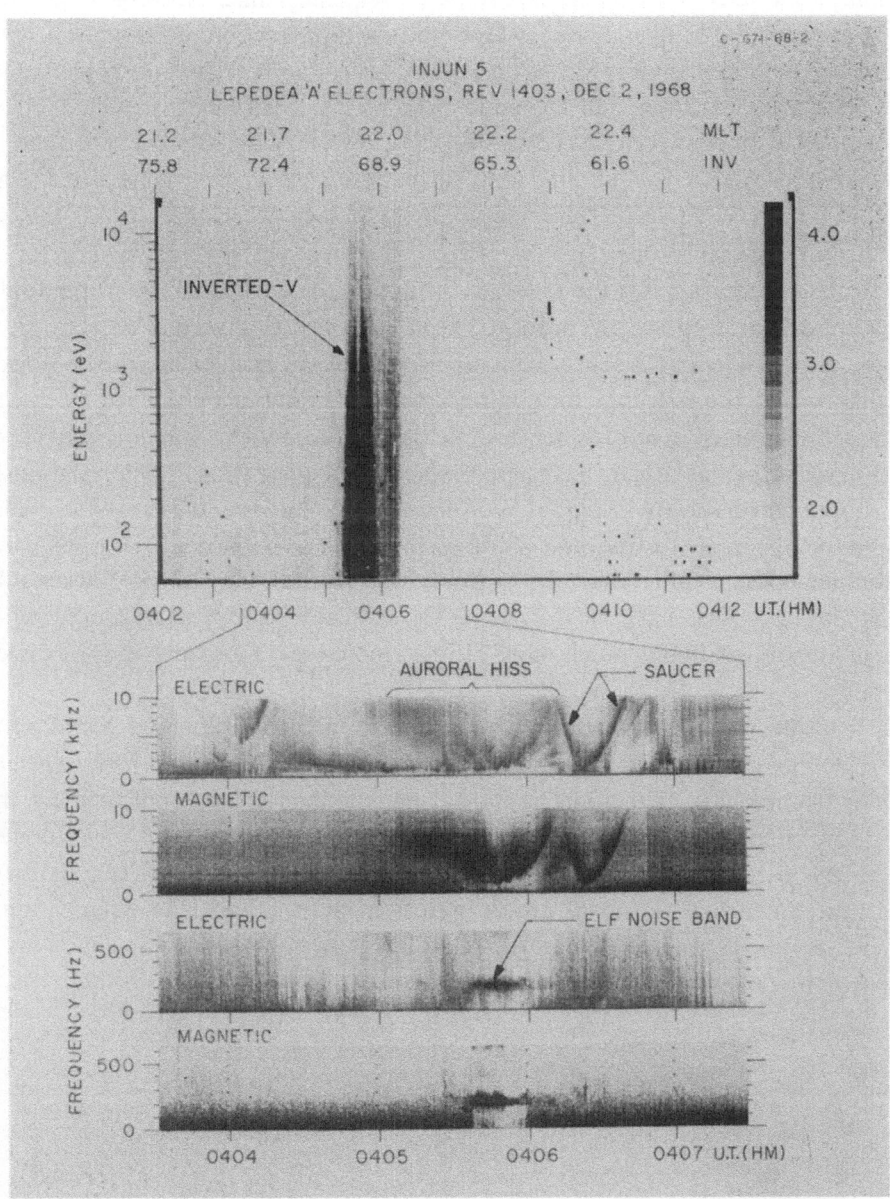

Fig. 3. High resolution frequency-time spectrograms of electric and magnetic fields detected
by the Injun 5 spacecraft at an altitude of about 2,500 km showing the occurrence of
auroral hiss, saucer emissions and an ELF noise band in close association with an
inverted-V electron precipitation event (from GURNETT and FRANK, 1972a).

occurs in direct association with an intense inverted-V electron precipitation event
of the type first discussed by FRANK and ACKERSON (1971). A VLF saucer emission
is also evident near the low-latitude boundary of the electron precipitation region.
Both the auroral hiss and the saucer emissions are characterized by a V-shaped

frequency-time structure. This characteristic frequency-time variation is a spatial effect caused by the frequency-dependent limiting ray direction of the whistler mode at large wave normal angles (MOSIER and GURNETT, 1969). Essentially the source illuminates a region along the magnetic field with a beam-width which increases with increasing frequency. As the spacecraft approaches the auroral field lines the highest frequencies are encountered first, since these rays can propagate at the largest angle to the magnetic field. The essential distinction between the auroral hiss and the saucer emissions is the direction of propagation, which is downward for the auroral hiss and upward for the saucers. Although the particles responsible for the downward propagating auroral hiss have been identified as inverted-V electrons, the particles responsible for the saucer emissions have not yet been established. It seems most likely that the upward propagating saucer emissions are produced by upward streaming ionospheric electrons of very low energy, $\sim$ few eV, which constitute the return current for the nearby inverted-V electron precipitation. This relationship is illustrated schematically in Fig. 4. For observing altitudes in the range from about 1,500 to 3,000 km the latitudinal width of the saucer emissions, 10 to 100 km, is usually somewhat smaller than the width of the auroral hiss, 100 to 500 km. The saucers also have very sharp spectral structure, indicating a very small source region, whereas the auroral hiss is much more diffuse, indicating a broader more extended source.

The event in Fig. 3 also illustrates the occurrence of ELF noise bands in the same region as the inverted-V electron precipitation and the auroral hiss. The electromagnetic character of these narrow band emissions is clearly indicated by their detection with both the electric and magnetic antennas. Poynting flux measurements indicate the presence of both upgoing and downgoing components. These emissions occur near but below the local proton gyrofrequency. Since the polarization of these waves has never been measured it is not known whether these emissions are propagating in the whistler mode or the ion cyclotron mode.

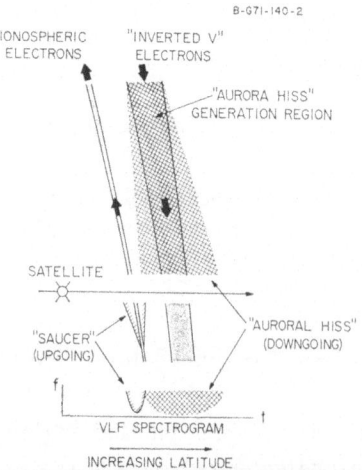

Fig. 4. A schematic illustration showing the spatial relationships between the downward propagating auroral hiss and the upward propagating saucer emissions for events such as in Fig. 3. The association of the saucer emissions with upward streaming ionospheric electrons has not yet been confirmed.

3.   Auroral Kilometric Radiation

The intense radio emissions escaping outward from the earth's auroral regions at frequencies above the local electron plasma frequency are characterized by a very intense peak in the frequency spectrum from about 100 to 500 kHz. Figure 5 shows the median power flux spectrums of this radiation at various local times around the earth, as measured from the IMP-6 spacecraft at radial distances greater than 25 R_E (KAISER and ALEXANDER, 1977). The intensity of this noise is highly variable and is closely correlated with the auroral electrojet index, AE. During geomagnetically quiet times the radiation intensity at 25 R_E is often completely below the galactic background, whereas at other times the intensity can be as much as six to eight orders of magnitude above the galactic background. Power fluxes as large as 10^{-14} watts $m^{-2} Hz^{-1}$ have been observed at 30 R_E, with even larger intensities closer to the earth (GURNETT, 1974). The occurrence of intense bursts of kilometric radiation is closely associated with the occurrence of auroral arcs. This association is illustrated in Fig. 6, which shows the occurrence of intense bursts of kilometric radiation at 178 kHz during periods (passes 1094 and 1096) when discrete auroral arcs are present

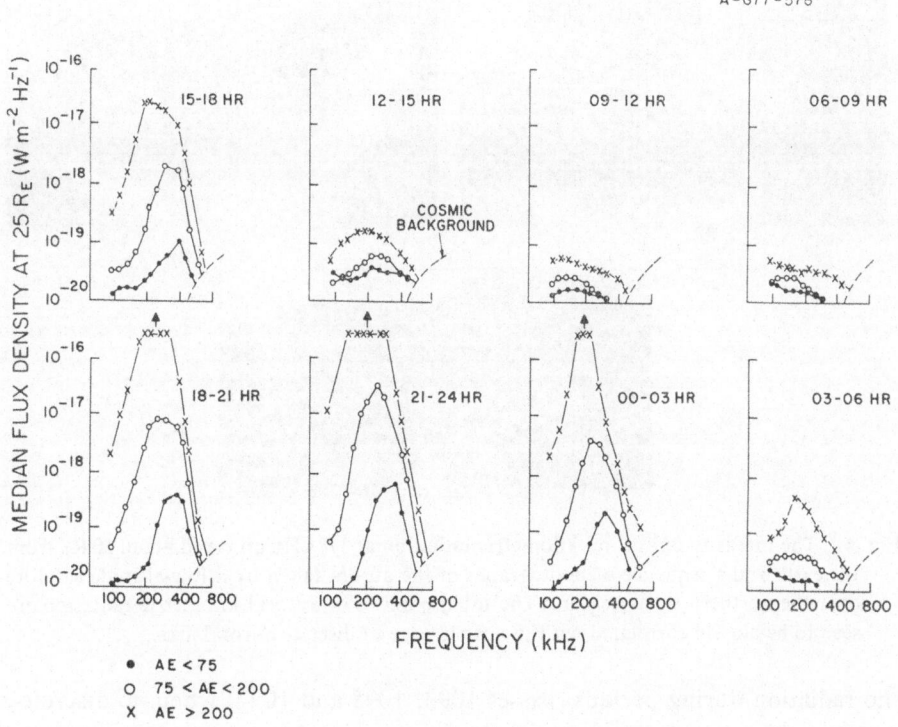

Fig. 5.   Median spectrums of auroral kilometric radiation observed at various local times around the earth by the IMP-6 spacecraft at radial distances $R > 25 R_E$. The three spectrums in each plot are for various ranges of the auroral electrojet, AE, index. The kilometric radio emissions from the earth are closely correlated with auroral zone currents as indicated by the AE index (from KAISER and ALEXANDER, 1977).

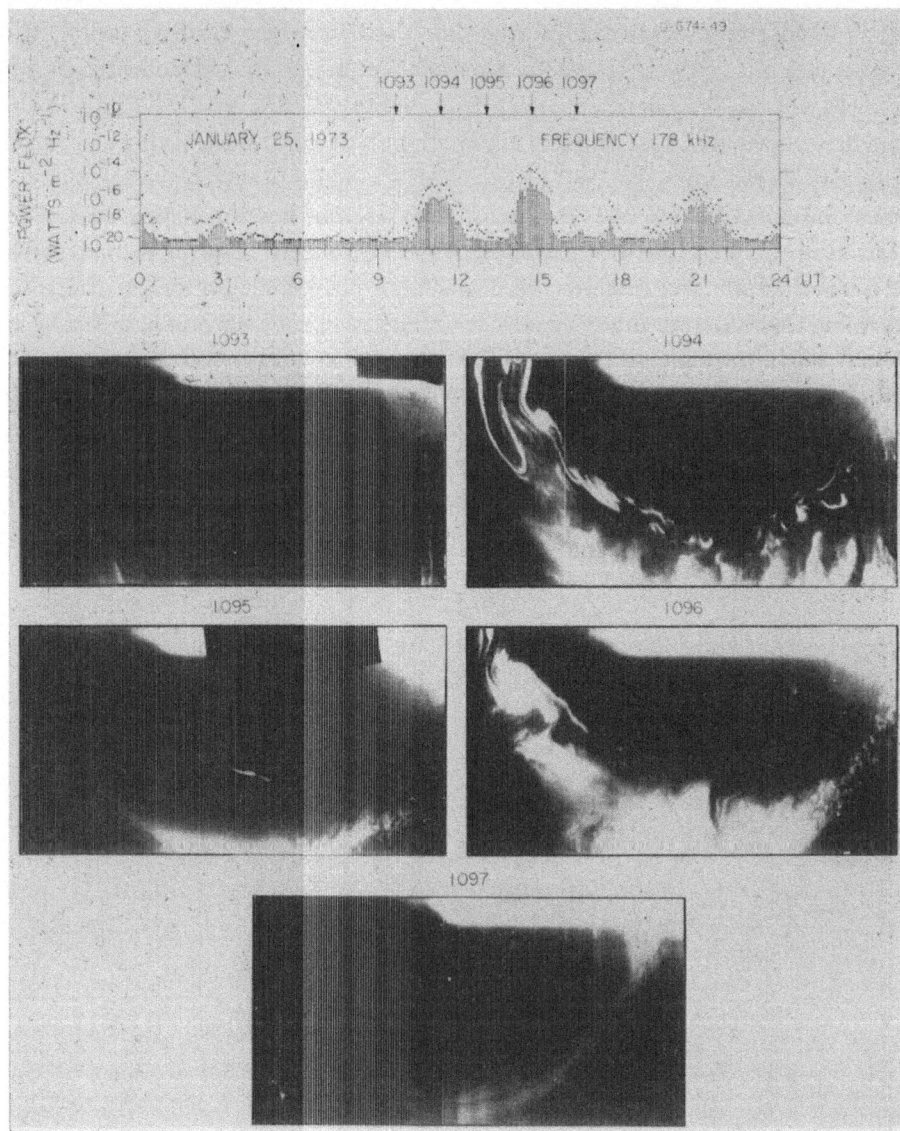

Fig. 6. The intensity of auroral kilometric radiation at 178 kHz observed about 30 R_E from the earth and a sequence of photographs of the aurora taken by a low-altitude satellite over the northern polar region. The intense bursts of auroral kilometric radiation are seen to be closely correlated with the occurrence of discrete auroral arcs.

and no radiation during periods (passes 1093, 1095 and 1097) when no discrete arcs are present. The association with discrete arcs provides substantial evidence that the generation of auroral kilometric radiation is closely associated with inverted-V events, since the inverted-V electron precipitation is associated with discrete arcs. Both direction finding measurements (KURTH *et al.*, 1975; KAISER and STONE, 1975) and lunar occultation measurements (ALEXANDER and KAISER, 1976) show that the most intense bursts of kilometric radiation come from the auroral field lines on the night side of the earth at radial distances ranging from 2 to 5 R_E. These results are illustrated in Fig. 7, which shows a series of source positions obtained by occultation measurements from the RAE-2 spacecraft in orbit around the moon. The moon in this case was in the dusk meridian plane (magnetic local time ~17.5 hr) so that the day-night position of the source can be resolved. The occultation measurements at 1255, 1635 and 2020 UT indicate the occurrence of multiple sources located at various points along a magnetic field line at about 70 to 75° invariant latitude. Occasionally day-side sources are also observed (ALEXANDER and KAISER, 1976, 1977), apparently associated with the day-side polar cusp region. The day-side sources occur less frequently and are less intense than the night-side sources. Angular distribution measurements by GREEN *et al.* (1977) are also consistent with a night-side, high-latitude source for the intense auroral kilometric radiation. A typical angular distribution is illustrated in Fig. 8, which shows the frequency of occurrence of auroral kilometric radiation intensities above a preset threshold at 178 kHz as a function of the magnetic latitude and magnetic local time of the observing point. A sharp low-latitude boundary is evident, varying from about 45° on the day side of the earth to near the equator on the night side of the earth. At large distances from the earth this cutoff forms a cone-shaped boundary, with the axis of the cone tipped toward local evening (~22 hr, magnetic local time) by about 20°. The radiation is confined

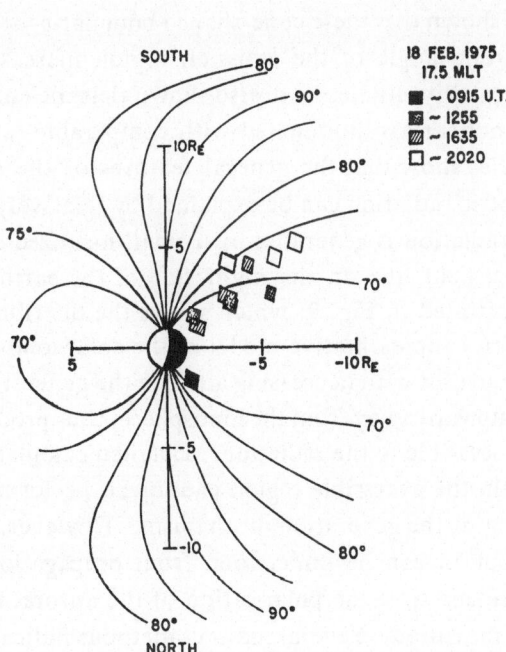

Fig. 7. Source positions for auroral kilometric radiation as determined by lunar occultation measurements with the RAE 2 spacecraft in orbit around the moon (from ALEXANDER and KAISER, 1976).

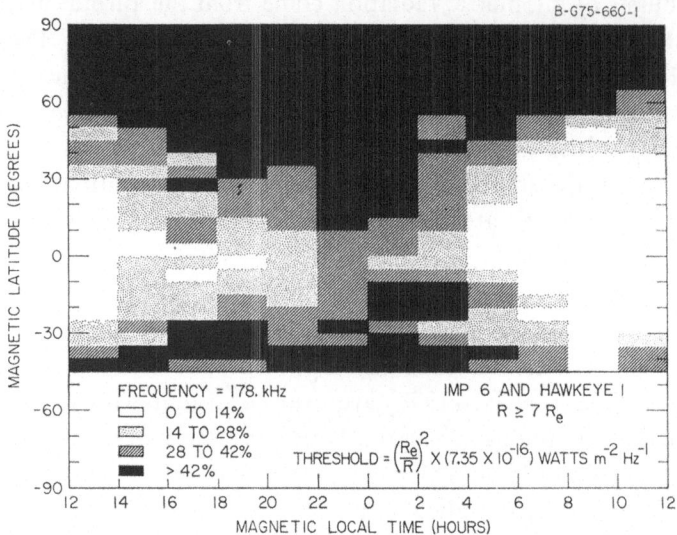

Fig. 8. The angular distribution of auroral kilometric radiation with intensities above a given threshold as a function of magnetic latitude and magnetic local time.

almost entirely to the poleward side of this boundary. Symmetrical cone-shaped boundaries are present in the northern and southern hemispheres, evidently corresponding to sources in the northern and southern auroral regions, respectively. Detailed studies have shown that these cone-shaped boundaries are strongly frequency dependent, with the solid angle of the emission region increasing with increasing frequency. Multiple satellite studies have also shown that the entire region poleward of the cone-shaped boundary is illuminated with comparable radiation intensities.

Ray tracing studies show that the general features of the angular distribution of the auroral kilometric radiation can be explained by relatively simple propagation considerations if the radiation is generated by a small localized source at about 2 to $3 R_E$ along an auroral field line on the night side of the earth. Some typical ray tracing results are illustrated in Fig. 9, which shows the distribution of ray paths at various frequencies for a representative model of the polar ionosphere. Because the index of refraction decreases with decreasing altitude the general effect is for the ray paths to be refracted upward away from the ionosphere, thus producing a low-latitude cutoff in the region accessible to the radiation. Detailed calculations of the distribution of intensity within the accessible region cannot be performed without a more detailed understanding of the generation mechanism. However, the overall features of the angular distribution can be understood from propagation considerations of this type. At the present time the polarization of the auroral kilometric radiation has not been directly measured. Several indirect methods indicate that the polarization is right hand with respect to the magnetic field in the generation region. For example, comparison of ray paths for the right and left hand modes of propagation, as in Fig. 9, with observed angular distributions, as in Fig. 10, give the best fit and

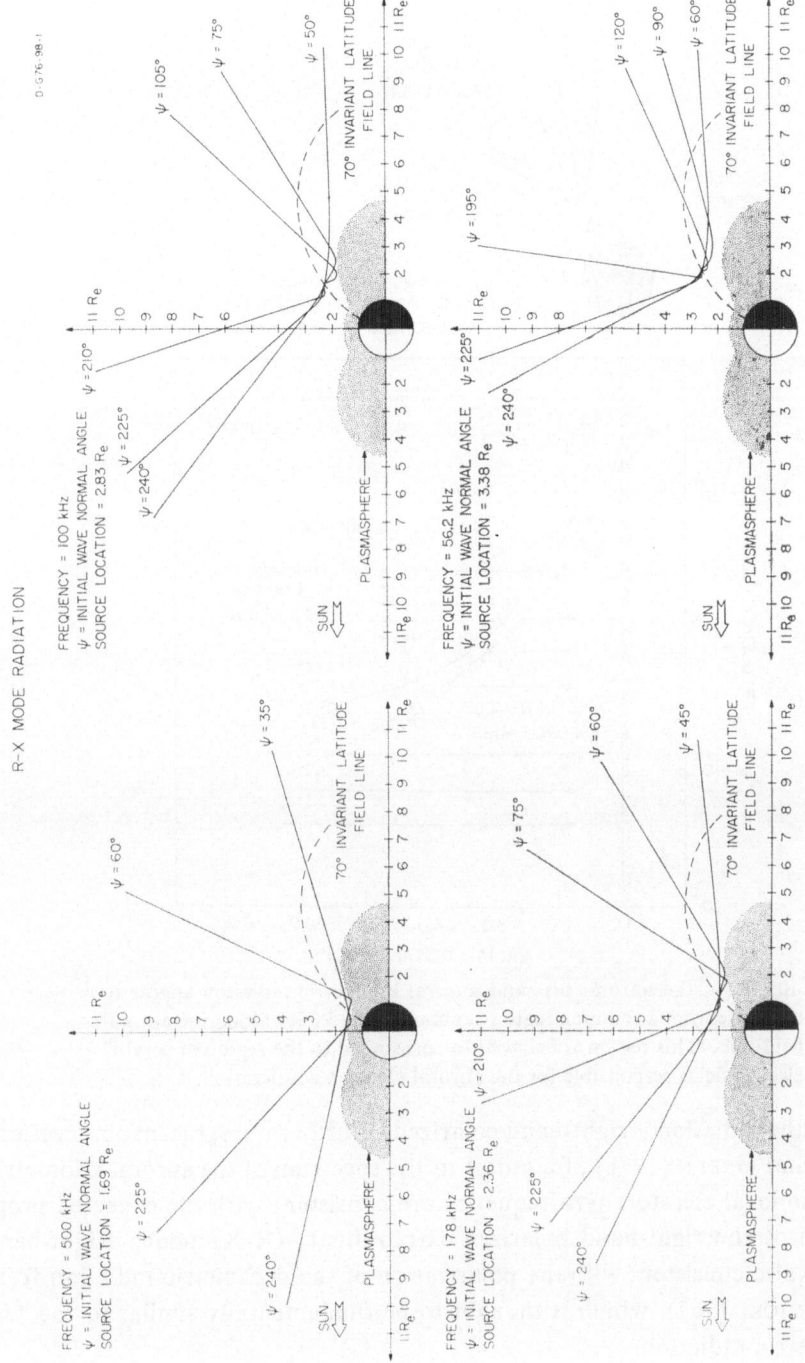

Fig. 9. Ray paths computed for various initial wave normal angles for a point source located at various points along an auroral field line at 70° invariant latitude. The cone shaped latitudinal cutoff evident in Fig. 8 is the result of refraction of the radiation upward away from the ionosphere in the region near the source (from GREEN et al., 1977).

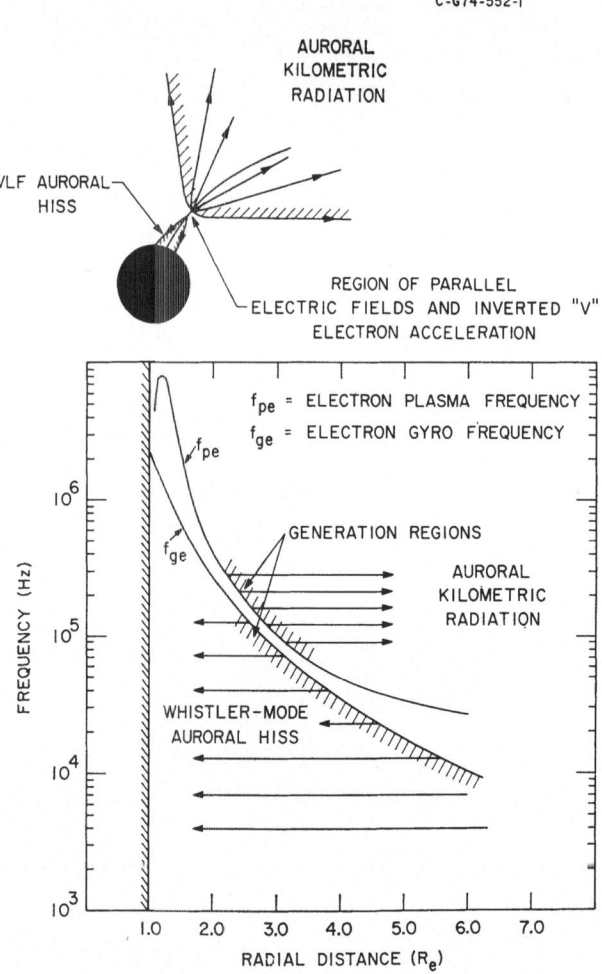

Fig. 10.   Both the auroral hiss and auroral kilometric radiation appear to
originate from a common region at about 2.0 to 5.0 R_E along the auroral
field line.   This region is believed to correspond to the region of parallel
electric fields responsible for the auroral electron acceleration.

consistency if the radiation is right-hand polarized.   Furthermore, recent observations
by GURNETT and GREEN (1978) of a cutoff in the spectrum of the auroral kilometric
radiation at the local electron gyrofrequency are consistent with the expected propagation cutoff of the right-hand polarized extraordinary (R-X) mode.   Right-hand
polarization is also consistent with the polarization of the decametric radiation from
Jupiter (WARWICK, 1967), which is thought to be fundamentally similar to the terrestrial kilometric radiation.

4.   Generation Mechanisms

When comparing these various auroral electromagnetic emissions one cannot
help but be impressed by the close similarity of the spatial regions within which the

auroral hiss and the auroral kilometric radiation are generated. In both cases the radiation is believed to be generated in a relatively small region along the auroral field lines at about $3.0\,R_E$ radial distance. Evidence that the source region is relatively small and localized is provided by the distinct V-shaped spectral features often observed in the auroral hiss and by the small angular size of the auroral kilometric radiation source obtained from the direction finding and lunar occultation measurements. Although no evidence exists showing that these two types of radiation originate from exactly the same region the evidence does strongly suggest that the two source regions overlap to a substantial extent. Since both the auroral hiss and the auroral kilometric radiation are closely associated with the inverted-V electron precipitation it seems most likely that the generation of these radio emissions is closely related to the acceleration of these electrons. Substantial evidence now exists (HAERENDEL et al., 1976; MOZER et al., 1977) showing that precipitating auroral electrons are accelerated by parallel electric fields at altitudes ranging from 5,000 to 15,000 km, in the same general region where these radio emissions are produced.

Although both the auroral hiss and the auroral kilometric radiation appear to originate from a common spatial region, distinctly different mechanisms are required to explain these two types of radiation since they are propagating in different plasma wave modes. As shown in Fig. 10, the whistler-mode auroral hiss must be generated at frequencies below the local electron gyrofrequency, whereas the auroral kilometric radiation must be generated at frequencies above the local electron plasma frequency. For many years it has been suggested that auroral hiss is produced by incoherent Cerenkov radiation from the precipitating auroral electrons (ELLIS, 1957; JØRGENSEN, 1968; LIM and LAASPERE, 1972; TAYLOR and SHAWHAN, 1974). Although the Cerenkov mechanism has many desirable features, it is generally concluded that this mechanism produces power fluxes which are several orders of magnitude too low to explain the observed auroral hiss intensities. At the present time the best possibility for explaining the observed auroral hiss intensities appears to be the mechanism proposed by MAGGS (1976), in which incoherent Cerenkov radiation generated by the precipitating electron beam is amplified to higher intensities by a whistler-mode instability. Similar mechanisms also appear to be responsible for the generation of saucer emissions (JAMES, 1976), although the details have not been investigated as thoroughly as for the auroral hiss. Cerenkov radiation has also been considered as a mechanism for generating the auroral kilometric radiation and again the computed power fluxes are much too small. Rough estimates indicate that an overall efficiency of 0.1 to 1% is required for the generation of this radiation. Such high conversion efficiencies can only be produced by a coherent plasma instability. Instability mechanisms which have been proposed include the generation of escaping electromagnetic radiation by interactions between electrostatic waves at the upper hybrid resonance (BENSON, 1975; BARBOSA, 1976), by interactions of the precipitating electron beam with ion turbulence (PALMADESSO et al., 1976), by electrostatic waves produced by energetic ions streaming outward from the auroral regions (BOSWELL, 1977), and by direct conversion via coherent electron cyclotron emission from the precipitating electrons (MELROSE, 1976). At the present time, no concensus exists about which, if any, of these mechanisms can account for the intense kilometric radiation from the earth's auroral regions.

5. Conclusions

It is evident that the auroral particle distributions produce a variety of complex and very interesting electromagnetic emissions. At the present time the mechanisms by which these electromagnetic emissions are generated are rather poorly understood and the full explanation of these radio emission processes represents a significant challenge to both the theorists and the experimentalists. It should, however, be possible to arrive at a reasonably clear understanding of how these radio emissions are generated since a great deal is already known about the charged particle distributions and processes which occur along the auroral field lines and our understanding of these processes is advancing rapidly. Because a comparable detailed knowledge of the charged particle distribution will probably never be known for most other radio sources in the universe, the study of these terrestrial radio emissions provides a unique opportunity to extend our understanding of similar electromagnetic plasma wave emissions from other planets and astronomical objects.

This research was supported in part by the National Aeronautics and Space Administration under Contracts NAS1-11257, NAS1-13129 and NAS5-11431 and Grants NGL-16-001-002 and NGL-16-001-043.

REFERENCES

ALEXANDER, J.K. and M.L. KAISER, Terrestrial kilometric radiation. 1. Spatial structure studies, *J. Geophys. Res.*, **81**, 5948–5956, 1976.

ALEXANDER, J.K. and M.L. KAISER, Terrestrial kilometric radiation. 2. Emission from the magnetospheric cusp and dayside magnetosheath, *J. Geophys. Res.*, **82**, 98–104, 1977.

BARBOSA, D.D., Electrostatic mode coupling at $2\omega_{UH}$: A generation mechanism for auroral kilometric radiation, Ph. D. dissertation, Dept. of Phys., Univ. of Calif., Los Angeles, 1976.

BENEDIKTOV, E.A., G.G. GETMANTSEV, Yu.A. SAZONOV, and A.F. TARASOV, Preliminary results of measurements of the intensity of distributed extraterrestrial radio-frequency emission at 725 and 1525-kHz frequencies by the satellite electron-2, *Kosm. Issled.*, **3**, 614–617, 1965.

BENEDIKTOV, E.A., G.G. GETMANTSEV, N.A. MITYAKOV, V.O. RAPOPORT, and A.F. TARASOV, Relation between geomagnetic activity and the sporadic radio emission recorded by the elektron satellites, *Kosm. Issled.*, **6**, 946–949, 1968.

BENSON, R.F., Source mechanism for terrestrial kilometric radiation, *Geophys. Res. Lett.*, **2**, 52–55, 1975.

BOSWELL, R.W., Energetic ions and electromagnetic radiation in auroral regions, preprint, Max-Planck-Institut für Extraterrestrische Physik, Garching, Germany, 1977.

BROWN, L.W., The galactic radio spectrum between 130 kHz and 2,600 kHz, *Astrophys. J.*, **180**, 359–370, 1973.

BURTON, E.T. and E.M. BOARDMAN, Audio-frequency atmospherics, *Proc. IRE*, **21**, 1476–1494, 1933.

DOWDEN, R.L., Low frequency (100 kc/s) radio noise from the aurora, *Nature*, **184**, 803, 1959.

DUNCAN, R.A. and G.R. ELLIS, Simultaneous occurrence of subvisual aurorae and radio noise bursts on 4.6 kc/s, *Nature*, **183**, 1618–1619, 1959.

DUNCKEL, N., B. FICKLIN, L. RORDEN, and R.A. HELLIWELL, Low frequency noise observed in the distant magnetosphere with OGO 1, *J. Geophys. Res.*, **75**, 1854–1862, 1970.

ELLIS, G. R., Low-frequency radio emission auroral, *J. Atmos. Terr. Phys.*, **10**, 302–306, 1957.

FRANK, L.A. and K.L. ACKERSON, Observations of charged-particle precipitation into the auroral zone, *J. Geophys. Res.*, **76**, 3612–3643, 1971.

FRANK, L.A. and K.L. ACKERSON, Local-time survey of plasma at low altitudes over the auroral zones, *J. Geophys. Res.*, **77**, 4116–4127, 1972.

GREEN, J.L., D.A. GURNETT, and S.D. SHAWHAN, The angular distribution of auroral kilometric radiation, *J. Geophys. Res.*, **82**, 1825–1838, 1977.

GURNETT, D.A., A satellite study of VLF hiss, *J. Geophys. Res.*, **71**, 5599–5615, 1966.

GURNETT, D.A. and L.A. FRANK, VLF hiss and related plasma observations in the polar magnetosphere, *J. Geophys. Res.*, **77**, 172–190, 1972a.

GURNETT, D.A. and L.A. FRANK, ELF noise bands associated with auroral electron precipitation, *J. Geophys. Res.*, **77**, 3411–3417, 1972b.

GURNETT, D.A., The earth as a radio source: Terrestrial kilometric radiation, *J. Geophys. Res.*, **79**, 4227–4238, 1974.

GURNETT, D.A. and J.L. GREEN, On the polarization and origin of auroral kilometric radiation, *J. Geophys. Res.*, **83**, 689–696, 1978.

HAERENDEL, G., E. RIEGER, A. VALENZUELA, H. FÖPPL, H.C. STENBAEK-NIELSEN, and E.M. WESCOTT, First observation of electrostatic acceleration of barium ions into the magnetosphere, in *European Programmes on Sounding-Rocket and Balloon Research in the Auroral Zone*, European Space Agency Report ESA-SP115, pp. 203–211, August 1976.

HARANG, L. and R. LARSEN, Radio wave emissions in the VLF band observed near the auroral zone. 1. Occurrence of emissions during disturbances, *J. Atmos. Terr. Phys.*, **27**, 481–497, 1964.

HARTZ, T.R., Low frequency noise emissions and their significance for energetic particle processes in the polar ionosphere, in *The Polar Ionosphere and Magnetospheric Processes*, edited by G. Skovli, pp. 151–160, Gordon and Breach, New York, 1970.

HELLIWELL, R.A., *Whistlers and Related Ionospheric Phenomena*, p. 207, Stanford University Press, Stanford, California, 1965.

HOFFMAN, R.A. and T. LAASPERE, Comparison of very-low-frequency auroral hiss with precipitating low-energy electrons by the use of simultaneous data from two OGO 4 experiments, *J. Geophys. Res.*, **77**, 640–650, 1972.

HUGHES, A.R.W., T.R. KAISER, and K. BULLOUGH, The frequency of occurrence of VLF radio emissions at high latitudes, in *Space Research*, Vol. XI, Akademie-Verlag, Berlin, 1323–1330, 1971.

JAMES, H.G., VLF saucers, *J. Geophys. Res.*, **81**, 501–514, 1976.

JØRGENSEN, T.S. and E. UNGSTRUP, Direct observation of correlation between aurorae and hiss in Greenland, *Nature*, **194**, 462–463, 1962.

JØRGENSEN, T.S., Interpretation of auroral hiss measured on OGO 2 and at Byrd Station in terms of incoherent Cerenkov radiation, *J. Geophys. Res.*, **73**, 1055–1069, 1968.

KAISER, M.L. and R.G. STONE, Earth as an intense planetary radio source: Similarities to Jupiter and Saturn, *Science*, **189**, 285–289, 1975.

KAISER, M.L. and J.K. ALEXANDER, Terrestrial kilometric radiation: 3. Average spectral properties, *J. Geophys. Res.*, **82**, 3273–3280, 1977.

KURTH, W.S., M.M. BAUMBACK, and D.A. GURNETT, Direction-finding measurements of auroral kilometric radiation, *J. Geophys. Res.*, **80**, 2764–2770, 1975.

LIM, T.L. and T. LAASPERE, An evaluation of the intesity of Cerenkov radiation from auroral electrons with energies down to 100 eV, *J. Geophys. Res.*, **77**, 4145–4157, 1972.

MAGGS, J.E., Coherent generation of VLF hiss, *J. Geophys. Res.*, **81**, 1707–1724, 1976.

MARTIN, L.H., R.A. HELLIWELL, and K.R. MARKS, Association between aurorae and very-low-frequency hiss observed at Byrd Station, Antarctica, *Nature*, **187**, 751–753, 1960.

MELROSE, D.B., An interpretation of Jupiter's decametric radiation and terrestrial kilometric radiation as direct amplified gyroemission, *Astrophys. J.*, **207**, 651–662, 1976.

Morozumi, H.M., Semi-diurnal auroral peak and VLF emissions observed at the south pole, 1960, *Trans. Am. Geophys. Union*, **44**, 798–806, 1963.

Mosier, S.R. and D.A. Gurnett, VLF measurements of the Poynting flux along the geomagnetic field with the Injun 5 satellite, *J. Geophys. Res.*, **74**, 5675–5687, 1969.

Mozer, F.S., C.W. Carlson, M.K. Hudson, R.B. Torbert, B. Parady, J. Yatteau, and M.C. Kelley, Observations of paired electrostatic shocks in the polar magnetosphere, *Phys. Rev. Lett.*, **38**, 292–295, 1977.

Palmadesso, P., T.P. Coffey, S.L. Ossakow, and K. Papadopoulos, Generation of terrestrial kilometric radiation by a beam-driven electromagnetic instability, *J. Geophys. Res.*, **81**, 1762–1770, 1976.

Stone, R.G., Radio physics of the outer solar system, *Space Sci. Rev.*, **14**, 534–551, 1973.

Taylor, W.W.L. and S.D. Shawhan, A test of incoherent Cerenkov radiation for VLF hiss and other magnetospheric emissions, *J. Geophys. Res.*, **79**, 105–117, 1974.

Warwick, J.W., Radiophysics of Jupiter, *Space Sci. Rev.*, **6**, 841–891, 1967.

Theory of Electromagnetic Waves on Auroral Field Lines

J.E. MAGGS

Institute of Geophysics and Planetary Physics,
University of California at Los Angeles,
Los Angeles, California, U.S.A.

(Received October 17, 1977)

Intense bursts of electromagnetic radiation generated in auroral regions during substorms radiate as much as one percent of the precipitated electron energy. The generation mechanism for the intense bursts of radio noise is apparently associated with the auroral electron beam. Some processes suggested for direct generation of electromagnetic noise by the auroral electron beam are briefly reviewed but it is suggested that the electromagnetic noise bursts are created indirectly due to radiation from strong electrostatic turbulence generated by the auroral electron beam. Due to refraction of ray paths out of the spatially limited auroral electron beam, beams below a certain intensity generate low levels of electrostatic noise as commonly observed by polar orbiting satellites. However, above a certain level, the electrostatic noise generated by the beam becomes intense and forms 'solitons' (large ion density cavities) or 'spiky turbulence.' Soliton turbulence can radiate efficiently enough to account for the observed levels of electromagnetic radiation.

1. Introduction

It has recently been established that the earth is, at times, a very powerful radio source (GURNETT, 1974). Strong bursts of electromagnetic radiation called auroral kilometric radiation (AKR) are emitted from the polar regions of the earth (KURTH *et al.*, 1975; ALEXANDER and KAISER, 1976) with radiated power levels that may be as high as one gigawatt (GURNETT, 1974). The radio frequency bursts of electromagnetic energy are apparently associated with auroral arcs because the source region of the bursts is on or near the field lines on which auroral arcs occur and the bursts are correlated with changes in the brightness of the arcs (GURNETT, 1974). It is likely, then, that the generation mechanism of the radio bursts is closely associated with the same fluxes of keV energy electrons precipitating into the polar ionosphere that generate the auroral arcs. In this paper I will briefly review some of the plasma physical processes that have been suggested for generating electromagnetic radiation from electron fluxes similar to those observed in auroral arcs.

While the study of the AKR generation mechanism is an interesting plasma physics problem it may eventually prove to be of greater astrophysical interest. There is some evidence that strong bursts of electromagnetic noise are emitted as a general rule when a magnetized rotating body interacts with a solar wind-like plasma flow (KENNEL and MAGGS, 1976). The decametric radio bursts from Jupiter (DESCH and Carr, 1974) and the hectometric radio bursts from Saturn (Brown, 1975) have power spectra and occurrence patterns similar to AKR and the radio bursts may represent a general phenomena that Kennel and Maggs (1976) call magnetospheric radio bursts (MRB). The process generating bursts of radio noise from the earth is then especially interesting because it could prove to be a general astrophysical process associated with MRB generation. The MRB generation mechanism could provide an alternative to synchrotron radiation as an explanation for some strong astrophysical radio sources.

2.   General Properties of AKR

The observed features of AKR place restrictions on the types of processes involved in generating the electromagnetic noise. Probably the most severe restriction results from the large power levels observed. A gigawatt of radiated power is about one percent of the power deposited into the auroral ionosphere by an intense flux of precipitating electrons (Axford, 1967; Akasofu, 1968). The generation process must involve instabilities capable of producing about 20 e-foldings of power above ambient noise levels in order to explain the observed power levels. All of the generation mechanisms discussed in this review rely on the precipitating auroral electron fluxes (which I will call the auroral electron beam) as the source of free energy for producing AKR. The velocity space distribution of the precipitating auroral elec-

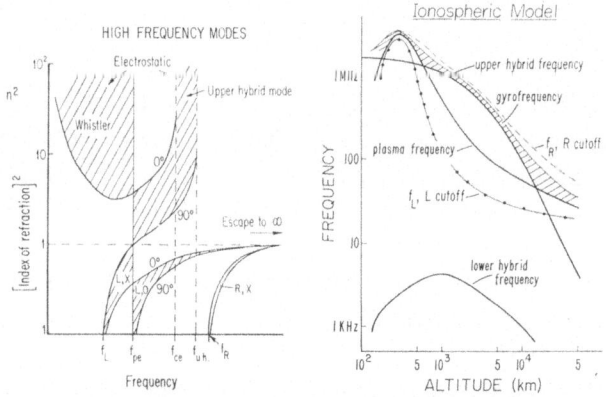

Fig. 1.   The square of the index of refraction of the cold plasma modes is shown as a function of frequency for the case that the electron plasma frequency is less than the electron gyrofrequency. Also shown is a model (after Gurnett, 1974) of the variation of the fundamental frequencies of the plasma with altitude in the undisturbed or quiet time auroral ionosphere. The whistler and upper hybrid modes are electrostatic at large values of the index of refraction and have slow phase and group velocities. These electrostatic modes are readily amplified by the auroral electron beam. However, they can not propagate out to large distances in the magnetosphere where kilometric radiation is observed. Either the electrostatic modes are converted to the electromagnetic (L, O) or (R, X) mode or the kilometric radiation is initially generated in the electromagnetic modes.

trons will be important in determining the wave mode involved in the instability mechanisms producing kilometric radiation. Ultimately, however, the wave noise must be in either the (L, O) or (R, X) electromagnetic modes because AKR has been observed in the magnetosphere at frequencies well above the local electron gyro-frequency (f_{ce}) and plasma frequency (f_{pe}). Figure 1 illustrates that only the (L, O) or (R, X) electromagnetic modes could have propagated to the space craft from the auroral regions.

In the source region the kilometric radiation may be produced directly in the L or R electromagnetic modes or it may be produced secondarily from other modes. As indicated in Fig. 1 the whistler and upper hybrid modes can be electrostatic when propagating at angles near the resonance cone angle. These electrostatic modes have slow phase velocities and can be strongly amplified by interacting with the electron beam through the Landau resonance (MAGGS, 1978a) but can not sub-sequently propagate into the outer magnetosphere without being converted to the L or R electromagnetic mode. On the other hand amplification of the L and R modes can not take place through the Landau resonance because these modes have phase velocities faster than the speed of light.

In reviewing production mechanisms of AKR I divide them into two categories. The first category includes those mechanisms in which the electromagnetic noise is produced directly in the (L, O) or (R, X) mode. In these mechanisms the L or R mode is ultimately amplified by the auroral electron beam but the amplification process may involve other plasma waves. The second category includes those pro-cesses which require the production of strong electrostatic noise by the auroral electron beam which is then partially converted by various processes into electro-magnetic radiation. In discussing the various production processes I will give the wave modes involved, the requirements on the plasma and beam parameters neces-sary for production and an estimate of the total power the process can generate.

3. Production Directly in (L, O) or (R, X)

The most direct mechanism for production of AKR involves the electromagnetic (R, X) mode. MELROSE (1976) suggests that an (R, X) mode with frequency near the right-hand cutoff frequency (f_R) propagating in the direction of the precipitating electrons can doppler shift to gyroresonance with the beam particles. The required resonance condition for propagation along the magnetic field is

$$\frac{f-f_{ce}}{f} = \frac{un}{c} \tag{1}$$

where u is the beam speed, c the speed of light, n the index of refraction and f the wave frequency. Since auroral electrons have energies of only a few keV, u/c is small and with $n<1$, f must be very close to f_{ce} to satisfy Eq. (1). Since the fre-quency of the R mode is above the right-hand cutoff, Eq. (1) requires that the plasma frequency be much smaller than the electron gyrofrequency $(f_{pe} \ll f_{ce})$. The mechanism will be very sensitive to the ratio of f_{pe} to f_{ce}. Strong amplification of the doppler shifted R mode occurs if the electron beam has an anisotropic temperature distribution such that

$$\frac{T_\perp}{T_{//}} > \frac{c^2}{u^2} \, . \tag{2}$$

For electrons of auroral energies of 1–10 keV Eq. (2) requires temperature anisotropy of 10^3–10^2 which is a severe restriction on the electron velocity distribution and not typically observed in auroral arcs (MENG 1976, ARNOLDY and CHOY, 1973). Melrose points out however that an initially isotropic electron beam will evolve into a distribution with $T_\perp/T_{//} > 1$ if it streams into regions of higher field strength. The growth rate of the process is very rapid and 20 e-foldings of power can be produced in 50–100 km so that the process is efficient enough to produce the observed power in AKR.

A process for amplifying the L mode has been suggested by PALMADESSO et al. (1976). If ion turbulence is present in the same region as the auroral electron beam an electromagnetic wave in the L mode can form an electrostatic beat wave in either the whistler or upper hybrid mode (see Fig. 1) that is Landau resonant with the beam particles and thus amplified. The amplification of the beat wave results in amplification of the electromagnetic wave because energy is transferred from the beat wave and the ion turbulence to the electromagnetic wave. In order for the processes to produce net growth the k-spectrum of ion turbulence must be such that beat waves are resonant mostly with the beam and not with electrons of several hundred eV energies or less. This constitutes a rather severe restriction on the k-spectrum of the ion turbulence. Due to the high group velocity of the electromagnetic wave the convective growth rate of the process is not large and distances on the order of 500 km or more are required to produce the power observed in AKR. Since this distance is apparently larger than the distance the upper hybrid mode could propagate in the topside ionosphere without reflection Palmadesso suggests that AKR is most likely produced in the (L, X) mode with the electrostatic whistler as the beat wave. In this case the propagation distances can exceed the required amplification distances but now the (L, X) mode must tunnel into the (L, O) mode to escape to infinity. For efficient tunneling the (L, X) mode must be propagating near parallel to the magnetic field and if the (L, X) mode is generated nearly isotropically the tunneling process greatly reduces the overall efficiency. However, Palmadesso estimates that, for strong ion turbulence and high auroral beam fluxes, power levels between 10^8 and 10^9 watts can be generated.

4.   Production of Electrostatic Noise by the Auroral Electron Beam

To evaluate the importance of processes that might generate AKR by converting electrostatic noise into electromagnetic radiation it is useful to have an estimate of the frequency spectrum and power levels of electrostatic noise produced by the auroral electron beam. The auroral beam produces strong electrostatic noise by convective amplification of the ambient noise levels in the plasma. The auroral electron beam can amplify noise because the beam density and speed are large enough to produce a region of positive slope in the electron velocity space distribution. The slow phase velocity of the electrostatic plasma modes leads to strong convective amplification. The power flux frequency spectrum of waves produced by the auroral beam can be found by integrating the wave kinetic equation (MAGGS, 1976) in a model of the auroral ionosphere and auroral electron beam. In particular, a procedure for calculating the power flux spectrum of electrostatic noise produced by a model of a quiet auroral arc has been developed by MAGGS (1978a). I will discuss the assumptions on which the quiet arc model is based and the results of the model to describe the predicted properties of auroral beam generated electrostatic noise.

The quiet arc model is based on the results of satellite surveys of quiet evening auroral arcs using electron differential energy analyzers and auroral imaging as reported by MENG (1976). Quiet-evening auroral arcs are discrete auroral forms that show little internal spatial structure. They are a few to several tens of kilometers in width and stretch for a thousand to several thousands of kilometers along the evening auroral oval. The most interesting feature of the quiet evening auroral arc is the form of the differential energy spectrum of electrons precipitating in the arc. The differential energy spectra in quiet evening auroral arcs almost invariably have a region of positive slope and correspond to velocity space distributions that can be modeled by drifting Maxwellian electron beams of a few keV energies (MAGGS, 1977). The quiet arc is then modeled as a sheet beam aligned along the magnetic field; narrow in the north-south direction, and extended in the east-west direction. The velocity distribution of electrons precipitating in the arc are modeled as a drifting Maxwellian with drift energy $E_b = 1.5$ keV and temperature $\bar{T}_b = 0.2 E_b$. The number

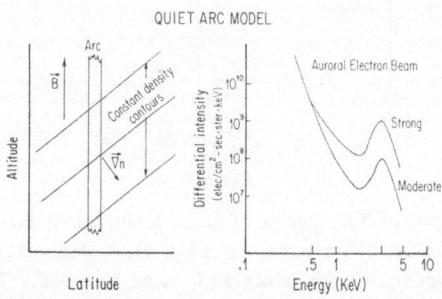

Fig. 2. A model of a quiet auroral arc and the auroral ionosphere are used in calculating the generation of electrostatic noise by the auroral electron beam. The arc is assumed to be field-aligned while the contours of constant ionospheric electron density lie at an angle to the magnetic field. The electron number density gradient is taken to lie in the magnetic meridian plane (i.e., the north-south direction). The electron distribution has a region of positive slope in velocity space. Two cases are considered: a strong beam with peak differential intensity of 10^9 elec/cm²-sec-ster-keV, and a moderate beam of intensity 10^8 elec/cm²-sec-ster-keV.

density of the beam is allowed to vary so that the effects of varying beam intensity on the power levels of the electrostatic noise produced can be studied.

The spatial structure of the auroral electron beam is important for determining the power flux of beam-amplified noise. The refraction of ray paths due to variations in the ionospheric number density and magnetic field strength limits the path length of rays passing through the beam. Noise with ray paths initially propagating in the amplification region can be refracted out of the beam and the noise is no longer amplified. Thus at a given frequency the power flux will be strongly influenced by the rate of refraction and spatial extent of the auroral beam. The rate of refraction is determined by the rate of variation of ionospheric number density and field strength. The model ionosphere is assumed to have a constant magnetic field and the number density is assumed to vary linearly in the direction along the magnetic field with scale height $L_{//}$ and in the north-south direction perpendicular to the field with scale height, $L_{\perp}$.

Figure 3 shows some power flux spectra of electrostatic noise predicted by the quiet arc model. In both cases illustrated the plasma frequency is taken to be less than the electron gyrofrequency and the ionospheric perpendicular and parallel scale heights are 1,000 km. A strong beam of peak differential intensity 10^9 (elec/cm²-sec-stev-keV) and 1 km in north-south thickness should produce strong fluxes of electrostatic noise in excess of 10^{-13} W/m²-Hz near the lower hybrid resonance frequency. Such intense hiss spectra are often observed in the auroral zone as illustrated by a spectrum taken from Alouette 2. On the other hand a 25 km thick beam of moderate

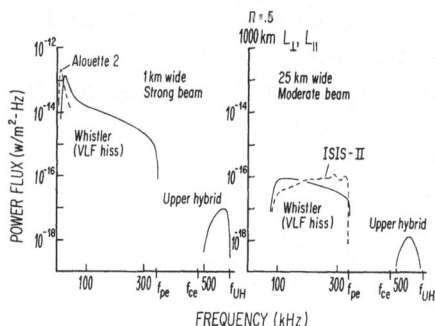

Fig. 3. The frequency power flux spectra of noise produced in the quiet arc model agree well with observed VLF hiss spectra. A 1 km thick strong beam produces a spectra peaked near the lower hybrid frequency as is often observed. The observation shown from Alouette 2 was teken in an ionosphere with a lower hybrid frequency smaller than that of the model ionosphere, but illustrates the tendency to peak near the lower hybrid frequency. A 25 km thick moderate beam produces a rather broad spectrum. The absolute power flux levels of the example taken from Isis-II is not known and so only the shapes should be compared. The spectrum taken from Isis-II was observed in conjunction with a beam-like velocity distribution in an auroral arc. Upper hybrid noise is also produced by the beam but with lower power fluxes than the whistler mode when $f_{pe} < f_{ce}$ as for this example ($R = f_{pe}^2 / f_{ce}^2 = 0.5$). However, the power flux of the upper hybrid noise can exceed the whistler when $f_{pe} > f_{ce}$.

intensity (10^8 elec/(cm²-sec-ster-keV)) produces a broad flat spectrum with intensities of only 10^{-16} W/cm²-Hz. Noise spectrum of this nature are also commonly observed in the auroral zone as illustrated by an example from Isis-II. In the examples of observed spectra only the shapes should be compared because the absolute intensities of the observed noise is not known. Note that the quiet arc model also predicts the presence of upper hybrid noise. In the examples illustrated the upper hybrid noise is weaker than the whistler noise. This is always the situation in the region $f_{pe} < f_{ce}$ but the power fluxes in the upper hybrid mode can exceed those in the whistler mode for the case $f_{pe} > f_{ce}$. Generally the agreement between the observed properties of VLF hiss and the whistler mode power flux predicted by the quiet arc model is quite good. This gives us confidence that beam associated electrostatic noise is generated by convective amplification of ambient noise and the electrostatic wave amplitude is limited by refraction out of the spatially limited auroral beam.

Observationally the generation of strong bursts of kilometric radiation does not appear to be associated with quiet auroral arcs but rather with intense breakup arcs. In these intense arcs the precipitating electron fluxes are over an order of magnitude higher than in quiet arcs (MENG *et al.*, 1978) but the differential intensities observed at 1–2 thousand kms altitude do not exhibit regions of positive slope. However, we can extend the quiet arc model to learn about the intense arc by introducing the concept of the beam persistence limit. The beam persistence limit is a level of beam intensity that distinguishes between the quiet and intense arc. It cannot be given as an absolute number because it depends on the beam thickness and the ionospheric gradients. In the quiet arc the power levels of beam amplified noise are low enough that nonlinear processes are not important. Convective amplification at the linear growth rate and limitation of power fluxes by ray refraction out of the amplification region are adequate to explain the observed power spectra. Since the beam produced power fluxes are low quasilinear diffusion is too weak to alter the velocity distribution of the precipitating electrons and a beam-like velocity distribution is thus observed in conjunction with the beam-generated noise. In the intense arc, on the other hand, refraction does not limit the noise to low levels and nonlinear processes dominate. Quasilinear diffusion is strong and the beam velocity distribution is plateaued— flattened by diffusion in velocity space. The beam no longer generates electrostatic noise once it is flattened so that the intense beam generates noise only from its source point to its plateau point, probably a relatively short distance along the field line. Whistler noise generated in this region propagates downward but is then not observed in conjunction with a beam-like velocity distribution. The geometry of the noise-producing region of the intense arc being relatively short along the field should favor amplification of frequencies near f_{pe} because rays at those frequencies propagate mostly across the magnetic field. Thus the intense arc probably produces intense electrostatic noise near the plasma frequency in either the whistler mode if $f_{pe} < f_{ce}$ or the upper hybrid mode if $f_{pe} > f_{ce}$. In considering the conversion of electrostatic noise into electromagnetic radiation we will take the electrostatic noise to have frequencies near f_{pe}.

5. Conversion of Electrostatic to Electromagnetic Noise

The most direct process for converting electrostatic noise to electromagnetic radiation is through linear conversion of the upper hybrid mode to the electromagnetic (L, O) mode due to propagation in the inhomogeneous ionosphere. OYA (1974) first suggested that Jupiter's decametric radiation could be produced in this manner. BENSON (1975) and JONES (1977) suggested that AKR could also be generated by this process. For a given ionospheric gradient there is a region in wave number, or k, space such that upper hybrid electrostatic noise in that region can propagate to a region of the ionosphere in which the wave frequency is below the plasma frequency. The wave then has the properties of the (L, X) mode (see Fig. 1) and the mode propagates to the left-hand cutoff (f_L) where it is reflected. The wave eventually propagates to the region $f > f_{pe}$ where it is again an upper hybrid mode. At two points along the ray path the wave frequency equals the plasma frequency and some of the wave energy is coupled into the (L, O) mode at these points. The (L, O) mode then escapes to infinity. The amount of wave energy converted to the (L, O) mode depends on the magnitude of the background density gradient and the angle between the wave k vector and the direction of the density gradient. If the change in number density occurs over a short distance (shorter than a wave length) so that the ionospheric density variation appears to the wave as a plane interface between two media of differing index of refraction the coupling to the (L, O) mode can be very efficient (OYA, 1974). Up to 10 % of the incident electrostatic energy can be reflected in the (L, O) mode. If the density gradient is not sharp the wave must propagate to the L cutoff, be reflected and tunnel through to the (L, O) mode. Only noise propagating near parallel to the magnetic field (k typically within 3° of B) efficiently tunnels into the (L, O) mode (BUDDEN, 1961); PALMADESSO et al., 1976). Since the noise in the (L, X) mode is isotropically distributed in angle only about $(3/90)^2 \sim 10^{-3}$ of (L, X) mode converts to (L, O). Further, the electrostatic noise is generated isotropically in k space with indices of refraction about equal to $c/u = ck_{//}/\omega$. Thus only $(u/c)^2 \sim 10^{-2}$–10^{-3} is propagated to regions in k space with indices of refraction on the order of unity; that is, to the (L, X) mode. Thus the overall efficiency of the conversion process in the absence of sharp gradients is $\sim 10^{-5}$–10^{-6}.

In the case of sharp gradients the (L, O) mode is radiated along the direction of the interface (OYA, 1974). To estimate the radiated power assume that many discontinuities in density exist along the arc and that its east-west vertical surface radiates (L, O) radiation. The surface area, S, is perhaps as much as 3,000 km in vertical extent by 10,000 km east-west extent or 3×10^{17} cm². Denote the ratio between the electrostatic energy density, $E^2/8\pi$, and the ionospheric energy density, $n_0 T_e$, by W_0 ($W_0 \equiv E^2/8\pi n_0 T_e$). Take $n_0 T_e$ for the ionosphere to be 3×10^{-9} ergs/cm³ then the total radiated power is

$$P = \varepsilon W_0 S n_0 T_e v_g = 3 \times 10^{11} W_0 \text{ watts}$$

where ε^{-1} is the ratio of incident electrostatic noise to radiated (L, O) noise and v_g is the group velocity of the (L, O) mode which is roughly the speed of light. This process could be efficient enough to produce kilometric radiation if $W_0 \sim 10^{-3}$. In the smooth gradient case the radiation is emitted along the field where the surface area of the source is smaller: the north-south distance 300 km times the east-west distance 10,000 km or 3×10^{16} cm^2. In this case then the total radiated power is only 10^5-$10^6 W_0$ watts and the mechanism is not efficient enough to account for the kilometric radiation. If this mechanism is operating the density gradients must be sharp and locally confined but this may be the case in a turbulent plasma. Two objections may be raised to this process. First the efficiency is very sensitive to the angle between k and the interface. Secondly, we expect strong upper hybrid turbulence with $f > f_{pe}$ only in the $R > 1$ region which may not occur over a large enough region of the field line to account for the observed bandwidth of AKR.

A nonlinear mechanism for converting strong upper hybrid noise to electromagnetic radiation has been suggested by BARBOSA (1976). Barbosa calculates the production of strong upper hybrid noise, primarily through cyclotron resonance growth in the $R < 1$ region. He assumes that the noise constitutes statistically homogeneous turbulence. One of the nonlinear processes occurring in the turbulence is the conversion of the electrostatic noise into electrostatic radiation from near head-on collisions of two electrostatic waves. In this process electromagnetic radiation is generated in both the (R, X) and (L, O) modes at twice the electrostatic wave frequency. The efficiency of the process depends on the square of the electrostatic energy density. Using the ionospheric model shown in Fig. 1 and the observed band width of auroral kilometric radiation Barbosa calculates a source volume of 5×10^{26} cm^3 in a dipole field. Using this source volume and his calculated k spectrum Barbosa concludes that 10^5-$10^9 W_0^2 (T/5 \text{ eV})^{7/2}$ watts of power can be radiated from statistically homogeneous upper hybrid turbulence. The process depends very strongly on the ionospheric temperature and Barbosa suggests that temperature increases due to heating from beam-generated electrostatic noise may turn on the kilometric radiation process. The process requires temperatures on the order of 5 eV and W_0 on the order of unity to explain the highest observed power levels of AKR. For $W_0 \sim 1$ the electric field is a few volts per meter and electric field strengths that large have not been observed in the auroral ionosphere.

6. Radiation from Soliton Turbulence

Observations of strong turbulence in laboratory plasmas has revealed that strong plasma turbulence tends to form localized regions of intense electric fields called solitons or cavitons (WONG and QUON, 1975). The strong turbulence called spiky turbulence is far from statistically homogeneous. Thus in the calculation of radiation from spiky turbulence the assumption of statistical homogeneity is not used. Before attempting to calculate the radiation from spiky turbulence it is necessary to determine the expected spatial configuration of the solitons in auroral beam generated spiky turbulence.

A soliton or caviton consists of a large density depression in the plasma con-

Soliton Turbulence

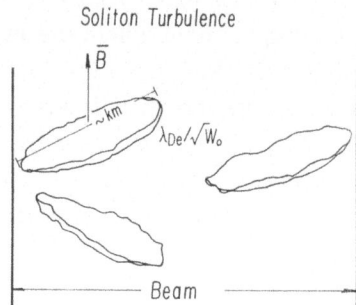

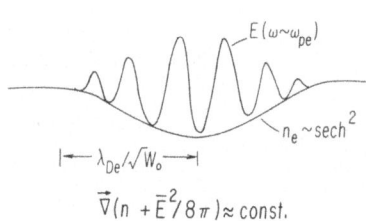

$$\vec{\nabla}(n + \bar{E}^2/8\pi) \approx const.$$

Fig. 4. Solitons formed in one-dimensional plasmas consist of high-frequency electrostatic noise trapped in a low-frequency density cavity. The radiation pressure and particle pressure are constant across the structure. The background plasma has a density depression that has a sech² dependence. The auroral beam is envisaged as forming many disk-like solitons aligned nearly perpendicular to the magnetic field.

taining trapped wave energy. The soliton is a self-consistent nonlinear structure in the plasma (KARPMAN 1975). The density depression traps wave energy and the pressure of the trapped wave energy prevents the density cavity from collapsing. The size of the cavity is determined by a balance between the tendency of wave energy trapped in the cavity to increase due to self focusing and the loss of wave energy by diffraction out of the cavity (KAW et al., 1975). Electrostatic plasma oscillations in a one dimensional plasma form solitons with a sech² density profile and thickness $\lambda_{De}/\sqrt{W_0}$ where λ_{De} is the electron Debye length and W_0 is based on the maximum trapped wave amplitude (ZAKHAROV and SHABAT, 1972). The one dimensional soliton is illustrated in Fig. 4.

It can be shown that in a magnetized plasma the electrostatic whistler and upper hybrid modes can form sheet-like solitons (MORALES and LEE, 1975, MAGGS, 1978b). In these soliton structures the density depression is a two dimensional planar structure. The axis of the sheet or the plane along which the density is minimum forms an angle to the magnetic field. The angle made by the soliton axis to the magnetic field, θ_A, depends on the frequency and amplitude of the electrostatic mode trapped in the soliton as illustrated by Fig. 5. For frequencies near the plasma frequency both the upper hybrid and whistler modes tend to form solitons nearly perpendicular to the magnetic field. The soliton axis angle is the complement of the resonance cone angle, θ_R (the angle at which a plane wave with the same frequency would propagate). If the amplitude of the trapped mode increases the whistler soliton axis angle, θ_A, shifts towards smaller values while the upper hybrid soliton angle shifts to larger values.

Large solitons have a tendency to collapse and form smaller solitons (ZAKHAROV,

SOLITON FORMATION

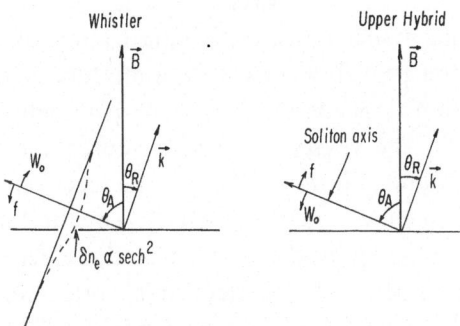

Fig. 5. In a magnetized plasma the electrostatic whistler or upper hybrid mode can form sheet-like solitons. The axis of the density depression makes an angle to the field, θ_A, that depends on the frequency and amplitude of the trapped radiation. The soliton axis angle is near the resonance cone angle, θ_R, for small amplitudes. At large amplitudes the whistler solitons tend towards smaller angles while the upper hybrid solitons tend toward larger angles. For frequencies near the plasma frequency both types of soliton are aligned nearly perpendicular to the magnetic field at small amplitudes.

1972, KAW et al., 1975). The sheet-like solitons described above probably collapse and form disk-like soliton structures (GALEEV and KRASNOSELSICK, 1978). However, the collapse of sheet-like solitons has not been studied and the assumption that disk-like structures are formed is based on considerations of symmetry. To calculate the radiation from solitons formed by auroral beam-generated electrostatic noise it is assumed that the beam or arc region of space is filled with disk-like soliton structures aligned nearly perpendicular to the magnetic field as illustrated in Fig. 4. Actually the beam-generated noise probably forms a nonlinear structure consisting of multiple solitons. This nonlinear structure is complicated and consists of a nonlinear superposition of basic soliton structures (HIROTA, 1972). However, it is considerably easier to calculate the radiation from soliton turbulence under the above assumptions.

The interaction of the trapped high-frequency electrostatic noise with the density gradients associated with the soliton produces nonlinear currents in the plasma (ZHELEZNYAKOV, 1970). As the high-frequency electrostatic mode propagates into the wall of the cavity the space charge density associated with the wave changes because the plasma density is changing. The high-frequency space charge density is largest at the point where the frequency of the trapped wave equals the local plasma frequency. The space charge associated with the wave interacts with the electric field of the wave to produce a nonlinear current at twice the trapped wave frequency along the wall of the cavity. Because of the finite spatial extent perpendicular to the magnetic field of the disk-like solitons the nonlinear current is not curl-free and can therefore be the source of electromagnetic radiation. The radiation has wave lengths on the order of the size of the disks.

The total power radiated by a soliton is calculated by assuming a two dimensional sheet-like soliton with a sech² profile in an unmagnetized plasma with trapped electrostatic noise at frequencies near the plasma frequency. The spatial structure of the eigen functions at twice the wave frequency are then calculated. Because an unmagnetized plasma model is used the polarization of the eigen functions is not found. The radiation probably consists of a mixture of R and L modes whose axial ratio of polarization depends on the plasma parameters in the source. With the nonlinear current in the cavity walls as a source the electric field intensity at infinity can then be found (MAGGS, 1978c). The electromagnetic power radiated at twice the trapped frequency (approximately $2f_{pe}$) by the three-dimensional beam-generated solitons is then given by the product of the electric field energy density at infinity, the group velocity of the electromagnetic radiation and the surface area of the soliton. This procedure is considerably different than that used by GALEEV and KRASNOSELSICK (1978) but gives similar results. The total radiated power of kilometric radiation produced by soliton turbulence is found by assuming 10^{10} disk-like solitons 1 km in radius are formed by the intense auroral beam. The largest total source volume is probably about 10^{25} cm³ (300 km N-S extent by 3,000 km vertical extent by 10^4 km E-W extent). For solitons about 10 Debye lengths thick (roughly 300 cm but the soliton thickness is actually given by $\lambda_{De}\sqrt{W_0}$) the total volume occupied by 10^{10} solitons is 1 % of the total volume. Figure 6 shows the relation between the total power radiated by beam-generated soliton turbulence and the ionospheric temperature for various values of W_0 and the spatially averaged value of W_0, $\bar{W}_0$. As in the case of homogeneous turbulence the amount of radiated power is a strong function of temperature but now the radiated

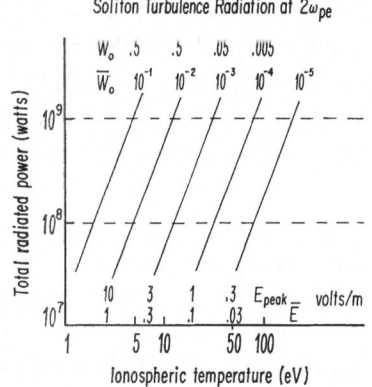

Fig. 6.   The total power radiated from turbulence consisting of many disk-like solitons is a strong function of the background ionospheric plasma temperature. The total power radiated is plotted for various values of the ratio of the maximum trapped electrostatic energy density to ionospheric energy density, W_0; along with its corresponding spatially averaged value, $\bar{W}_0$. The electric field in volts per meter associated with the values of W_0 and $\bar{W}_0$ are given as well. Soliton turbulence can radiate between 10^8–10^9 watts of power in an ionosphere of less then 1 eV temperature only if $\bar{W}_0 \approx 1$. In a warm ionosphere soliton turbulence can radiate up to 10^9 watts for modest values of spatially averaged electric field (100–300 mV/m).

power is proportional to $T^{5/2}$. Radiation from soliton turbulence is strong enough to account for observed levels of kilometric radiation. Furthermore, if the ionospheric temperature in the source region is tens of electron volts, a gigawatt of total radiated power can be produced by fields with a spatially averaged value of a few 100 mV/m.

In summary, my view of the radiation produced by the auroral electron beam is the following. In the quiet auroral arc the beam intensity and thickness are small enough that the beam-generated whistler or upper hybrid electrostatic noise levels are not high enough to quasilinearly diffuse the beam. In this case VLF hiss and upper hybrid noise is observed in conjunction with a beam-like velocity distribution. If the source size or strength increases above a certain limit, however, the beam generates intense electrostatic noise near the plasma frequency. The intense electrostatic noise forms spiky turbulence consisting of disk-like soliton structures aligned nearly perpendicular to the magnetic field. The solitons formed by the intense beam radiate electromagnetic energy at $2f_{pe}$ in the R and L modes. The beam interacts with the soliton structures and is quasilinearly diffused.

7. Diagnostic Measurements

I've discussed five theories for the production of kilometric radiation and now I will discuss what experimental observations and measurements can be made to determine which, if any, are operative along the auroral field lines. Surely, a measurement of the electromagnetic wave polarization will help distinguish between the various mechanisms. Two of the mechanisms suggested produce pure (L, O) radiation, one exclusively (R, X) and two a mixture of R and L. While the polarization measurement can easily be made, care should be taken to sample the polarization of the radiation over the entire emission cone. Furthermore the polarization as a function of frequency should be studied because the polarization may depend on the ratio of the gyrofrequency and plasma frequency at the altitude of emission in the source.

Altitude profiles in the source region of the basic plasma parameters, such as number density and temperature, are also needed to determine if the conditions required for operation of some mechanisms are met. Sharp density gradients are required in one mechanism while high ionospheric temperature leads to greater radiation efficiency in others. Also the velocity space distribution in the source region is important. Melrose's mechanism requires a beam distribution with a large temperature anisotropy while the generation of strong electrostatic noise requires a beam-like distribution. Furthermore, the evolution of the velocity distribution in the source region might yield information about the strength and types of wave particle interactions occurring in the source region.

If electrostatic turbulence is involved in the production of kilometric radiation it will be necessary to distinguish between homogeneous and spiky turbulence. This distinction can be made using high spatial and temporal resolution measurements of the high- and low-frequency electric fields. Also peak field detection is desirable because the peak field of spiky turbulence is much larger than the average field. Correlation studies between low- and high-frequency turbulence can also be used to determine if the trapped high-frequency wave configuration of spiky turbulence is present in the plasma.

Further theoretical progress on some of the proposed mechanisms requires lengthy and complicated calculations. Experimental observations in the earth's auroral ionosphere are badly needed to provide incentive and guidance in carrying out these calculations. The understanding of the generation mechanism of auroral kilometric radiation has the potential to greatly increase our knowledge of other planetary magnetospheres, and perhaps, other astrophysical objects as well.

REFERENCES

AKASOFU, S.I., *Polar and Magnetospheric Substorms*, p. 222, D. Reidel, Dordrecht, Holland, 1968.

ALEXANDER, J.K. and M.L. KAISER, Terrestrial kilometric radiation. 1. Spatial structure studies, *J. Geophys. Res.*, **81**, 5948–5956, 1976.

ARNOLDY, R.L. and L.W. CHOY, Auroral electrons of energy less than 1 keV observed at rocket altitudes, *J. Geophys. Res.*, **78**, 2187–2200, 1973.

AXFORD, W.I., Magnetic storm effects associated with the tail of the magnetosphere, *Space Sci. Rev.*, **7**, 149–157, 1967.

BARBOSA, D.D., Electrostatic mode coupling at $2\omega_{UH}$: A generation mechanism for auroral kilometric radiation, Ph.D. Thesis, UCLA, Los Angeles, California, 1976.

BENSON, R.F., Source mechanism for terrestrial kilometric radiation, *Geophys. Res. Lett.*, **2**, 52–55, 1975.

BROWN, L.W., Saturn radio emission near 1 MHz, *Astrophys. J. Lett.*, **198**, L89–L92, 1975.

BUDDEN, K.G., *Radio Waves in the Ionosphere*, p. 424, Cambridge University Press, New York, New York, 1961.

DESCH, M.D. and T.D. CARR, Decametric and hectometric observations of Jupiter from the RAE-1 satellite, *Astrophys. J. Lett.*, **194**, L57–L59, 1974.

GALEEV, A.A. and V.V. KRASNOSELSICK, Cylindrical Lanmuir turbulence in the earth's magnetosphere as a source of kilometric radiation, *Sov. Phys.*, **II**, 1978 (in press).

GURNETT, D.A., The earth as a radio source: Terrestrial kilometric radiation, *J. Geophys. Res.*, **79**, 4227–4238, 1974.

HIROTA, R., Exact solution of the modified Korteweg-de Vries Equation for multiple collisions of Solitons, *J. Phys. Soc. Japan*, **33**, 1456–1458, 1972.

JONES, D., Mode-coupling of Z-mode waves as a source of terrestrial kilometric and Jovian decametric radiations, *Astron. Astrophys.*, **55**, 245–252, 1977.

KARPMAN, V.I., *Nonlinear Waves in Dispersive Media*, p. 141, Pergamon Press, New York, 1975.

KAW, P.K., K. NISHIKAWA, Y. YOSHIDA, and A. HASEGAWA, Two dimensional and three dimensional envelope solitons, *Phys. Rev. Lett.*, **35**, 88–91, 1975.

KENNEL, C.F. and J.E. MAGGS, Possibility of detecting magnetospheric radio bursts from Uranus and Neptune, *Nature*, **261**, 299–301, 1976.

KURTH, W.S., M.M. BAUMBACK, and D.A. GURNETT, Direction finding measurements of auroral kilometric radiation, *J. Geophys. Res.*, **80**, 2764, 1975.

MAGGS, J.E., Coherent generation of VLF hiss, *J. Geophys. Res.*, **81**, 1707–1724, 1976.

MAGGS, J.E., Electrostatic noise generated by the auroral electron beam, *J. Geophys. Res.*, **83**, 3173–3188, 1978a.

MAGGS, J.E., Formation of electrostatic upper hybrid and whistler wave solitons, 1978b (in preparation).

MAGGS, J.E., Radiation from soliton turbulence as a source of kilometric radiation, 1978c (in preparation).

MELROSE, D.B., An interpretation of Jupiter's decametric radiation and the terrestrial kilometric radiation as direct amplified gyroemission, *Astrophys. J.*, **207**, 651–662, 1976.

MENG, C.I., Simultaneous observations of low-energy electron precipitation and optical auroral arcs in the evening sector by the DMSP 32 satellite, *J. Geophys. Res.*, **81**, 2771–2785, 1976.

MENG, C.I., A.L. SNYDER, and H.W. KROEHL, Observations of auroral westward traveling surges and electron precipitations, *J. Geophys. Res.*, **83**, 575–585, 1978.

MORALES, G.J. and Y.C. LEE, Nonlinear filamentation of lower-hybrid cones, *Phys. Rev. Lett.*, **35**, 930–933, 1975.

OYA, H., Origin of Jovian decametric wave emissions—Conversion from the electron cyclotron plasma wave to the ordinary mode electromagnetic wave, *Planet. Space Sci.*, **22**, 687–708, 1974.

PALMADESSO, P., T.P. COFFEY, S.L. OSSAKOW, and K. PAPADOPOULOS, Generation of terrestrial kilometric radiation by a beam-driven electromagnetic instability, *J. Geophys. Res.*, **81**, 1762–1770, 1976.

WONG, A.Y. and B.H. QUON, Spatial collapse of beam-driven plasma waves, *Phys. Rev. Lett.*, **34**, 1499–1502, 1975.

ZAKHAROV, V.E., Collapse of Langmuir Waves, *Soviet Phys. JETP*, **35**, 908–914, 1972.

ZAKHAROV, V.E. and A.B. SHABAT, Exact theory of two-dimensional self focusing and one dimensional self-modulation of waves in nonlinear media, *Soviet Phys. JETP*, **34**, 62–69, 1972.

ZHELEZNYAKOV, V.V., *Radio emission of the Sun and Planets*, p. 376, Pergamon Press, New York, 1970.

RAPID AURORAL FLUCTUATIONS
AND
ASSOCIATED PHENOMENON

Observations of Rapid Auroral Fluctuations

Takasi OGUTI

Geophysics Research Laboratory, University of Tokyo,
Tokyo 113, Japan

(Received February 7, 1978)

Rapid auroral fluctuations are categorized into several groups according to their deformation modes, and a composite distribution of these modes is given with respect to the auroral oval. The distribution could impose some restrictions on the nature of plasma in the magnetosphere involved in the auroral precipitations.

1. Introduction

The rapid fluctuation of the aurora, in its luminosity and in its pattern as well, is a fairly common condition of auroral displays and has long been studied, since as early as the end of the 9th century. The early studies have inevitably been based on visual observations so that these have mostly been qualitative descriptions on the most remarkable features with a rather limited range in periodicity. These observations have been summarized by CAMPBELL and REES (1961).

The second phase of the research was opened by the introduction of photomultipliers. Since then, a quantitative discussion of the temporal variations has become possible. Subsequently the research activity on this subject has greatly increased and great progress has been made. That progress has been summarized in several comprehensive papers, e.g., CAMPBELL and REES (1961), SHEPHERD and PEMBERTON (1968) and KVIFTE and PETERSEN (1969) and books such as *Atmospheric Emissions* (OMHOLT et al., 1969), *The Optical Aurora* (OMHOLT, 1971) and *Aurora* (VALLANCE-JONES, 1974).

These studies, in spite of various locations of observation, seem to approach general agreement on several points regarding the rapid auroral fluctuations such as latitude dependence and local time dependence of the occurrence (HEPPNER, 1954; CAMPBELL and REES, 1961; AKASOFU, 1964, 1968; VICTOR, 1965; JOHANSEN and OMHOLT, 1966; OMHOLT and BERGER, 1967; OMHOLT et al., 1969; and KVIFTE and PETERSEN, 1969) spectral feature of temporal variation (e.g., CAMPBELL and REES, 1961; VICTOR, 1965; JOHANSEN and OMHOLT, 1966; PAULSON and SHEPHERD, 1966; PAULSON et al., 1967; SHEPHERD and PEMBERTON, 1968), and phase differences between several characteristic emissions (PAULSON and SHEPHERD, 1965; EVANS and VALLANCE-JONES, 1965). The relation to the variation in temperature in aurora (HILLIARD and SHEPHERD, 1966), relation to geomagnetic pulsations (CAMPBELL and REES, 1961; VICTOR, 1965; PAULSON et al., 1967; CAMPBELL, 1970), coherency between the fluctuations from separated fields (Shepherd and Pemberton, 1968; Campbell, 1970) have been studied also during this phase of the research.

However, photometers alone are not suitable for the determination of rapid deformation modes of aurora. These also give important information on the physical processes of acceleration and/or pitch angle scattering of electrons in the magnetosphere.

The third phase of research thus is indebted to the introduction of high sensitive TV cameras (Davis, 1966, 1967; Scourfield and Parsons, 1969a) that enabled us to record the rapid fluctuations with a sampling rate as high as 60 frames per second. Cresswell and Davis (1966) have confirmed the occurrence region of pulsating aurora to be the equatorward boundary of the auroral activity, Beach et al. (1968) have visualized the flickering aurora, defining it as a group of small patches (1–5 km) that usually move very rapidly (~ 5 km/sec), and Hallinan and Davis (1970) have found the so-called "auroral curl." Fast auroral waves noted by Cresswell and Belon (1966) and by Cresswell (1968) on the basis of the data from separated field photometers have been confirmed by TV imaging by Scourfield and Parsons (1971a). Scourfield and Parsons (1971b) also found the expansion mode of pulsating patches, and Cresswell and Davis (1966) and Scourfield et al. (1972) have shown the independence of fluctuations between different patches, and in contrast to this, a certain coherency between the different parts of a single patch. Oguti (1975a), and Oguti and Watanabe (1976) have shown the existence of a poleward propagation mode and its close relation to geomagnetic pulsations.

In spite of these increased observations and several morphological studies (e.g., Cresswell, 1972; Thomas et al., 1973; Oguti, 1975a) on the basis of TV observation, the classifications of variation modes seem still not to be in general agreement, and even a simple morphological picture is still missing as to the distribution of each deformation mode. In this paper, the author tentatively proposes a simple classification by re-examining previous results, and subsequently synthesizes distribution of rapid auroral fluctuations with respect to the global auroral pattern.

2.   Classifications of Fluctuation Modes

While not everywhere the same at a given level of activity the aurora is generally understood to follow time sequence involving the substorm growth phase, expansion phase and recovery phase. This idea suggests that the terms "expansion" and "recovery," at least, define the nature of regions rather than of the temporal phases as a whole. Furthermore this means that the aurora left in the wake or in at lower latitudes could be already in a state of recovery or constitute after effects, while the auroras at the front of an expansion are still in the expansion state.

Following this idea, the aurora is first placed into three categories according to three typical regions, namely the region along and near the expansion front, in the wake, both of them in the dusk to midnight sectors, and in the equatorward region in the dawn sector. Thus the tentative classifications of rapid auroral fluctuations are proposed as follows:

1) In and near discrete forms (Expansion Region)
   Splitting-unfolding or folding over (including outward streaming rotation or leading-rotation)
   Flickering
2) In smoke-like forms (Wake Region)
   Rotational variations (curl or trailing rotation, fluttering)
   Folding and meandering (including some of fluttering)
3) In pulsating forms (Recovery or After Effect Region)
   Pulsation
   Expansion (including core expansion, streaming and flaming)
   Poleward propagation of arcs (including streaming)
   Equatorward propagation of arc fragments (auroral waves)
   Flash

Examples are shown below.

3. Splitting-Unfolding and/or Folding-over

The left half of Fig. 1 shows an example of what the author calls "splitting-unfolding" of a discrete band. The sequence of photographs is taken every 1/2 sec. The right half schematically illustrates how the splitting and unfolding occur. This also indicates how the protrusion to the upper side (high-latitude side) propagates toward the right (eastward) and how it reverses on the lower side (low-latitude side) during a course of a splitting-unfolding. The length of the arc shown in these pictures is about 80 km, so that the characteristic scale of the deformation here is about 50–60 km.

The splitting-unfolding or folding-over can occur in various scales in the fashion shown (Oguti, 1975c). Among them, the smallest (1–5 km) and the fastest (several to 20 Hz) often occurs along the discrete arc prior to and during the active phase of auroral expansion; the expansion not necessarily of global scale but of local and small scale. An example is shown in Fig. 2, where the left hand side sequence of photographs, reproduced at the rate of 60 frames per second, indicates deformations of a very thin arc. The direction of this arc is roughly east to west; the left is toward the east, and the top is toward low latitude. The length of arc covered in the pictures here is about 100 km. In the right half of the diagram, the solid lines are traces of the shift of identified peeled-off structures that protrude on the upper side (low latitude side) and the dotted lines are those on the high latitude side, both deduced from the sequential pictures at left. The direction of propagation evidently is reversed beyond the main band, and the trends reveal a strong sheer along the arc clockwise viewed along the magnetic field and consistent with clockwise unfolding shown in Fig. 1.

This mode is apt to be regarded as the same group as flickering, but is classified

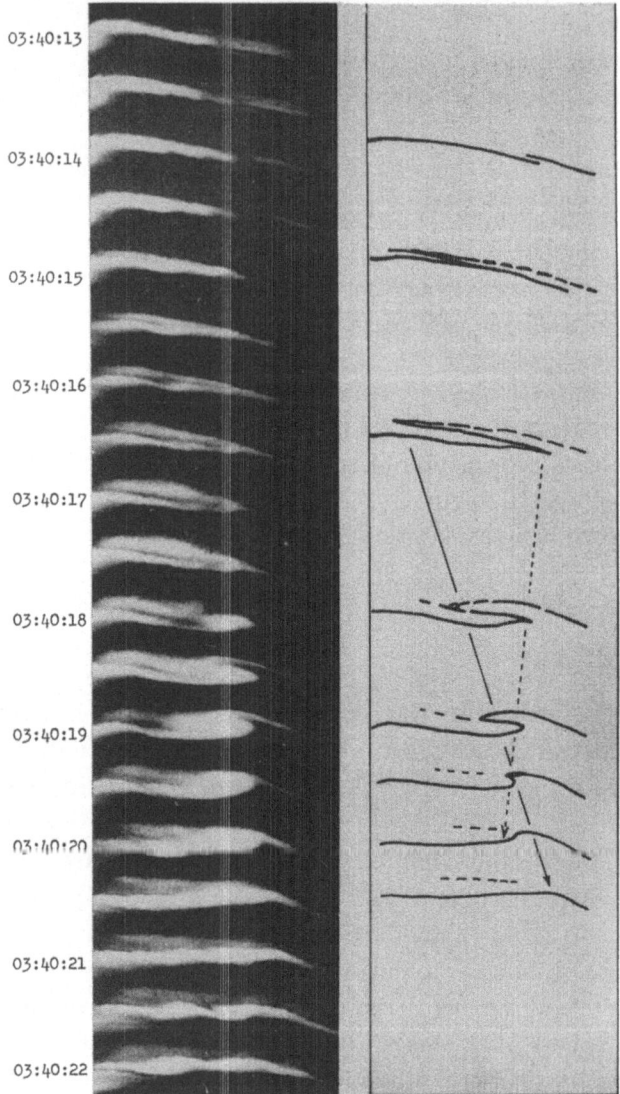

Fig. 1. Typical "splitting-unfolding" deformation of an arc seen near
magnetic zenith; 03:40:13–03:40:22 UT on May 18, 1971, at Syowa
Station. Pictures on the left hand side are sequential reproductions
from video tape with a time interval of 1/2 sec and with an exposure
of 1/60 sec each. The direction of the arc is roughly in east-west; left
is to west and the top is to high latitude. The length of the arc covered here is about 80 km. The right half schematically illustrates
how the splitting-unfolding proceeds along this arc. Protrusions on
each side of the arc propagate in reverse direction as shown by solid
and dotted arrows respectively.

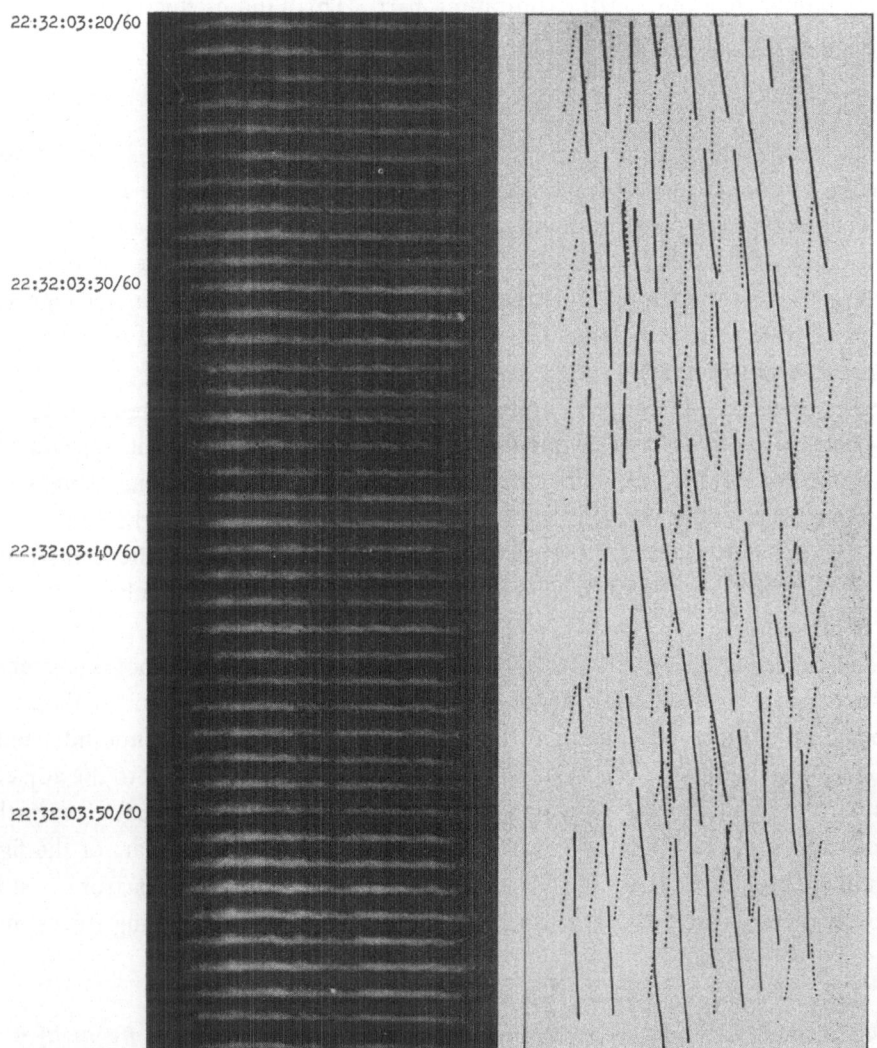

Fig. 2. An example of rapid splitting-unfolding along a discrete arc seen near the zenith.
The pictures on the left are sequential reproductions from video tape at the rate of 60
fields/sec; 22:32:03:20/60–22:32:03:59/60 UT on Aug. 25, 1971 at Syowa Station. The direction of the arc is roughly in east-west; left is to east and the top is to equatorward. The
right half indicates the shifts of protrusions (split ends) in the same fashion as in Fig. 1.
Solid lines are the traces of upper (equatorward) protrusions and the dotted lines are of
lower (poleward) protrusions.

here otherwise because the mode is different from the flickering defined by BEACH
et al. (1968). The periodic splitting here always appears to be a feature of the discrete band itself whereas the flickering of patches appears mostly in the remnants
just behind the expansion aurora or behind the fluttering of discrete band. This
category could include also outward streaming rotation or leading rotation (OGUTI,
1974) because it can be a kind of splitting-unfolding process where the original arc

is much fainter than the splitting-unfolding part. This kind of fluctuation usually is
related to auroral hiss emissions (Oguti, 1975b).

3.1   Flickering

The condition of flickering is well defined by Beach et al. (1968), such that the
flickering aurora consists of a number of interconnected, rapidly moving diffuse
spots with scale of 1–5 km in diameter. This condition of the display is often found
in the remnant diffuse arc behind the expanding discrete band or behind a rapid
fluttering of discrete band. The frequency of the flickering aurora is roughly the
same or a little less than that of rapid splitting-unfolding described in the previous
section (several to 10 Hz).

Sometimes the spot appears to rotate clockwise around a fixed point or to rotate
with coherent periodic change in ellipticity. On other occasions, it appears to be
illuminated by fainter, wave-like structures periodically. Sometimes bright spots
can even be stationary and exist without change in luminosity, and the fainter wave-
like structures only represent the periodic variations. In any case flickering reveals
itself as a vibrating column along field line when viewed perpendicular to the magnetic field.

The coherency of the fluctuation is usually good so far as one spot is concerned.
The spot changes periodically not only in its luminosity but also in its shape, size
and position. The brightening and the motion of different spots are not independent
of each other and sequential brightenings usually occur from one spot to the adjacent.
The situation is seen in Fig. 3 which indicates the temporal and spatial variations
of flickering spots in a form of time-position display. The upper part of the figure
is obtained through a slit in NE-SW direction and the lower comes from a slit in
NW-SE direction, both 40 km in length. The several upward moving traces in the

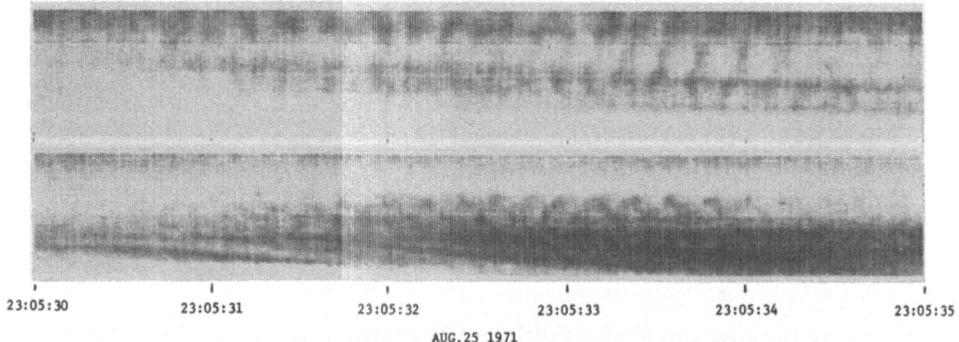

23:05:30   23:05:31   23:05:32   23:05:33   23:05:34   23:05:35

AUG. 25 1971

Fig. 3.   Time-position display of a flickering aurora seen near the magnetic zenith, 23:05:30–
23:05:35 UT on Aug. 25, 1971 at Syowa Station. The top and bottom figures are obtained
through slits in front of monitor TV in NE-SW and in NW-SE directions respectively.
The ordinates of both pictures cover about 40 km. The darkness represent the auroral
luminosity. Sequential upward shifts of brightness from spot to spot are seen during the
period from 23:05:33 to 23:05:34 in the top figure.

upper part from 23:05:33 to 23:05:34 represent 4 different spots; the upward motion of the traces indicates the sequential brightening of these spots, from NE to SW.

3.2 Rotational variations (curls or trailing rotations)

The curl (HALLINAN and DAVIS, 1970) or trailing rotations (OGUTI, 1974) some-

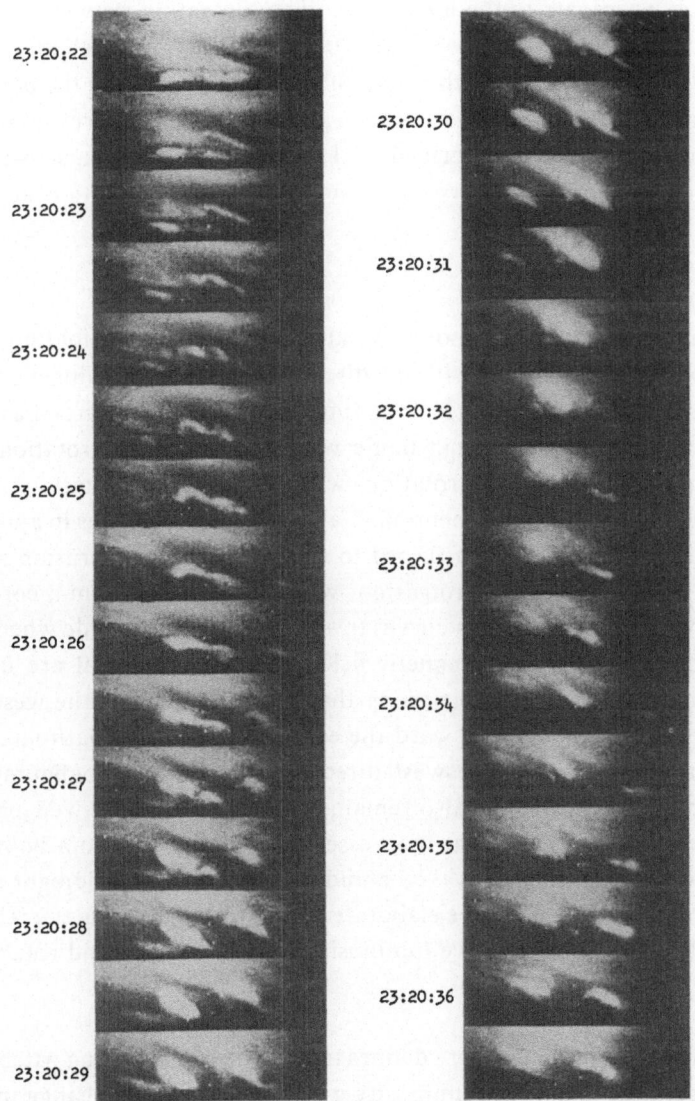

Fig. 4. An example of q-p occurrence of curls (or trailing rotations) with protrusions seen near the magnetic zenith, 23:20:22:0–23:20:36:5 UT on Aug. 25, 1971, at Syowa Station. The pictures are reproductions from video tape at the rate each 1/2 sec with exposure of 1/60 sec. Length covered here is about 80 km, the left is to west and the top is to poleward. q-p generations and leftward (westward) propagations of the rotational structures are evident.

times show a characteristic quasi-periodic condition both in space and in time. The structure usually emerges from a main arc as a protrusion that consists of one or a few rotational cores, An example of such a repetition of the trailing rotation is shown in Fig. 4. The arc here is roughly in east-west direction; left is toward west and the top is to high latitude; the length is about 80 km. Several groups of small trailing rotations are found to repeatedly appear along an arc in a form of protrusions and they propagate to the left (west). The separation between each group of trailing rotations is found to be about 30 km, the period is about 5 sec and the speed here is about 8 km/sec in this particular example. During the period of occurrence of bright and rapid trailing rotations, structureless q-p chorus emission (similar to the post midnight chorus described by TSURUTANI and SMITH, 1974) often can be observed, but the one-to-one correspondence between the rotation and the chorus emission is not yet confirmed.

3.3   Sheer folding and meandering

Another typical mode in smoke-like aurora in the wake is folding likely due to sheer motion, or meandering, without discernible rotations. This mode sometimes is periodic in space and quasi-periodic in time. This may be regarded as a less active and a larger "splitting-unfolding" mode with less prominent rotational unfolding motion, or propagating trailing rotations with no prominent rotations.

Although the four modes mentioned above show differences in appearance, two common features are found with regard to the sense of the protrusion and the sense of rotation. That is to say, a protrusion, when it generates from a core or from an arc and grows outwards, always curves toward the right-hand side when viewed parallel (not anti-parallel) to the magnetic field. Hence the auroral arc usually forms a saw-tooth structure with protrusions that protrude toward the west and toward high latitude or that protrude toward the east and toward low latitude. Even when the main arc is not in the east-west direction, the relative configuration is as described. The sense of rotation also remains clockwise when viewed parallel to the magnetic field. These two are the most essential features of aurora both in the poleward expanding front and the wake behind it in the dusk to midnight sectors. The varieties of appearances could be elaborated upon by considering the relative speeds, relative scale of sizes and relative luminosities of protrusions and rotations.

3.4   Pure pulsation

The term pure pulsation here defines independently pulsating patches and bands with quasi-periodic change in luminosity without remarkable change in pattern and in position, except the effect of slow drift as a whole. This is the typical condition of auroral activity in the dawn sectors near the low latitude border of the auroral activity, as noted by CRESSWELL and DAVIS (1966). It has a broad spectral peak around 10 seconds, distributed in a range of periods from a few seconds to a few tens of second, and, more importantly, the period of this mode seems not to depend much upon geomagnetic latitude, although the latitude of its occurrence undoubtedly varies according to the general activity. The persistence of this 10-sec band is evident in a fairly wide range of latitudes, e.g., from Churchill (IMS observation on Feb. 19, 1977) to Saskatoon (e.g., SHEPHERD and PEMBERTON, 1968). Another important point is that it often includes a period other than and much shorter than 10 sec. That is to say, the 'switch-on' level of luminosity of a pulsating patch often fluctuates with period from 0.1 to 1 sec. This has been noted by OMHOLT et al. at Tromsö (1969) and by OGUTI at Syowa Station (1976).

The pulsating patches and arc fragments usually form several groups that may be calld "core of pulsations" (see Fig. 6). The cores usually occur aligned along the low latitude border of the auroral activity, and are separated some hundreds of kilometers from each other. This is the most active part of the pulsating aurora. This mode usually relates to geomagnetic pulsations observed under or near the patches, but the correlation between them is usually poor because a number of independently pulsating patches can incoherently affect geomagnetic variations.

3.5 Expansion mode

Although very few authors have reported on the expanding patch of aurora (e.g., SCOURFIELD and PARSONS, 1971b) this mode really exists in the core region mentioned above. This might be placed into a pulsation category since the pulsating patch usually shows more or less expansion tendency during the period of its increasing luminosity, but the association of chorus groups with this mode, as seen in Fig. 5, could indicate that it is qualitatively different from the pulsation that usually does not relate to chorus emissions. This mode could appear also as a streaming mode when expansion occurs along the elongated direction of patches or along the interconnection structures between patches or cores.

Another expansion is the expansion of the core. While the individual patches pulsate with about 10-sec period, a core can expand by increasing the number of the member patches or by expansion of individual patches with a longer period, about 10–50 sec in a quasi-periodic fashion. The period of the core expansion likely depends on latitude, being longer in higher latitudes, in contrast to the persistense of the period of 10-sec band. Radar aurora observed from Ottawa (MCDIARMID and MCNAMARA, 1972) also indicates the existence of non-propagating, periodic echoes with the same period range.

3.6 Poleward propagating mode

The poleward propagating mode of aurora is fairly common condition at Syowa Station (OGUTI, 1975a; OGUTI and WATANABE, 1975) although few have been reported at other locations. This mode is a common feature of radar aurora, too (KANEDA et al., 1964; MCDIARMID and MCNAMARA, 1972). The poleward propagating aurora at Syowa (OGUTI and WATANABE, 1975) shows characteristic linear relations between repetition frequency and propagation speed similar to that of radar aurora found by MCDIARMID and MCNAMARA (1972). The similarity of the characteristics seems to be sufficient to conclude the identity of the two poleward propa-

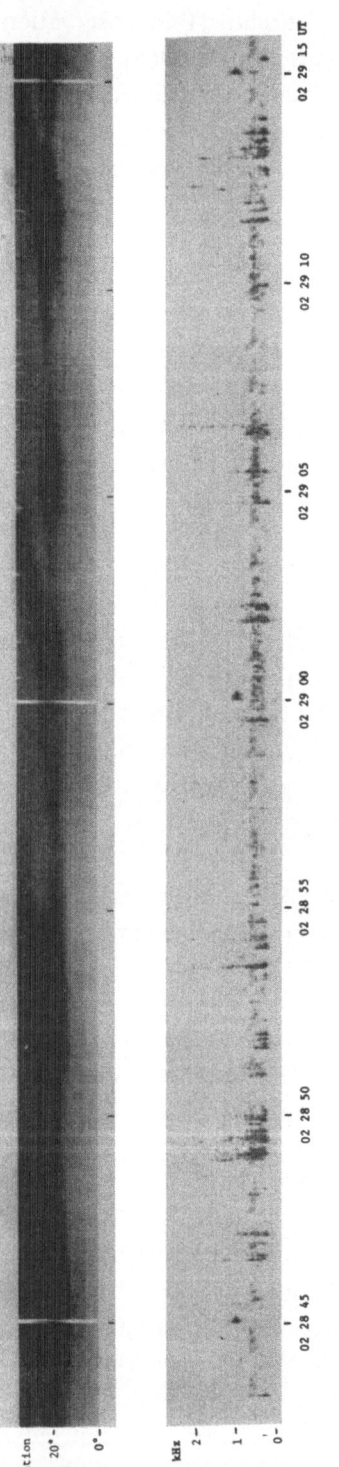

Elevation

20° —

0° —

kHz

2 —

1 —

0 —

02 28 45   02 28 50   02 28 55   02 29 00   02 29 05   02 29 10   02 29 15 UT

July 22 1971

Fig. 5

Fig. 5.   Time-position display of "core" expansions and poleward propagating modes; 02:28:43–02:29:16 UT on July 22, 1971 at Syowa Station (top panel). Ordinate covers about 40° of elevation from the equatorward horizon. The core expansion is seen in a horizontal trace and poleward propagating mode is seen in up-going traces, both of them being in q–p fashion. The chorus emissions in a time-frequency spectrogram in the bottom panel indicate that they are related to expansion mode but not to the poleward propagation mode.

Fig. 6.   Distributions of rapid auroral fluctuations. Splitting-unfolding and flickering are common conditions in and near an active discrete arc, and pulsation and flash near low latitude borders of active diffuse aurora in the dawn quadrant as well. The propagating modes (both of poleward and equatorward) occur quite frequently but not always.

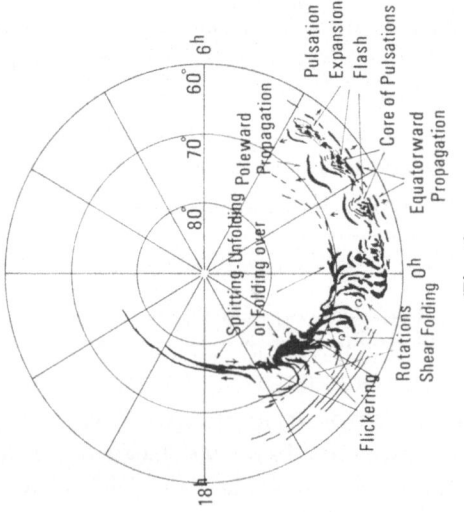

Fig. 6

gating phenomena. The poleward propagating mode appears in the region somewhat higher in latitude than the region of pure pulsation (core), usually in a shape that envelops the poleward side of the cores. The period of the propagating mode is longer than both the period of pulsating patches and the core expansion occurring at the same time. Very close association with geomagnetic pulsations is an important point of this mode (OGUTI and WATANABE, 1976). When the propagation has east-west components rather than poleward, the streaming condition appears along auroral structures usually elongated in east-west direction.

Figure 5 (top) shows the typical poleward propagating aurora along with the q-p expansion of a core in a form of time-position display again; where the top is to high latitude and the slit covers 40° of elevation from the equatorward horizon. At the middle of the figure, the q-p expansion (q-p increase in number of patches in a core) of a core is seen as q-p thickening of the traces without change in its position, while the propagating mode is seen as traces moving upward with time.

The existence of the poleward propagating mode indicates that the quasi-periodic variations of the plasma in the magnetosphere, some of them at least, are generated in the magnetosphere, in contrast to the belief that all come from outside.

3.7 Equatorward propagating mode

This mode is called "fast auroral wave" by CRESSWELL and BELON (1966). As stated by CRESSWELL (1968), it is really a fast wave-like propagation of faint arc fragments usually elongated in east-west direction, with horizontal scale several tens of km in length and about 10 km in width. The propagation speed is in a range from several tens of km/sec to 200 km/sec, and the repetition time is in a range from 1 to several seconds. The horizontal distance of the traverse of one wave is roughly the same to the poleward propagating mode in the previous section, but the mode of the propagation of the wave here is always smooth with a constant speed during its traverse, in contrast to the stepwise propagation of the poleward propagating mode. The occurrence of this mode seems to be limited to the region below geomagnetic latitude 65°. No evidence of this mode has been found at Syowa Station whereas it is often found at College (e.g., CRESSWELL and BELON, 1966; CRESSWELL, 1968) and at Cold Lake (SCOURFIELD and PARSONS, 1971a). The wave appears to be generated near the low-latitude boundary of or sometimes within the core of pulsating aurora.

3.8 Flash mode

Toward dawn, the pulsating aurora and the expanding aurora, seem to be gradually replaced by the flash mode. The temporal characteristics of this mode are similar to the pulsating aurora but the faster modulation dominates over the 10-sec band in the flash mode. The structure of the aurora involved is diffuse, faint and widespread within surfaces over 100 km in horizontal scale, in contrast to the fairly discrete, small-scale structure of the pulsating patches.

A strong impression exists that the fast modulation of pulsating patches originally is the nature of the flash mode and that it modulates the switch on level of the pulsating patches by overlapping the patches, whereas the 10-sec band is characteristic of the pulsating patches themselves. This mode is usually associated with VLF chorus emissions, but one to one correspondence between each flash and each riser of chorus is quite hard to find because so many chorus risers and flash auroras usually occur at the same time. The aurora studied by UNGSTRUP (1966), although he called it flickering, must be this mode. Satellite observations of precipitation associated with chorus (e.g., OLIVEN and GURNETT, 1968; HOLZER et al., 1974) likely relate also to this. Most of X-ray bursts so far studied (e.g., ANDERSON et al., 1966; PARKS et al., 1968; VENKATESAN et al., 1968; FOSTER and ROSENBERG, 1976) are likely related to this mode too. Because the change in luminosity of the flash mode is usually so coherent over a fairly widespread area, coherent pulses of X-rays are to be expected. By contrast pulsating patches are smaller and temporally independent so would possibly result in noise-like X-ray variations due to overlapping of effects from many patches (SCOURFIELD et al., 1970).

3.9 Propagation along an arc (streaming)

This mode likely appears associated with the expansion mode or poleward propagation mode, as mentioned before, and hence could be classified into either category. Its period is in a range roughly the same as the poleward propagation mode.

3.10 Propagation along the magnetic field (flaming)

Although quite many have reported on this mode (e.g., CRESSWELL, 1969; SCOURFIELD and PARSONS, 1969b) it could not be found during the observations of several years at Syowa Station, Antarctica and in Canada, except for an appearance of flaming due to a perspective effect of horizontal motions. It seems to be a very rare condition of aurora and its distribution is not as clear as the other modes.

4. Summary and Discussion

The rapid fluctuation of aurora shows typical modes of variation according to location with respect to the auroral oval. The characteristics of auroral modes in the expansion region and in the wake just behind the expansion front are a combination of a right hand side splitting (or right hand side curving extension of an arm) and of clockwise rotation viewed along the magnetic field. The variations can occur in a wide range of horizontal scale sizes. The smallest and fastest of them are the splitting-unfolding process in the expansion front and flickering in the wake, whereas the largest and the slowest may be the auroral bulge. The middle scale and less active examples are the curl or trailing rotations where the rotational motion dominates, and shear folding where the right hand side splitting dominates.

In the region of recovery (or in the region of after effects), namely in low-latitude part of auroral oval, especially near dawn, the typical condition of the auroral display is pulsation. Here several cores of pulsating aurora that consist of a number of pulsating or expanding patches appear in an alignment along the oval. The structure of the core gradually becomes diffuse, or from another point of view, the patches are replaced or overlapped by diffuse surfaces that flash with period from 0.1–1 sec near dawn. Much to envelop the core regions to the poleward side of them, the poleward propagating auroras appear. The equatorward propagating mode (auroral wave), on the other hand, appears on the equatorward side of the core region. Thus, the distribution of each mode along an oval can be shown as in Fig. 6.

One of the most important characteristics of the auroral rapid variation is that the energy of the corresponding electrons is higher than that in more common aurora (EVANS, 1976; WHALEN et al., 1971; BRYANT et al., 1971; COURTIER et al., 1974; BRYANT et al., 1975; BROWN et al., 1976). The energy range of the corresponding electrons, in conjunction with the association of VLF chorus emissions and the association of geomagnetic pulsation as well, suggests that the coupling between energetic auroral electrons, VLF waves and HM waves in the magnetosphere plays an essential role in the region of after-effect. Since the core region of pulsating aurora is the most active region of pulsation, this region possibly links to the inner edge of the injected energetic electrons where an inward steep gradient of the plasma pressure is expected, and hence where instabilities such as due to drift-Alfvén and drift-bounce coupling tend to occur. Hence it is possible that the HM waves are first generated in this region and then modulate the pitch angle of the electrons, which subsequently results in an enhancement of cyclotron coupling between energetic electrons and VLF waves under a possible influence of cold plasma. Another possibility that the coupling between VLF waves and energetic electrons, resulting in the rapid escape of electrons in some regions in the magnetosphere, subsequently produces HM waves in this region, could not be overlooked. The propagating modes are understood to be due to compressional HM-waves generated here. HM waves could also be produced near the ionosphere as a result of q-p precipitation of electrons through q-p change in ionospheric electric current, possibly being fed back to the magnetosphere.

On the other hand, the precipitation in the expansion region must be understood in another way. The clockwise rotation suggest that a charge excess is one of the main reason of the precipitation here. Here, the relation between auroral hiss emissions and splitting-unfolding and flickering (the latter appears to relate to stationary hiss emission although this is not confirmed yet) is understood in terms of coherent Cerenkov emission that might be generated from some corrective motions of electrons at the steep boundary of small structures in the electron beams such as that corresponding to the splitting-unfolding and flickering. The fact that auroral hiss is mostly observed only when the corresponding aurora shows rapid variation supports this mechanism. The possibility of a wave conversion mechanism from electrostatic to electromagnetic near or in the auroral precipitations also can not be excluded. In any case, the close relation between auroral rapid variation and burst-like enhancement of auroral his emission not only in time (OGUTI, 1975b) but also in space (HAYASHI et al., 1978 (this issue)) suggests that the two phenomena are coupled, not in a wide-spread region in the magnetosphere, but in a space relatively near to the electron precipitation.

REFERENCES

AKASOFU, S.-I., The development of the auroral substorm, *Planet. Space Sci.*, **12**, 273–282, 1964.

AKASOFU, S.-I., *Polar and Magnetospheric Substorms*, Reidel Publ. Co., Dordrecht, Holland, 1968.

ANDERSON, K.A., L.M. CHASE, H.S. HUDSON, M. LAMPTON, D.W. MILTON, and G.K. PARKS, Ballon and rocket observations of auroral-zone microbursts, *J. Geophys. Res.*, **71**, 4617–4629, 1966.

BEACH, R., G.R. CRESSWELL, T.N. DAVIS, T.J. HALLINAN, and L.R. SWEET, Flickering, a 10-cps fluctuation within bright auroras, *Planet. Space Sci.*, **16**, 1525–1529, 1968.

BROWN, N.B., T.N. DAVIS, T.J. HALLINAN, and H.C. STENBAEK-NIELSEN, Altitude of pulsating aurora determined by a new instrumental technique, *Geophys. Res. Lett.*, **3**, 403–404, 1976.

BRYANT, D.A., G.M. COURTIER, and G. BENNETT, Equatorial modulation in a pulsating aurora, in *The Radiating Atmosphere*, edited by B.M. McCormac, pp. 170–175, Reidel Publ. Co., Dordrecht, Holland, 1971.

CAMPBELL, W.H., Rapid auroral luminosity fluctuations and geomagnetic field pulsations, *J. Geophys. Res.*, **75**, 6182–6208, 1970.

CAMPBELL, W.H. and M.H. REES, A study of auroral coruscations, *J. Geophys. Res.*, **66**, 41–55, 1961.

COURTIER, G.M., M.J. SMITH, and D.A. BRYANT, Auroral electrons observed using a mother-daughter rockets, in *Magnetospheric Physics*, edited by B.M. McCormac, pp. 207–211, Reidel Publ. Co., Dordrecht, Holland, 1974.

CRESSWELL, G.R., Fast auroral waves, *Planet. Space Sci.*, **16**, 1453–1464, 1968.

CRESSWELL, G.R., Flaming auroras, *J. Atmos. Terr. Phys.*, **31**, 179–183, 1969.

CRESSWELL, G.R., The morphology of displays of pulsating auroras, *J. Atmos. Terr. Phys.*, **34**, 549–554, 1972.

CRESSWELL, G.R. and A.E. BELON, Observations of fast auroral waves, *Planet. Space Sci.*, **14**, 299–301, 1966.

CRESSWELL, G.R. and T.N. DAVIS, Observations on pulsating auroras, *J. Geophys. Res.*, **71**, 3155–3163, 1966.

DAVIS, T.N., The application of image orthicon techniques to auroral observation, *Space Sci. Rev.*, **6**, 222–247, 1966.

DAVIS, T.N., Cinematographic observations of fast auroral variations, in *Aurora and Airglow*, edited by B.M. McCormac, pp. 133–141, Reinhold Publ. Co., New York, 1967.

DAVIS, T.N., Auroral morphology, *Planet. Space Sci.*, **20**, 1369–1373, 1972.

EVANS, D.S., A 10-cps periodicity in the precipitation of auroral-zone electrons, *J. Geophys. Res.*, **72**, 4281–4291, 1967.

EVANS, W.F.J. and A. VALLANCE-JONES, Some observations of type B aurora with a multichannel photometer, *Can. J. Phys.*, **43**, 697–704, 1965.

FOSTER, J.C. and T.J. ROSENBERG, Electron precipitation and VLF emissions associated with cyclotron resonance interactions near the plasmapause, *J. Geophys. Res.*, **81**, 2183–2192, 1976.

HALLINAN, T.J. and T.N. DAVIS, Small-scale auroral arc distortions, *Planet. Space Sci.*, **18**, 1735–1744, 1970.

HAYASHI, K., K. TSURUDA, S. KOKUBUN, T. OGUTI, R.E. HORITA, and T. WATANABE, Arrival directions of VLF emissions and corresponding activities in aurora and geomagnetic pulsations, *J. Geomag. Geoelectr.*, **30**, 361–362, 1978.

HEPPNER, J.P., Time sequences and spatial relations in auroral activity during magnetic bays at College, Alaska, *J. Geophys. Res.*, **59**, 329–338, 1954.

HILLIARD, R.I. and G.G. SHEPHERD, Upper atmospheric temperatures from Doppler line width-IV, A detailed study using the OI 5577 Å auroral and nightglow emission, *Planet. Space Sci.*, **14**, 383–406, 1966.

HOLZER, R.E., T.A. FARLEY, R.K. BURTON, and M.C. CHAPMAN, A correlated study of ELF waves and electron precipitation on Ogo 6, *J. Geophys. Res.*, **79**, 1007–1013, 1974.

JOHANSEN, O.E. and A. OMHOLT, A study of pulsating aurora, *Planet. Space Sci.*, **14**, 207–215, 1966.

KANEDA, E., S. KOKUBUN, T. OGUTI, and T. NAGATA, Auroral radar echoes associated with Pc-5, *Rep. Ionosph. Space Res. Japan*, **18**, 165–172, 1964.

KVIFTE, G.J. and H. PETERSEN, Morphology of the pulsating aurora, *Planet. Space Sci.*, **17**, 1599–1607, 1969.

MCDIARMID, D.R. and A.G. MCNAMARA, Periodically varying radio aurora, *Ann. Geophys.*, **28**, 433–441, 1972.

OGUTI, T., Rotational deformations and related drift motions of auroral arc, *J. Geophys. Res.*, **79**, 3861–3865, 1974.

OGUTI, T., Metamorphoses of aurora, *Mem. Natl. Inst. Polar Res.*, *Ser. A, No. 12*, 1–101, 1975a.

OGUTI, T., Hiss emitting auroral activity, *J. Atmos. Terr. Phys.*, **37**, 761–768, 1975b.

OGUTI, T., Similarity between global auroral deformations in DAPP photographs and small scale deformations observed by a TV camera, *J. Atmos. Terr. Phys.*, **37**, 1413–1418, 1975c.

OGUTI, T., Recurrent auroral patterns, *J. Geophys. Res.*, **81**, 1782–1786, 1976.

OGUTI, T. and T. WATANABE, Quasi-periodic poleward propagation of on-off switching aurora and associated geomagnetic pulsations in the dawn, *J. Atmos. Terr. Phys.*, **38**, 543–551, 1976.

OLIVEN, M.N. and D.A. GURNETT, Microburst phenomena, 3. An association between microbursts and VLF chorus, *J. Geophys. Res.*, **73**, 2355–2362, 1968.

OMHOLT, A., *The Optical Aurora*, Springer-Verlag, New York, 1971.

OMHOLT, A. and S. BERGER, The occurrence of auroral pulsations in the frequency range 0.01–0.1 c/s over Tromsö, *Planet. Space Sci.*, **15**, 1075–1080, 1967.

OMHOLT, A., G.J., KVIFTE, and H. PETERSEN, Pulsating aurora, in *Atmospheric Emissions*, edited by B.M. McCormac and A. Omholt, pp. 131–143, Van Nostrand Reinhold Co., New York, 1969.

PARKS, G.K., F.V. CORONITI, R.L. MCPHERRON, and K.A. ANDERSON, Studies of magnetospheric substorm, 1. Characteristics of modulated electron precipitation occurring during auroral substorms, *J. Geophys. Res.*, **73**, 1685–1696, 1968.

PAULSON, L.V. and G.G. SHEPHERD, A cross spectral method for determining the mean lifetime of metastable oxygen atoms from photometric observations of quiet auroras, *J. Atmos. Terr. Phys.*, **27**, 831–841, 1965.

PAULSON, K.V. and G.G. SHEPHERD, Short lived brightness oscillations in active aurora, *Can. J. Phys.*, **44**, 921–924, 1966.

PAULSON, K.V., G.G. SHEPHERD, and P. GRAYTONE, A note on auroral-type fluctuations in the earth's electromagnetic field, *Can. J. Phys.*, **45**, 2813–2821, 1967.

SCOURFIELD, M.W.J. and N.R. PARSONS, An image intensifier-vidicon system for auroral cinematography, *Planet. Space Sci.*, **17**, 75–81, 1969a.

SCOURFIELD, M.W.J. and N.R. PARSONS, Auroral pulsations and flaming, *Planet. Space Sci.*, **17**, 1141–1147, 1969b.

SCOURFIELD, M.W.J. and N.R. PARSONS, Television imaging of fast auroral waves, *Planet. Space Sci.*, **19**, 437–442, 1971a.

SCOURFIELD, M.W.J. and N.R. PARSONS, Pulsating auroral patches exhibiting sudden intensity-dependent spatial expansion, *J. Geophys. Res.*, **76**, 4518–4524, 1971b.

SCOURFIELD, M.W.J., W.F. INNES, and N.R. PARSONS, Spatial coherency in pulsating aurora, *Planet. Space Sci.*, **20**, 1843–1848, 1972.

SCOURFIELD, M.W.J., G.R. CRESSWELL, G.R. PILKINGTON, and N.R. PARSONS, Auroral pulsations-television image and X-ray correlations, *Planet. Space Sci.*, **18**, 495–499, 1970.

SHEPHERD, G.G. and E.V. PEMBERTON, Characteristics of auroral brightness fluctuations, *Radio Sci.*, **3**, 650–658, 1968.

THOMAS, I.L., M.W.J. SCOURFIELD, and N.R. PARSONS, Classification of optical auroral pulsations, *Can. J. Phys.*, **51**, 2209–2215, 1973.

TSURUTANI, B.T. and E.J. SMITH, Post midnight chorus: A substorm phenomenon, *J. Geophys. Res.*, **79**, 118–127, 1974.

UNGSTRUP, E., Association between VLF emissions and flickering aurora, *J. Geophys. Res.*, **71**, 2395–2396, 1966.

VALLANCE-JONES, A., *Aurora*, D. Reidel Publ. Co., Dordrecht, Holland, 1974.

VENKATESAN, D., M.N. OLIVEN, P.J. EDWARDS, K.G. MCCRACKEN, and M. STEINBOCK, Microburst phenomena, 1. Auroral zone X-rays, *J. Geophys. Res.*, **73**, 2333–2343, 1968.

VICTOR, L.J., Correlated auroral and geomagnetic micropulsations in the period range 5 to 40 seconds, *J. Geophys. Res.*, **70**, 3123–3130, 1965.

WHALEN, B.B., J.R. MILLER, and I.B. MCDIARMID, Energetic particle measurements in a pulsating aurora, *J. Geophys. Res.*, **76**, 978–986, 1971.

Highlights in the Studies of the Relationship of Geomagnetic Field Changes to Auroral Luminosity

W.H. CAMPBELL

U.S. Geological Survey, Denver, Colorado 80225, U.S.A.

(Received February 7, 1978)

In the middle of the 18th century Celsius observed that there was a correspondence between a great aurora in Europe and the extreme motion of his observed compass needle. By the nineteenth century it was well established that the geomagnetic field always fluctuated violently at the height of an auroral display, that the high latitude zones of peak field disturbance and luminosity had similar locations, and that there were concurrent solar cycle changes in activity levels for the two phenomena. After the International Geophysical Year of 1957 to 1959 the correspondence of the ionospheric electrojet currents and auroral forms became a focus of observational programs. In these recent times the studies of a relationship between short-period pulsations of the geomagnetic field and luminosity pulsations (or the pulsations of bombarding, low energy electrons causing the auroral intensity changes) appeared prominently in the literature. A summary of the scientific progress in the study of these phenomena is presented in this paper.

1. Introduction

About two and one-half centuries have passed since the first discovery of a relationship between the earth's magnetic field variations and the Northern Lights. During that period the instrumentation for recording geomagnetic and auroral phenomena has improved greatly in comparison to those early years when observers watched angular variations of foot-long compass needles and shivered, out in the cold polar winter nights, while making notes on the auroral forms dancing overhead. This report will try to point out the highlights of scientific discovery over that long period. The emphasis of this paper will be on the recent studies of concurrent rapid fluctuations of field and luminosity. To develop that focus I will mention the historical milestones of discovery of a relationship between auroras and the geomagnetic field before the International Geophysical Years of 1957 to 1959, and I will refer to some more-recent selected papers that disclose the relationships between ionospheric current and luminosity. A more extensive review of the early history of geomagnetic phenomena can be found in CHAPMAN and BARTEL's (1940) book "Geomagnetism."

2. Early Discoveries

In the first half of the 18th century evidence of a relationship between the aurora and magnetic field began to appear in the literature. In 1716 HALEY reported a spectacular auroral display and described the characteristic ray alignment that was later found to be parallel to the earth's magnetic field lines. Celsius is credited (see HANSTEEN, 1819) with the first recognition of an association between the aurora and geomagnetic disturbances at Upsala, Sweden, in 1741 during a period of joint field observations with Graham of London. In the first detailed treatise on aurora de MAIRAN (1733, 1754) described the recognized connection of the two phenomena. Aurora in the southern hemisphere (at what we later learned to be a high-latitude magnetic dip location comparable to the auroral borealis) was observed in 1773 by that great world explorer Captain James COOK (1961). Enough auroral occurrences had been reported by late in the 18th century that WILCKE (1777) produced a catalog of corresponding auroral and magnetic events and he was able to deduce that auroral rays were surely aligned with the earth's magnetic dip direction. Finally, in 1790 CAVENDISH produced the first accurate determination of auroral intensity maximum at 84 to 114 km—a region later understood to be the current-carrying ionospheric level of our atmosphere.

Scientific progress was initiated the next century by a letter, in 1836, from Humbolt (see SCHERING, 1889) to the Royal Society of London. Excited by his successful scientific expedition to the New World he encouraged placement of magnetic observatories in the British Colonies. Humbolt's efforts resulted in the establishment of several stations with one at Toronto, Canada. By 1841 (GAUSS and WEBER, 1841) a special European "Magnetic Union" had obtained 10 to 20 sec interval measurements of several concurrent magnetic storms and auroras. SABINE (1851, 1852) analyzed the records from the new Toronto observatory and discovered a relationship between magnetic disturbances and sunspots. SABINE (1857) was also able to instruct the officers of the relief ship, "Plover," in magnetic and auroral observations which they subsequently carried out from the summer of 1852 to the summer of 1854 while at Point Barrow, Alaska (then Russian Territory), awaiting the possible arrival of a ship attempting a northwest passage across northern Canada. At this most valuable observatory, the best correspondence of the two phenomena were found under conditions of 24-hr darkness; a postmidnight maximum and a midday minimum in activity was reported.

PRESCOTT (1860), HANSTEEN (1860), and others described the strong induced currents on telegraph lines during a great aurora that extended across Europe in 1859. Although LOOMIS (1860) first proposed the existence of a latitude region for maximum luminosity occurrence, full credit for the first polar map of auroral frequency should go to FRITZ (1881) who plotted iso-occurrence lines that were concentric about the geomagnetic pole location.

The First Polar Year was established in 1882 to 1883 for an international effort of high-latitude research. Then, in 1898, BOLLER detailed a relationship between aurora and sunspot numbers which, considering SABINE's (op. cit.) magnetic observations, tied the auroral and magnetic phenomena even closer together.

BIRKLAND (1901) proposed that speeding elementary extraterrestrial particles followed geomagnetic field lines to bombard the atmosphere and cause auroras. He constructed laboratory models to give substance to his ideas while STÖRMER (1907) developed the appropriate mathematical background. Störmer's subsequent observational program showing a geomagnetic latitude alignment of auroral arcs, accurate height determinations, etc. are summarized in his later book (STÖRMER, 1955). Störmer is credited with the first formal report of pulsating auroras, with periods of several seconds to several minutes and measured heights of pulsation emission near 100 km.

Early in the 20th century, just before the Second Polar Year of 1932–1933, several additional theories about the origin of concurrent magnetic storms and auroras (MARIS and HULBURT, 1929; CHAPMAN and FERRARO, 1930) were published. These theories were surprisingly accurate in their picture of the solar-terrestrial environment and the formation of a magnetosphere, considering the lack of space-probe information at that time. Comprehensive reviews of these theories were given by HULBURT in 1937. Then in 1940 CHAPMAN and BARTELS published a textbook that summarized the world knowledge of geomagnetism and related phenomena such as auroras. The meticulous detail of this book made it a cornerstone of research on that subject for the subsequent 30 years.

In 1934 ROONEY had made an extensive study of the correspondence of earth-induced current fluctuations and moving auroral forms at College, Alaska. Later, CURRIE and EDWARDS (1936) used the Polar Year observations to show a correspondence between the nightly mean intensity of auroras and magnetic activity at Chesterfield, a location not far from what was to become the Baker Lake observatory of Canada.

In the last few years preceding the International Geophysical Years (IGY) of 1957 to 1959, several interesting discoveries occurred. VESTINE (1943) was the first to find a peak-for-peak correspondence between auroral luminosity pulsations and geomagnetic pulsations. In 1945 HARANG (1956) made a study of the variation of magnetic field vectors beneath active auroral forms in Tromso, Norway, using specially designed recording auroral-light cameras. Later, MEEK (1953) reported that the increase in intensity of an aurora is related to the rate of decrease of the magnetic northward field component, H, and that the equatorward motion of the aurora depends upon the maximum amplitude of H.

The pre-IGY period ended with two works of considerable importance. MEINEL et al. (1954) produced an excellent 55-year analysis showing that auroral and magnetic variations are at their maximum during the declining years of the 11-year solar cycle and that the auroras show evidence of a 27-day recurrence tendency which was already reported for magnetic storms. In 1954 HEPPNER published a greatly improved picture of the type of auroral changes associated with the progress of a magnetic storm in order to establish the morphological relationship between the two phenomena.

3.   Auroral Electrojet Current and Luminosity

With the advent of the IGY an extensive use of all-sky cameras was introduced and the first space measurements within the magnetosphere were made. The study of a relationship between auroral and field changes soon became only a small part of the newly developed science of solar-terrestrial disturbances. I will avoid this larger topic so that this section can be restricted to those selected publications representative of the direct relationship between the auroral electrojet and luminosity. A review of these publications should provide the necessary background to the understanding of the pulsation phenomena that occur within the common ionospheric region.

In the southern hemisphere EVANS and THOMAS (1959) detailed the geomagnetic latitude position of the southern auroral zone, whereas DENHOLM and BOND (1961) established that the southern auroral forms lie along lines of flow of the southern hemisphere geomagnetic disturbance-current system.

Investigating the great aurora of February, 1958, DAVIS and KIMBALL (1962) found a consistent alignment of the geomagnetic disturbance field perpendicular to the auroral arcs. Changes in the auroral location with varying levels of geomagnetic disturbances and time of day were given in AKASOFU (1962), AKASOFU and CHAPMAN (1962, 1963), FELDSTEIN (1963), and FELDSTEIN and STARKOV (1967). Later, AKASOFU (1968) described the process by which the magnetospheric substorm affected the aurora and concurrent geomagnetic field changes on a global scale. This author also showed a close relationship between the auroral surges and specific surface magnetic signatures.

The midday auroral intensity location is considered to be at the foot of earth field lines that connect to the sun-side neutral-point, or cusp, of the magnetopause. The motion of this part of the auroral oval should provide information on the magnetic flux transfer from the dayside of the magnetospheric boundary to the magnetotail during a solar-terrestrial substorm. Thus, in the early 1970's there was a burst of publications on the relationship between geomagnetic field levels and the midday auroral location (AKASOFU, 1972a, b; AKASOFU and CHAPMAN, 1972; AKASOFU and KIMBALL, 1973; KANEDA, 1973; and STARKOV et al., 1973).

As the newly formed space-science community focused their attention on models of sun, solar wind, and magnetosphere interaction, the workhorse all-sky cameras provided a continuing source of information about the auroral formation during geomagnetic substorms (AKASOFU et al., 1973; FELDSTEIN, 1974; GERARD, 1974; KAMIDE and AKASOFU, 1974; VOROBJEV, 1974; NAGATA et al., 1975; VOROBJEV et al., 1975; and VOROBJEV et al., 1976). Of particular interest is HORWITZ and AKASOFU's (1977) study of the response of dayside aurora and the interplanetary magnetic field to substorms. FUKUNISHI (1975) described the dynamic relationship of proton and electron auroras with respect to the auroral electrojet current.

Satellite measurements of the auroras gave science an entirely new perspective of the global changes in auroral forms and location at times of geomagnetic activity.

Representative results obtained from this new technique of observation are seen in the papers of CHUBB and HICKS (1970), PIKE and WHALEN (1974), SNYDER and AKASOFU (1974), KAMIDE and AKASOFU (1975), LUI et al. (1975), SHEPHERD et al. (1976), and BERKEY and KAMIDE (1976).

Specialized papers relating the aurora and its associated current to either the auroral electron precipitation patterns or the radar reflection from the special ionization were discussed in BROWN and CAMPBELL (1962), GOSLING (1966), HOFFMAN and BURCH (1973), KAMIDE and BREKKE (1975), TSUNODA et al. (1976), and McDIARMID et al. (1976). REES et al. (1976) modeled the electron density, conductivity, currents, and auroral emissions from the electron flux obtained with rocket-borne detectors.

Following an early study of BOSTROM (1964), the field-aligned currents, feeding the ionospheric electrojet system in the region of the auroras, became a subject of considerable interest by the middle 1970's (CLOUTIER et al., 1970; CHOY et al., 1971; PARK and CLOUTIER, 1971; WHALEN and McDIARMID, 1972; AKASOFU et al., 1974; ANDERSON and CLOUTIER, 1975; ARMSTRONG et al., 1975; CASSERLY and CLOUTIER, 1975; PAZICH and ANDERSON, 1975; SPIGER and ANDERSON, 1975; DAVIS and HALLINAN, 1976; HALLINAN, 1976; KAMIDE et al., 1976a, b; and MALTSEV et al., 1977).

4.   Luminosity and Field Pulsations

The greatly improved photometric and particle detection technology developed during the IGY period led to specialized studies of the rapid fluctuation in auroral light and its relationship to precipitating magnetospheric particles, ionospheric electron density changes, auroral radio wave absorption, and the geomagnetic field micropulsations (sometimes called the Ultra Low Frequency, ULF, part of the field). With the introduction of a new wide-angle photometer and sensitive induction magnetometer, improved records of these interdependent phenomena were obtained by CAMPBELL (1960), CAMPBELL and REES (1961), CAMPBELL and LEINBACH (1961), and CAMPBELL and MATSUSHITA (1962). The auroral pulsations were found to be only a few percent of the total light, to have a dominant period near 6 to 10 sec, and to obtain a maximum in occurrence and amplitude in the predawn hours. The accompanying magnetic field changes often tracked with luminosity in both amplitude and period. KAZAK et al. (1972) also studied the peak-for-peak correspondence between the two phenomena and CAMPBELL (1970) related the optical pulsations and the change in luminosity region to the field appearances; he also observed a similarity of pulsations at conjugate locations in opposite hemispheres.

CRESSWELL and DAVIS (1966) introduced the use of an image-orthicon television (TV) system to the specialized study of auroral dynamics and found the luminosity pulsations restricted to small field-aligned volumes. OGUTI (1975) and OGUTI and WATANABE (1976) also used the TV method at Syowa, Antarctica, to show that geomagnetic pulsations of a few seconds to several minutes are coherently related to auroral pulsations and occur during a typical quasi-periodic poleward propagation of the on-off switching auroral forms. Recent studies are by ROYRVIK and DAVIS (1977).

Special high-latitude, balloon-borne detectors were sometimes used to measure the X-rays from incoming electrons responsible for the formation of an auroral display. A number of these studies found pulsating events that would account for the concurrent auroral and field changes (cf. ANGER et al., 1963; BARCUS and CHRISTENSEN, 1965; BARCUS and ROSENBERG, 1966; ROSENBERG et al., 1971; HOFMANN and GREENE, 1972; GASSET et al., 1972). In support of the earlier publications JOHANSEN and OMHOLT (1966) found that observations do not show semi-trapped electrons bouncing on the field lines to be the source of the periodicity in the pulsations.

A high frequency part of the auroral fluctuations was investigated by PARKS et al. (1965); JOHANSEN (1966); EVANS (1967), OMHOLT and PETTERSEN (1967), BEACH et al. (1968); LIPOV and ROLDUGIN (1974); and PEMBERTON and SHEPHERD (1975). Specific peaks in the spectra were found near the 2.4, 1.2, 0.3, 0.1 sec period. However, HIRASAWA and KAMINUMA (1970) found the correlation between auroral and magnetic pulsations to be highest near the 6 to 10 sec period. OMHOLT (1971) and VALLANCE-JONES (1974) summarized the form and movement of pulsating auroras.

PARKS and WINCKLER (1969) showed that similar pulsations of 5 to 15 sec period occurred at the equatorial plane in the magnetosphere and at connecting conjugate areas on the earth's surface. A later study of incoming particles and resulting luminosity by WHALEN et al. (1971) indicated that the auroral pulsations result primarily from increases in pitch-angle diffusion causing large electron precipitation enhancements in the energy range of 15 to 80 keV.

Lower latitude effects were described by SAKURAI and SAITO (1976) who found that low latitude Pi2 geomagnetic pulsations occurred almost simultaneously with the sudden brightening of an emission arc at the onset of an auroral substorm. HALL (1974) described 6 to 10 sec pulsating aurora observed at middle latitudes during a magnetic bay event.

Theoretical models were contributed by several authors in addition to WHALEN et al. (op. cit.). CORONITI and KENNEL (1970) attributed the electron pulsation and resulting auroral effects to increased pitch-angle diffusion resulting from the modulation (by geomagnetic micropulsations) of equatorial-magnetosphere electron fluxes. ROLDUGIN et al. (1971) proposed that the electric fields, introduced into the ionosphere by the flux of auroral particles, produce variations in ionospheric currents responsible for the geomagnetic pulsations. MALTSEV et al. (1974) suggested that the appearance of a region of enhanced conductivity (associated with the auroral brightening) in the ionosphere leads to the origin of an Alfvén impulse propagating along field lines; this impulse, being reflected between hemispheres, was thought to produce standing Alfvén waves registered as Pi2 geomagnetic pulsations on the ground. Most recently, HASEGAWA (1976) proposed a model of the phenomena involving an incompressible MHD surface-wave that accelerated particles along the field lines.

Presently, researches still need to explain, more fully, the early morning auroral region preference of the pulsations, to find what occurs at polar-cap locations, to separate the HM wave, ionospheric current, absorption, and induced parts of the observed surface field, and to be able to predict the onset and location of this auroral phenomenon. There is still some dispute in the literature on three questions: (1) whether the bombarding electrons responsible for the luminosity obtain their characteristic pulsations periodicity, a) at ionospheric levels or b) at a far equatorial magnetospheric location; (2) how the field pulsations are transmitted to low latitude observatories; and (3) how well the pulsations may be used as direct indicators of the physical properties within the magnetosphere. From the very apparent increase in publications on the topic of concurrent rapid fluctuations in the geomagnetic field and auroral luminosity it is clear that more discoveries of the interrelationship of the two phenomena may be expected soon.

REFERENCES

AKASOFU, S.I., Large-scale auroral motions and polar magnetic disturbances-II: The changing distribution of the aurora during large magnetic storms, *J. Atmos. Terr. Phys.*, **24**, 723–727, 1962.

AKASOFU, S.I., Auroral substorms and associated magnetic disturbances, in *Polar and Magnetospheric Substorms*, pp. 22–48, D. Reidel Pub. Co., Dordrecht, 1968.

AKASOFU, S.I., Midday auroras and magnetospheric substorms, *J. Geophys. Res.*, **77**, 244–247, 1972a.

AKASOFU, S.I., Middy auroras at the South Pole during magnetospheric substorms, *J. Geophys. Res.*, **77**, 2303–2308, 1972b.

AKASOFU, S.I. and S. CHAPMAN, Large-scale auroral motions and polar magnetic disturbances-III: The aurora and magnetic storm of 11 February 1958, *J. Atmos. Terr. Phys.*, **24**, 785–796, 1962.

AKASOFU, S.I. and S. CHAPMAN, The lower limit of latitude (US sector) of northern quiet auroral arcs, and its relations to Dst(H), *J. Atmos. Terr. Phys.*, **25**, 9–12, 1963.

AKASOFU, S.I. and S. CHAPMAN, Magnetospheric storms, in *Solar Terrestrial Physics*, pp. 561–716, Oxford at the Clarendon Press, London, 1972.

AKASOFU, S.I. and D.S. KIMBALL, The behavior of midday auroras during substorms, *Planet. Space Sci.*, **21**, 696–698, 1973.

AKASOFU, S.I., S. DeFOREST, and C. McILWAIN, Auroral displays near the foot of the field line of the ATS-5 satellite, *Planet. Space Sci.*, **22**, 25–40, 1974.

AKASOFU, S.I., P.D. PERREAULT, F. YASUHARA, and C.I. MENG, Auroral substorms and the interplanetary magnetic field, *J. Geophys. Res.*, **78**, 7490–7508, 1973.

ANGER, C.D., J.R. BARCUS, R.R. BROWN, and D.S. EVANS, Auroral zone X-ray pulsations in the 1- to 15-second period range, *J. Geophys. Res.*, **68**, 1023–1028, 1963.

ANDERSON, H.R. and P.A. CLOUTIER, Simultaneous measurements of auroral particles and electric currents by a rocket-borne instrument system: Introductory Remarks, *J. Geophys. Res.*, **80**, 2146–2151, 1975.

ARMSTRONG, J.C., S.I. AKASOFU, and G. ROSTOKER, A comparison of satellite observations of Birkeland currents with ground observations of visible aurora and ionospheric currents, *J. Geophys. Res.*, **80**, 575–586, 1975.

BARCUS, J.R. and A. CHRISTENSEN, A 75-second periodicity in auroral-zone X rays, *J. Geophys. Res.*, **70**, 5455–5459, 1965.

BARCUS, J.R. and T.J. ROSENBERG, Energy spectrum for auroral-zone X rays, *J. Geophys. Res.*, **71**, 803–834, 1966.

BEACH, R., G.R. CRESSWELL, T.N. DAVIS, T.J. HALLINAN, and L.R. SWEET, Flickering, a 10-cps fluctuation within bright auroras, *Planet. Space Sci.*, **16**, 1525-1529, 1968.

BERKEY, F.T. and Y. KAMIDE, On the distribution of global auroras during intervals of magnetospheric quiet, *J. Geophys. Res.*, **81**, 4701-4714, 1976.

BIRKELAND, K., Expedition Norvegienne de 1899-1900, Resultats magnetiques, *Vidensk. Skrifter*, **I**, 1-80, 1901.

BOLLER, W., Das Sudlicht, *Beitr. Geophys.*, **3**, 56-130 and 550-609, 1898.

BOSTROM, R., A model of the auroral electrojet, *J. Geophys. Res.*, **69**, 4983-4999, 1964.

BROWN, R.R. and W.H. CAMPBELL, An auroral-zone electron precipitation event and its relationship to a magnetic bay, *J. Geophys. Res.*, **67**, 1357-1366, 1962.

CAMPBELL, W.H., Magnetic micropulsations, pulsating aurora, and ionospheric absorption, *J. Geophys. Res.*, **65**, 1833-1834, 1960.

CAMPBELL, W.H., Rapid auroral luminosity fluctuations and geomagnetic field pulsations, *J. Geophys. Res.*, **75**, 6182-6208, 1970.

CAMPBELL, W.H. and H. LEINBACH, Ionospheric absorption at times of auroral and magnetic pulsations, *J. Geophys. Res.*, **66**, 25-34, 1961.

CAMPBELL, W.H. and S. MATSUSHITA, Auroral-zone geomagnetic micropulsations with periods of 5 to 30 seconds, *J. Geophys. Res.*, **67**, 555-573, 1962.

CAMPBELL, W.H. and M.H. REES, A study of auroral coruscations, *J. Geophys. Res.*, **66**, 41-55, 1961.

CASSERLY, R.T., Jr. and P.A. CLOUTIER, Rocket-based magnetic observations of auroral Birkeland currents in association with a structured auroral arc, *J. Geophys. Res.*, **80**, 2165-2171, 1975.

CAVENDISH, H., On the height of luminous arch which was seen on February 23, 1784, *Philos. Trans. R. Soc.*, **80**, 101-105, 1790.

CHAPMAN, S. and J. BARTELS, *Geomagnetism*, 1049 pp., Oxford at the Clarendon Press, London, 1940.

CHAPMAN, S. and V.C.A. FERRARO, A new theory of magnetic storms, *Nature*, **126**, 129-130, 1930.

CHOY, L.W., R.L. ARNOLDY, W. POTTER, P. KINTNER, and L.J. CAHILL, Jr., Field-aligned particle currents near an auroral arc, *J. Geophys. Res.*, **76**, 8279-8298, 1971.

CHUBB, T.A. and G.T. HICKS, Observations of the aurora in the far ultraviolet from OGO 4, *J. Geophys. Res.*, **75**, 1290-1311, 1970.

CLOUTIER, P.A., H.R. ANDERSON, R.J. PARK, R.R. VONDRAK, R.J. SPIGER, and B.R. SANDEL, Detection of geomagnetically aligned currents associated with an auroral arc, *J. Geophys. Res.*, **75**, 2595-2600, 1970.

COOK, J., *The Journals of Captain James Cook*, II, p. 95, Cambridge University Press, Cambridge, U.K., 1961.

CORONITI, F.V. and C.F. KENNEL, Electron precipitation pulsations, *J. Geophys. Res.*, **75**, 1279-1289, 1970.

CRESSWELL, G.R. and T.N. DAVIS, Observations on pulsating auroras, *J. Geophys. Res.*, **71**, 3155-3163, 1966.

CURRIE, B.W. and H.W. EDWARDS, On the auroral spectrograms taken at Chesterfield, Canada, during 1932-1933, *Terr. Mag. Atmos. Electr.*, **41**, 265-278, 1936.

DAVIS, T.N. and T.J. HALLINAN, Auroral spirals: 1. Observations, *J. Geophys. Res.*, **81**, 3953-3958, 1976.

DAVIS, T.N. and D.S. KIMBALL, The auroral display of February 13-14, 1958, *Geophys. Inst. Rept. UAG-R120*, University of Alaska, 1-52, 1962.

de MAIRAN, J., Traite physique et historique de l'aurore boreale, *Suite des Mem. Acad. Roy. Sci.*, 1st ed., Paris, 1733 (2nd ed. 1754). (Reported in Chapman and Bartels, 1940)

DENHOLM, J.V. and F.R. BOND, Orientation of polar auroras, *Austral. J. Phys.*, **14**, 193-195, 1961.

EVANS, D.S., A 10-cps periodicity in the precipitation of auroral electrons, *J. Geophys. Res.*, **72**, 4281-4291, 1967.

EVANS, S. and G.M. THOMAS, The southern auroral zone in geomagnetic longitude sector 20°E, *J. Geophys. Res.*, **64**, 1381-1388, 1959.

FELDSTEIN, Y.I., Some problems concerning the morphology of auroras and magnetic disturbances at high latitudes, *Geomag. Aeron.*, **3**, 183-192, 1963.

FELDSTEIN, Y.I., Night-time aurora and its relation to the magnetosphere, *Ann. Geophys.*, **30**, 259–272, 1974.

FELDSTEIN, Y.I. and G.V. STARKOV, Dynamics of auroral belt and polar geomagnetic disturbances, *Planet. Space Sci.*, **15**, 209–229, 1967.

FRITZ, H., *Das Polarlicht*, Brockhaus Pub., Leipzig, 1881.

FUKUNISHI, H., Dynamic relationship between proton and electron auroral substorms, *J. Geophys. Res.*, **80**, 553–574, 1975.

GASSET, G., I.A. ZHULIN, F. CAMBOU, K.D. KANONIDI, N.G. KLEYMENOVA, O.M. RASPOPOV, A. SAINT-MARC, and J.P. TREILHOU, Modulation of auroral electron fluxes and geomagnetic pulsations during the storm of March 8, 1970, *Geomag. Aeron.*, **12**, 919–924, 1972.

GAUSS, C.F. and W. WEBER, *Resultate aus den Beobachtungen des Magnetishen Vereine in June 1841*, *Hefte 6*, Göttingen und Leipzig, 1841. (Reported in Chapman and Bartels, 1940)

GERARD, J.C., Ground-based photometric observations of the type A aurora of 17–18 December 1971, *Ann. Geophys.*, **30**, 387–396, 1974.

GOSLING, J.T., Localization and motion of energetic electron precipitation during magnetic bays, *J. Geophys. Res.*, **71**, 835–848, 1966.

HALEY, E., An account of the extraordinary meteor seen all over England on the 6th day of March 1716, *Philos. Trans. R. Soc.*, **6**, 406, 1716.

HALL, W.N., Mid-latitude pulsating auroras, *Planet. Space Sci.*, **22**, 1315–1321, 1974.

HALLINAN, T.J., Auroral spirals: 2. Theory, *J. Geophys. Res.*, **81**, 3959–3965, 1976

HANSTEEN, C., *Untersuchungen über den Magnetismus der Erde*, Christiania, 148 pp., 1819. (Reported in Chapman and Bartels, 1940)

HANSTEEN, C., The great auroral exhibition of Aug. 28th to Sept. 4th, 1859, 4th Article, 1. Observations at Christiania, *Am. J. Sci. Arts*, **29**, 386–388, 1860.

HARANG, L., The mean field of disturbance of polar geomagnetic storms, *J. Geophys. Res.*, 353–380, 1956.

HASEGAWA, A., Particle acceleration by MHD surface wave and formation of aurora, *J. Geophys. Res.*, **81**, 5083–5090, 1976.

HEPPNER, J.P., Time sequences and spatial relations in auroral activity during magnetic bays at College, Alaska, *J. Geophys. Res.*, **59**, 329–338, 1954.

HIRASAWA, T. and K. KAMINUMA, Space-time variation of aurora and magnetic disturbances—Auroral observations at Syowa station in Antarctica—1967–1968, Japanese Antarctic Research Expedition, *Sci. Rep. Ser. A, No. 8*, 1–29, 1970.

HOFFMAN, R.A. and J.L. BURCH, Electron precipitation patterns and substorm morphology, *J. Geophys. Res.*, **78**, 2867–2884, 1973.

HOFMANN, D.J. and R.A. GREENE, Balloon observations of simultaneous auroral X-ray and visible bursts, *J. Geophys. Res.*, **77**, 776–780, 1972.

HORWITZ, J.L. and S.I. AKASOFU, The response of the dayside aurora to sharp northward and southward transitions of the interplanetary magnetic field and to magnetospheric substorms, *J. Geophys. Res.*, **82**, 2723–2734, 1977.

HULBURT, E.O., Terrestrial magnetic variations and aurorae, *Rev. Mod. Phys.*, **9**, 44–68, 1937.

JOHANSEN, O.E., A possible relation between pulsations in the auroral luminosity and the energy spectrum of the primary particles, *Planet. Space Sci.*, **14**, 217–219, 1966.

JOHANSEN, O.E. and A. OMHOLT, A study of pulsating aurora, *Planet. Space Sci.*, **14**, 207–215, 1966.

KAMIDE, Y. and S.I. AKASOFU, Latitudinal cross section of the auroral electrojet and its relation to the interplanetary magnetic field polarity, *J. Geophys. Res.*, **79**, 3755–3790, 1974.

KAMIDE, Y. and S.I. AKASOFU, The auroral electrojet and global auroral features, *J. Geophys. Res.*, **80**, 3585–3602, 1975.

KAMIDE, Y. and A. BREKKE, Auroral electrojet current density deduced from the Chatanika radar and from the Alaska meridian chain of magnetic observatories, *J. Geophys. Res.*, **80**, 587–594, 1975.

KAMIDE, Y., S.I. AKASOFU, and G. ROSTOKER, Field-aligned currents and the auroral electrojet in the morning sector, *J. Geophys. Res.*, **81**, 6141–6147, 1976a.

KAMIDE, Y., F. YASUHARA, and S.I. AKASOFU, A model current system for the magnetospheric substorm, *Planet. Space Sci.*, **24**, 215–222, 1976b.

KANEDA, E., Dayside auroral activity and its relation to substorm, *Rep. Ionos. Space Res. Japan*, **27**, 209–212, 1973.

KAZAK, B.N., V.K. ROLDUGIN, and S.A. CHERNOUS, Simultaneity of the appearance of geomagnetic and auroral pulsations, *Geomag. Aeron.*, **12**, 817–819, 1972.

LIPOV, O.S. and V.K. ROLDUGIN, Auroral pulsations in the 1- to 10-Hz range, *Geomag. Aeron.*, **14**, 313–315, 1974.

LOOMIS, E., On the geographical distribution of auroras in the northern hemisphere, *Am. J. Sci. Arts*, **30**, 89–94, 1860.

LUI, A.T.Y., C.D. ANGER, and S.I. AKASOFU, The equatorward boundary of the diffuse aurora and auroral substorms as seen by the Isis 2 auroral scanning photometer, *J. Geophys. Res.*, **80**, 3603–3614, 1975.

MALTSEV, Y.P., S.V. LEONTYEV, and W.B. LYATSKY, Pi-2 pulsations as a result of evolution of an Alfvén impulse originating in the ionosphere during a brightening of aurora, *Planet. Space Sci.*, **22**, 1519–1533, 1974.

MALTSEV, Y.P., W.B. LYATSKY, and A.M. LYATSKAYA, Currents over the auroral arc, *Planet. Space Sci.*, **25**, 53–57, 1977.

MARIS, H.B. and E.O. HULBERT, Comets and terrestrial magnetic storms, *Phys. Rev.*, **33**, 1046–1060, 1929.

McDIARMID, D.R., F.R. HARRIS, and A.G. McNAMARA, Relationships between radio aurora, visual aurora and ionospheric currents during a sequence of magnetospheric substorms, *Planet. Space Sci.*, **24**, 717–725, 1976.

MEEK, J.H., Correlation of magnetic, auroral, and ionospheric variations at Saskatoon, *J. Geophys. Res.*, **58**, 445–456, 1953.

MEINEL, A.B., B.J. NEGAARD, and J.W. CHAMBERLAIN, A statistical analysis of low-latitude aurorae, *J. Geophys. Res.*, **59**, 407–413, 1954.

NAGATA, T., T. HIRASAWA, and M. AYUKAWA, Discrete and diffuse auroral belts in Antarctica, *Rep. Ionos. Space Res. Japan*, **29**, 149–156, 1975.

OGUTI, T., Metamorphoses of aurora, *Memoirs of National Institute of Polar Research*, Ser. A, No. 12, 101 pp., 1975.

OGUTI, T. and T. WATANABE, Quasi-periodic poleward propagation of on-off switching aurora and associated geomagnetic pulsations in the dawn, *J. Atmos. Terr. Phys.*, **38**, 543–551, 1976.

OMHOLT, A., *The Optical Aurora*, Springer-Verlag, New York, 1971.

OMHOLT, A. and H. PETTERSEN, Characteristics of high frequency auroral pulsations, *Planet. Space Sci.*, **15**, 347–355, 1967.

PARK, R.J. and P.A. CLOUTIER, Rocket-based measurement of Birkeland currents related to an auroral arc and electrojet, *J. Geophys. Res.*, **76**, 7714–7733, 1971.

PARKS, G.K. and J.R. WINCKLER, Simultaneous observations of the 5 to 15-sec period modulated energetic electron fluxes at the synchronous altitude and the auroral zone, *J. Geophys. Res.*, **74**, 4003–4017, 1969.

PARKS, G.K., H.S. HUDSON, D.W. MILTON, and K.A. ANDERSON, Spatial asymmetry and periodic time variations of X-ray microbursts in the auroral zone, *J. Geophys. Res.*, **70**, 4976–4978, 1965.

PAZICH, P.M. and H.R. ANDERSON, Rocket measurement of auroral electron fluxes associated with field-aligned currents, *J. Geophys. Res.*, **80**, 2152–2160, 1975.

PEMBERTON, E.V. and G.G. SHEPHERD, Spatial characteristics of auroral brightness fluctuation spectra, *Can. J. Geophys.*, **53**, 504–513, 1975.

PIKE, C.P. and J.A. WHALEN, Satellite observations of auroral substorms, *J. Geophys. Res.*, **79**, 985–1000, 1974.

PRESCOTT, G.B., The great auroral exhibition of August 28th to September 4th, 1859, 1. Observations made at Boston, Mass., and its vicinity, *Am. J. Sci. Arts*, **29**, 92–95, 1860.

PEES, M.H., G.J. ROMICK, H.R. ANDERSON, and R.T. CASSERLY, Jr., Calculation of auroral emissions from measured electron precipitation: Comparison with observation, *J. Geophys. Res.*, **81**, 5091–5096, 1976.

ROLDUGIN, V.K., B.N. BELENKAYA, and N.F. MALTSEVA, Electric fields in the ionosphere during pulsating auroras, *Geomag. Aeron.*, **11**, 691–694, 1971.

ROONEY, W.J., Aurorae and earth-currents, *Terr. Mag. Atmos. Electr.*, **39**, 103–109, 1934.

ROSENBERG, T.J., J. BJORDAL, H. TREFALL, G.J. KVIFTE, A. OMHOLT, and A. EGELAND, Correlation study of auroral luminosity and X-rays, *J. Geophys. Res.*, **76**, 122–132, 1971.

ROYRVIK, O. and T.N. DAVIS, Pulsating aurora: Local global morphology, *J. Geophys. Res.*, **82**, 4720–4740, 1977.

SABINE, E., On periodic laws discernible in the mean effects of larger magnetic disturbances I and II, *Philos. Trans. R. Soc., London*, 123–139 and 103–124, 1851 and 1852.

SABINE, E., On hourly observations of the magnetic declination, made by Captain Rochfort Maguire, R.N., and the Officers of H.M. Ship 'Plover,' in 1852, 1853 and 1854, at Point Barrow, on the shores of the Polar Sea, *Proc. R. Soc., London*, **8**, 610–614, 1857.

SAKURAI, T. and T. SAITO, Magnetic pulsation Pi2 and substorm onset, *Planet. Space Sci.*, **24**, 573–575, 1976.

SCHERING, K., Berichte über die Fortschritte unserer Kenntnisse vom Magnetismus der Erde, *Geogr. Jahrbuch*, **13**, 1889. (Reported in Chapman and Bartels, 1940)

SHEPHERD, G.G., F.W. THIRKETTLE, and C.D. ANGER, Topside optical view of the dayside cleft aurora, *Planet. Space Sci.*, **24**, 937–944, 1976.

SNYDER, A.L., Jr. and S.I. AKASOFU, Major auroral substorm features in the dark sector observed by a USAF DMSP satellite, *Planet. Space Sci.*, **22**, 1511–1517, 1974.

SPIGER, R.H. and H.R. ANDERSON, Electron currents associated with an auroral band, *J. Geophys. Res.*, **80**, 2161–2164, 1975.

STARKOV, G.V., Y.I. FELDSTEIN, and N.F. SHEVNINA, Auroras on the dayside of the oval during substorms, *Geomag. Aeron.*, **13**, 72–75, 1973.

STÖRMER, C., Sur les trajectories des electrises dans l'espace sons l'action du magnetisme terrestre avec application aux aurores boreales, *Arch. Sci. Phys.*, Geneve, **24**, 5, 113, 221, and 317, 1907.

STÖRMER, C., *The Polar Aurora*, Oxford Press, 1955.

TSUNODA, R.T., R.I. PRESNELL, Y. KAMIDE, and S.I. AKASOFU, Relationship of radar aurora, visual aurora, and auroral electrojets in the evening sector, *J. Geophys. Res.*, **81**, 6005–6015, 1976.

VALLANCE-JONES, A., *Aurora*, D. Reidel Pub. Co., New York, 1974.

VESTINE, E.H., Remarkable auroral forms, Meanook Observatory, Polar Year 1932–1933, *Terr. Mag. Atmos. Electr.*, **48**, 233–236, 1943.

VOROBJEV, V.G., SC-associated effects in auroras, *Geomag. Aeron.*, **14**, 72–74, 1974.

VOROBJEV, V.G., G. GUSTAFSSON, G.V. STARKOV, Y.I. FELDSTEIN, and N.F. SHEVNINA, Dynamics of day and night aurora during substorms, *Planet. Space Sci.*, **23**, 269–278, 1975.

VOROBJEV, V.G., G.V. STARKOV, and Y.I. FELDSTEIN, The auroral oval during the substorm development, *Planet. Space Sci.*, **24**, 955–965, 1976.

WHALEN, B.A. and I.B. McDIARMID, Observations of magnetic-field-aligned auroral-electron precipitation, *J. Geophys. Res.*, **77**, 191–202, 1972.

WHALEN, B.A., J.R. MILLER, and I.B. McDIARMID, Energetic particle measurements in a pulsating aurora, *J. Geophys. Res.*, **76**, 978–986, 1971.

WILCKE, J.C., Von den Jarlichen und Taglichen Bewegungen der Magnetnadel in Stockholm, *Svensk. Vet. Acad. Handl.*, pp. 273–300, 1777. (Reported in Chapman and Bartels, 1940)

Microburst Precipitation Phenomena

George K. PARKS

Space Sciences Division, Geophysics Program,
University of Washington, Seattle,
Washington 98195, U.S.A.

(Received February 7, 1978)

Auroral zone microburst electron precipitation is reviewed. It can be shown from existing data that microbursts represent a major precipitation form on the dayside magnetosphere. Balloon X-ray data indicate that microbursts are periodic and they are small-scaled. Rocket and sattellite data indicate that microbursts may consist of substructures and that microburst precipitation is fractionally more enhanced along the magnetic field direction. The source of electrons appears to be nearby, $\leq 4R_e$ from the surface of the earth. Microbursts are correlated to VLF and magnetic micropulsations. Study of microburst phenomena should lead to better understanding of wave-particle interaction mechanisms responsible for electron precipitation.

1. Introduction

This article reviews microburst precipitation phenomena. "Microburst" is a term used by ANDERSON and MILTON (1964) to describe the short, impulsive auroral zone electron precipitation of durations ~0.25 second. ANDERSON and MILTON (1964) have shown by means of a systematic study of auroral X-rays from precipitated energetic Van Allen electrons that during daylight hours the electron precipitation occurs primarily in the form of microbursts. Thus, microburst precipitation represents a major perturbation in magnetospheric electron population.

Direct detection of precipitation bursts with ≤ 1 second duration of electrons with ≥ 40 keV energies was made by an Injun 3 experiment (O'BRIEN, 1964). These bursts have subsequently been identified as microburst electrons (MILTON and OLIVEN, 1967; OLIVEN *et al.*, 1968). Microburst electron precipitation events have also been detected by a rocket experiment from Fort Churchill ($L=8$) (LAMPTON, 1967). The Injun 3 and the rocket data provide the only data available concerning microburst pitch-angle distributions and the primary electron energy spectra.

The rapid precipitation bursts of microbursts could be indicating that the first and second adiabatic invariants are violated. The study of microbursts could lead to important clues on the fundamental wave-particle and plasma interaction processes of relevance to understanding the particle loss mechanism in the magnetosphere. It is worth noting that impulsive electron precipitation with time scales ≤ 1 second is not limited to auroral latitudes. WINCKLER *et al.* (1962) have shown that X-rays detected over Minneapolis ($L=3.5$) during a storm consisted of similar rapid precipitation structures. Microburst-like precipitation structures have been detected subsequently in subauroal latitudes by Rosenberg *et al.* (1971), Zuhlin *et al.* (1972), Treilhou *et al.* (1973), and Foster and Rosenberg (1976). Electron precipitation with $\lesssim 1$ second structures is also observable in visual auroras (Cresswell and Davis, 1965) and those precipitation forms which are confined to local hours near midnight (Evans, 1967; Parks *et al.*, 1967).

2. Observations

2.1 Temporal characteristics

Examples of microbursts in electron precipitation phenomena are shown in Figs. 1 and 2. The temporal properties of microbursts can be summarized as follows:

2.1.1 Microburst widths

As shown in the top panel of Fig. 1, which is taken from Anderson and Milton (1964), microbursts are characterized by a ~ 0.25 second duration. Subsequent studies by Anderson *et al.* (1966) indicate that microbursts have durations from ~ 0.1–0.6 seconds. This result has been substantiated by Brown (1973). Statistical study of X-ray data obtained by omnidirectional detectors indicate that nearly 85% of all microbursts detected have durations between 0.1 and 0.3 second. Note that microbursts occur in singles, doubles, and in trains of more than 10. A central theoretical question is concerned with the meaning of microburst duration.

2.1.2 Microbursts and Pulsations

When groups of two or more microbursts occur, the groups tend to be separated by 5–15 seconds. Individual microbursts in these groups are spaced ~ 0.6 second apart. The microbursts are superposed on 5–15 second periodic X-ray pulsations. An example showing this is in the middle panel of Fig. 1 (taken from Parks, 1967). An interesting theoretical question here concerns the relationship between microbursts and the 5–15 second period pulsations (the X-ray pulsations are correlated with similar periodic magnetic micropulsations and visual auroral pulsations).

2.1.3 Substructures

Microbursts may consist of substructure as indicated in the rocket data of Lampton's (1967) which is shown in the bottom panel of Fig. 1. The existence of these substructures that are only a few tens of millisecond in duration indicates that microbursts may be generated locally in the ionosphere (more on this will be discussed in a later section). Additional evidence of microburst substructures is given in Parks (1974).

2.1.4 Injun 3 data

Microburst structures are observable at small and large pitch-angles as indicated in Fig. 2 (from Oliven *et al.*, 1968). This figure shows that the structures are more prominent at small pitch-angles (180° detector). Whereas only a factor of two increase in fluxes is seen in the perpendicular direction (90° detector), the fluxes along the magnetic field (180° detector) increase by an order of magnitude.

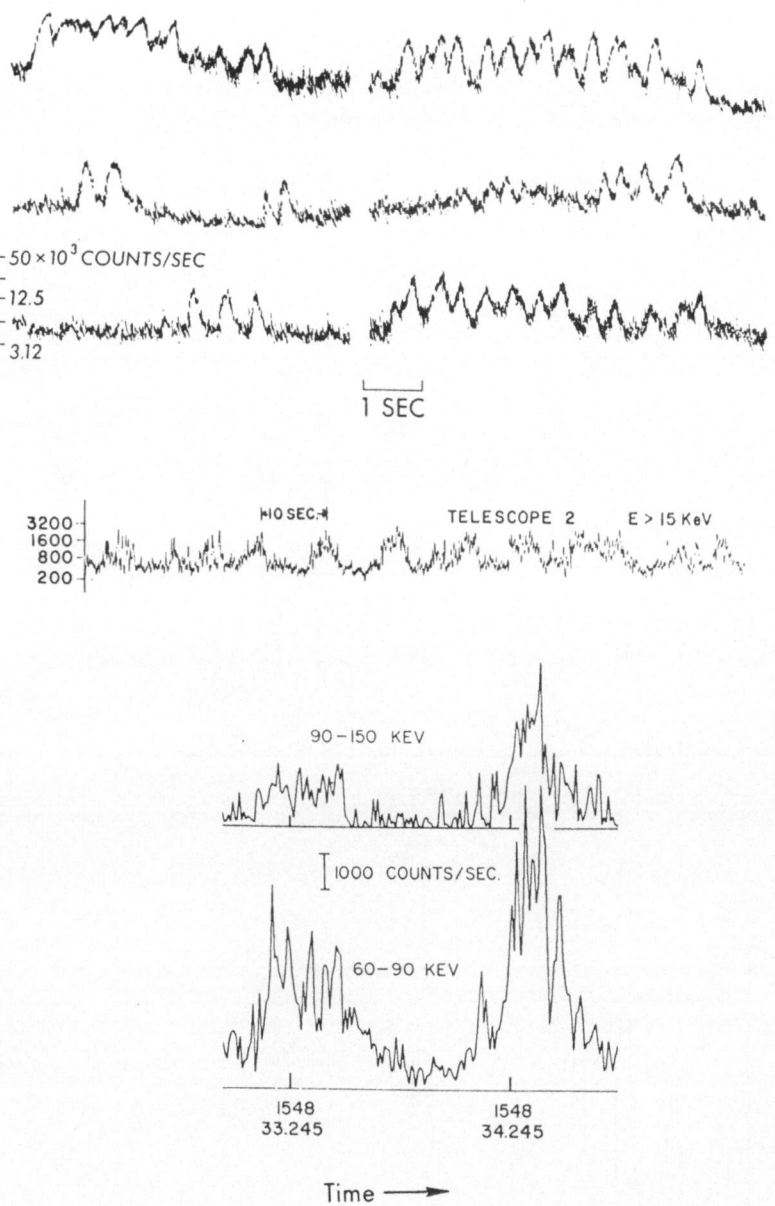

Fig. 1. Microburst precipitation structures as revealed in the X-ray and rocket electron
data.

2.1.5 *Periodicities*

It can be seen in the top panel of Fig. 1 that microbursts occur nearly periodically. Figure 3 shows some Fourier analysis results of microburst trains. The top
left panel shows the results originally shown in ANDERSON and MILTON (1964). A
weak but significant peak in observed at 0.6 second period (compare this spectrum

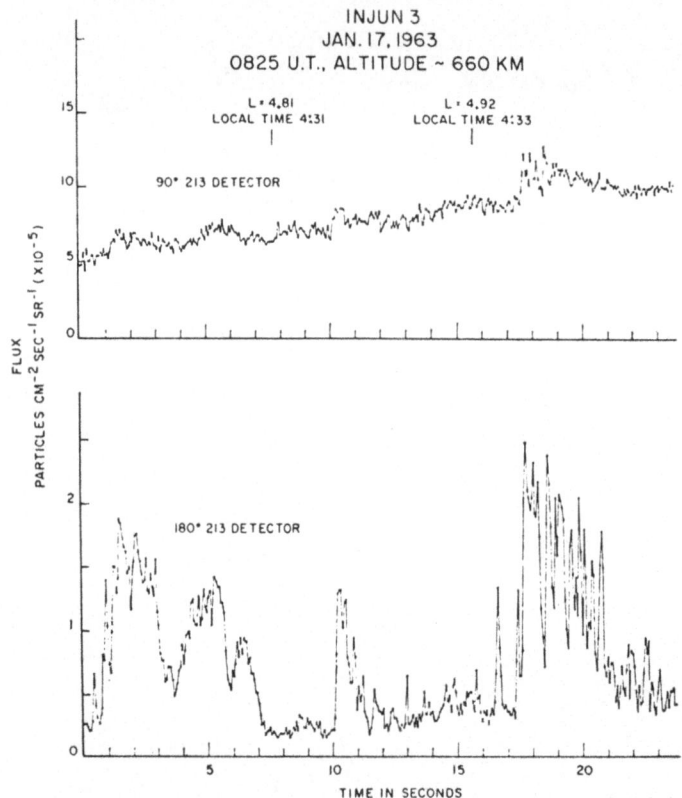

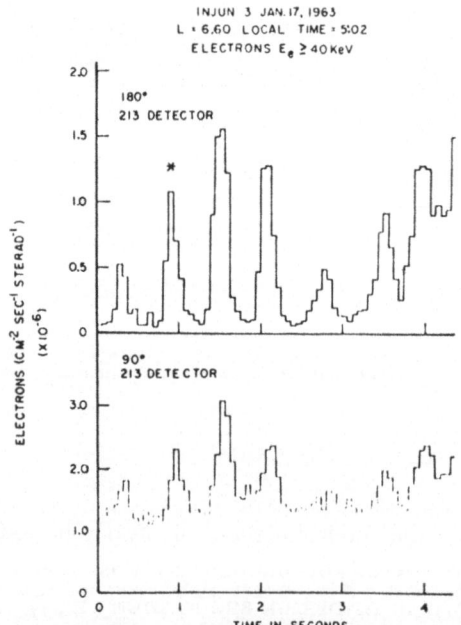

Fig. 2. Microburst electron precipitation structures observed by the Injun 3 satellite.

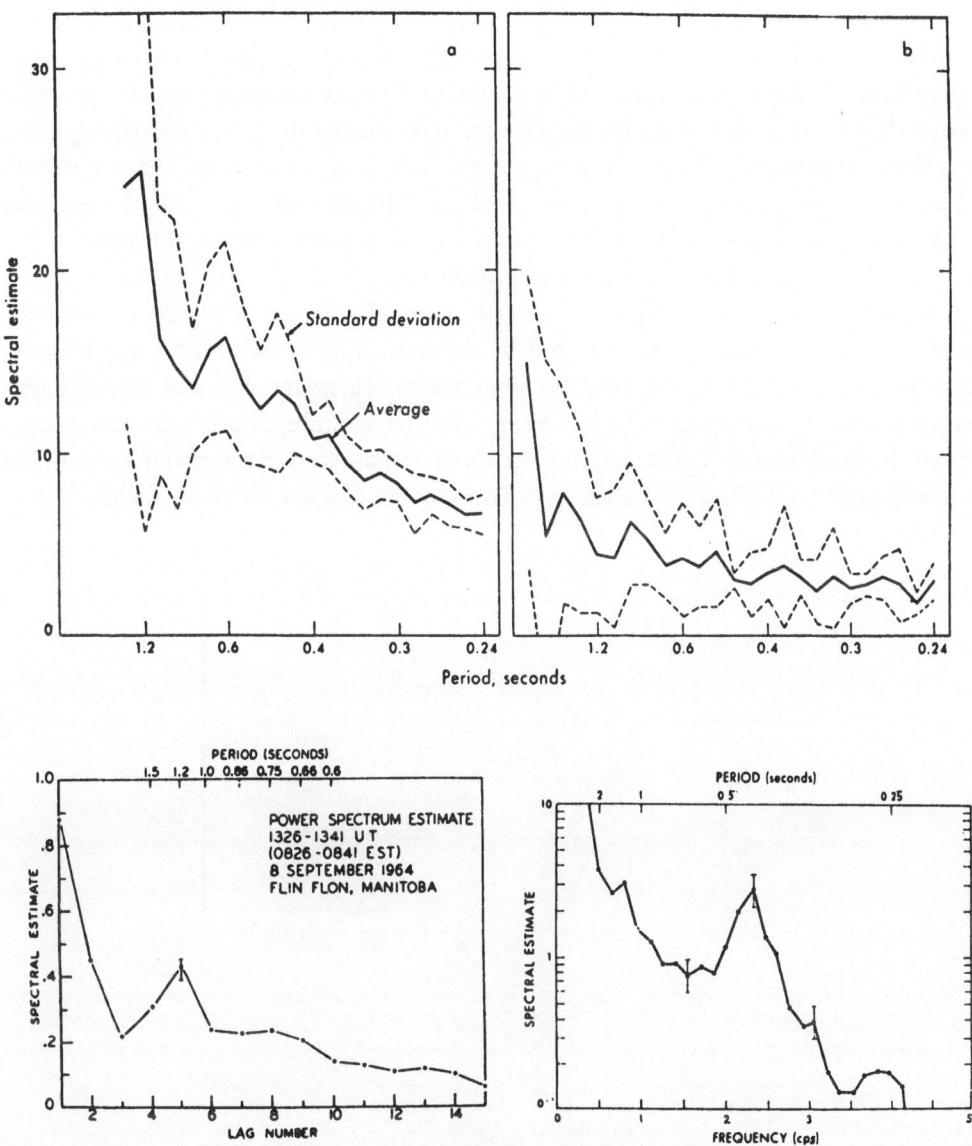

Fig. 3. Fourier analysis results of microburst trains to show the periodic properties.

to the one on the right which was obtained from analyzing random count-rates of a radioactive source). The bottom left panel shows results from analyzing microburst trains from highly collimated detectors (from PARKS *et al.*, 1965). Here a significant period was detected at ∼1.2 seconds. The bottom right hand panel shows the results taken from BROWN (1973) and represents "very intense" microburst events. A significant period was found around 0.6 second.

2.2 Spatial characteristics

2.2.1 Occurrence in local time

Microbursts are primarily detected in the daytime, from ∼0600–1800 local time.

The top panel of Fig. 4 shows the percentage of time when X-ray fluxes detected on balloons showed microburst structure (the data sample here represents more than 1,000 hours of X-ray data obtained in Alaska and Canada during 1961–1974). This result shows a high detection probability for microbursts in the early morning hours just after local dawn. It is interesting to note that X-ray microbursts have not been observed before local dawn and after local dusk (except for an isolated event described by VENKATESAN et al., 1968, that occurred around 0430 local time).

2.2.2   Occurrence in longitude and latitude

The bottom panel of Fig. 4 shows the latitudinal region where microbursts were detected by Injun 3 as reported by OLIVEN et al. (1968). This figure shows that microbursts occur even at 0200 local hours. However, it is not obvious that data obtained in a moving vehicle can be used to unambiguously sort out microbursts from other rapid precipitation forms that characterize the precipitation phenomena near midnight and the early morning hours (see PARKS et al., 1968).

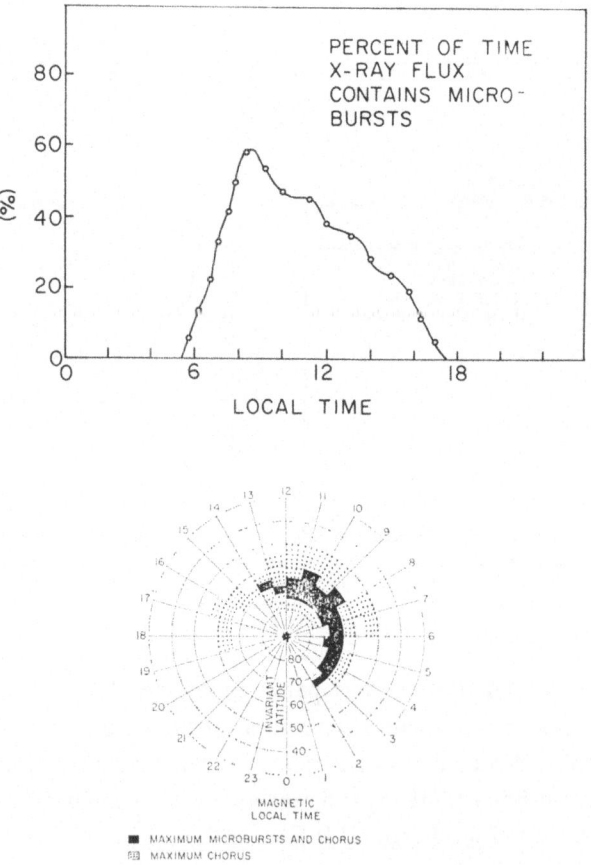

Fig. 4. Diurnal, latitudinal, and longitudinal information on microburst precipitation. Note that the VLF wave activity has similar occurrence pattern.

2.2.3 Spatial extent of microbursts

Auroral electron precipitation phenomena are substorm associated and therefore the precipitation region can be world-wide. Microburst precipitation is only one of the many varieties of forms that occur in smaller regions within the large precipita-

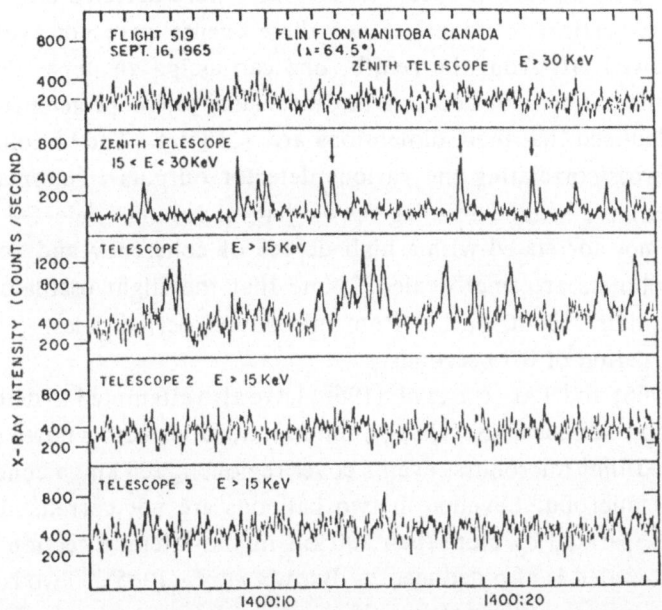

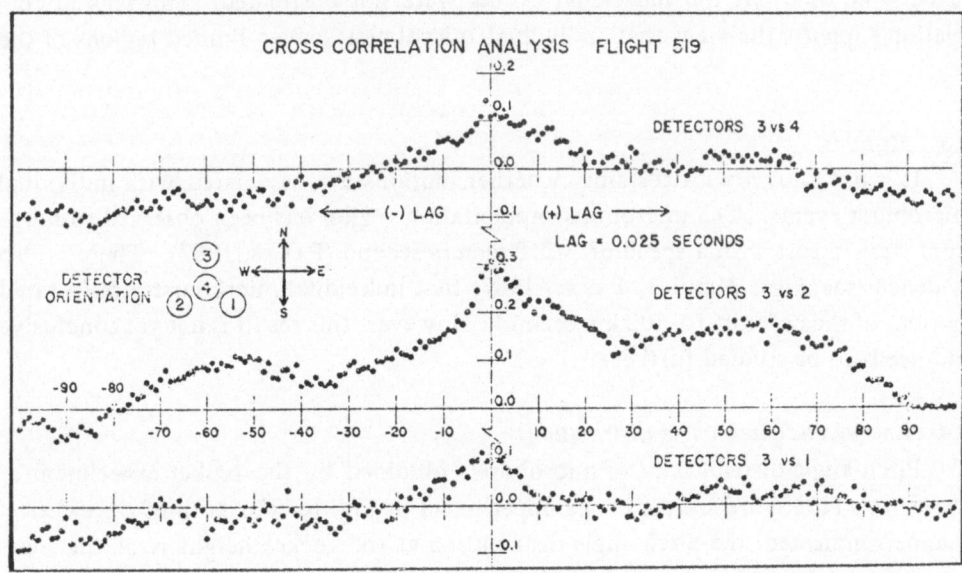

Fig. 5. Microbursts observed by narrow directional X-ray telescopes. Those events that have arrows over them cover 18 km in the lateral direction at 100 km electron stopping heights. The small cross-correlation coefficients between various telescopes indicate that microbursts are small-scaled.

tion region. Microburst precipitation is restricted to a few-hour region centered around 0700 local hours. Individual microbursts are even smaller, covering regions ≤ 80 km at 100 km heights. Sometimes microbursts cover regions smaller than about 18 km (PARKS, 1967).

The top panel of Fig. 5 shows microbursts detected by directional telescopes (PARKS, 1967). The events that are marked with an arrow were detected only within the field of view of the vertical telescope. Since these events were not evident in the detectors that viewed 30° from the zenith, one can assign an upper limit dimension to these events taking into account the individual telescope geometry factors. PARKS (1967) deduced that their dimensions are ≤ 18 km. The bottom panel shows the results of cross correlating the various detector outputs. The small correlation coefficient centered around zero lag shows that microbursts detected by the various detectors are not correlated with a high degree of coherency and confirms that individual microbursts are small-scaled. Note that the slight oscillatory behavior in the coefficient is a contribution from microburst periodicities. (Here 30 lags corresponds to a period of 0.75 seconds.)

BARCUS et al. (1966) and TREFALL et al. (1966) have also attempted to determine the individual microburst dimensions. Using omnidirectional detectors flown on two balloons, they deduced that microburst events cover regions ≤ 200 km, a conclusion based on the fact that microburst events on two balloons are not correlated when the two balloons are separated by more than 200 km in the lateral direction. That microbursts are small-scaled is also deduced by BROWN et al. (1965). Two balloons flown from conjugate regions of the northern hemisphere detected microburst events at the same time but the individual events were not correlated. This lack of correlation supports the view that individual microbursts cover limited regions of the sky.

2.3   Motions

It is not known with certainty whether motions are associated with individual microburst events. The microburst precipitation region has been observed to move from west to east with a speed of ∼ 250 meters/second (PARKS, 1967). There is also evidence (see Figs. 2 and 3, PARKS, 1967) that individual microbursts are in rapid motion, of the order of 10–100 km/second. However, this result is not yet conclusive and needs to be studied further.

2.4   Energy and pitch-angle distributions

Pitch-angle distributions of microbursts obtained by the rocket experiment of LAMPTON's (1967) are shown in the upper panel of Fig. 6. For both of the electron channels indicated, the pitch-angle distribution at the rocket height is of the "loss cone" shape. Note that the pitch-angles sampled are from about 30°–90°. The pitch-angle distribution taken at time 230 seconds represents data before enhancement of precipitation fluxes. Data at 270 and 274 seconds are for enhanced precipitation periods and for microbursts, respectively. According to these results, the electron

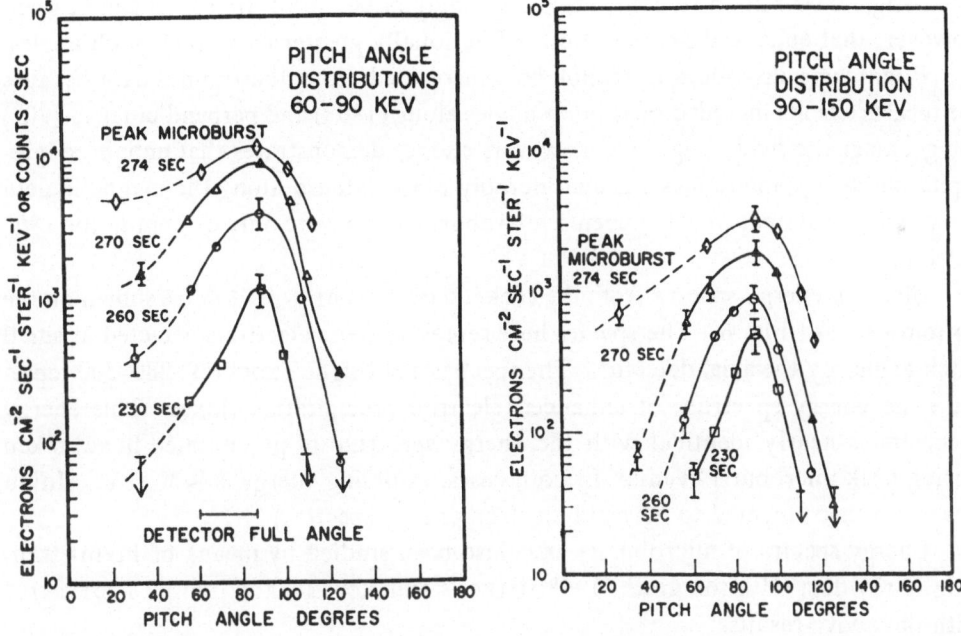

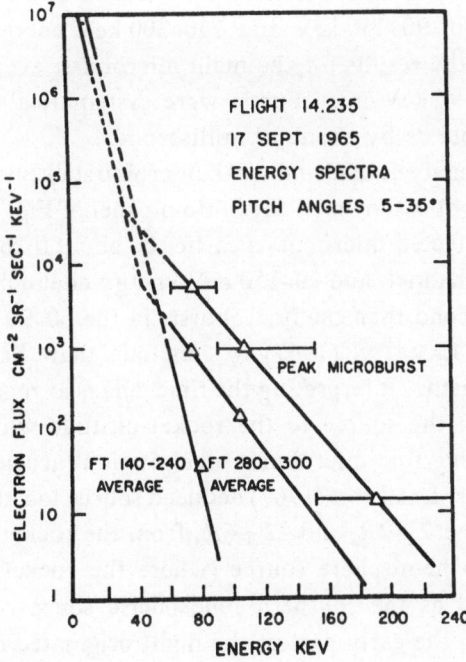

Fig. 6. Pitch-angle distribution of electron precipitation during microburst events. Primary electron energy spectra are shown as well.

pitch-angle distribution remains peaked strongly at 90° at all times. It can be seen, however, that enhanced precipitation is fractionally greater at smaller pitch-angles.

Pitch-angle distribution cannot be constructed from the Injun 3 data because the measurements included only pitch-angles along (180°) and perpendicular to (90°) the geomagnetic field. Figure 2, however, clearly demonstrates that enhanced precipitation during microbursts is considerably more intense along the magnetic field direction. As noted, in some events microburst fluxes were only evident in the 180° detector.

Electron energy spectra from the rocket data (LAMPTON, 1967) are shown on the bottom panel of Fig. 6. The spectra here represent those electrons detected at small pitch-angles by the axial detector. The spectrum at Flight Time (FT) 280–300 represents the energy spectrum of enhanced electron precipitation fluxes. This energy spectrum is nearly identical with the energy spectrum of precipitated fluxes taken during peak microburst events. In both cases, e-folding energy is ~ 30 keV. (Injun 3 was not instrumented to measure electron energy spectra.)

Energy spectra of microbursts have also been studied by means of bremsstrahlung X-ray data (HUDSON et al., 1965; BARCUS and ROSENBERG, 1966). These agree with the above results.

2.5 Cross-correlation analysis

LAMPTON (1967) has cross-correlated the microbursts in the 60–90 keV energy channel against those in 90–150 keV and 150–300 keV energy channels. The top panel of Fig. 7 shows the results for the main microburst events. It is evident that in this analysis the 60–90 keV microbursts were systematically detected later than the 90–150 keV microbursts by about 10 milliseconds.

Cross-correlation analysis made on weak microburst events that were detected in the early part of the flight is shown on the bottom panel of Fig. 7. Here, the 150–300 keV energy channel detected microbursts earlier by about 0.26 seconds than those in the 60–90 keV energy channel, and 90–150 keV energy channel detected microbursts earlier by about 0.1 second than the microbursts in the 60–90 keV energy channel.

These results led LAMPTON (1967) to conclude that the microburst electron source is close to the earth. Interpreting the time delays as representing transit delay times of electrons from the source to the rocket altitude where the measurements were made, and assuming that microburst precipitation (acceleration) occurs simultaneously at all energies, LAMPTON (1967) deduced source locations for the two cross-correlation results to be $2\pm 2R_e$ and $22\pm 6R_e$ from the rocket. The first was interpreted as the northern hemisphere source (where the rocket was flown), and the second was interpreted as the southern ionospheric souce. Evidently, the weak microbursts detected in the early part of the flight originated in the southern hemisphere and propagated into the northern hemisphere (on closed field lines), where they were detected.

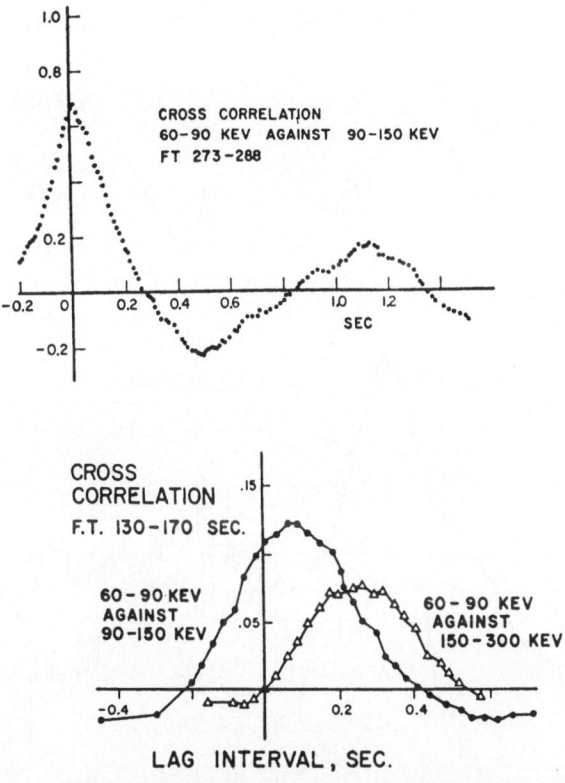

Fig. 7. Cross-correlation results of microburst electrons in the different energy channels.

2.6 Microburst correlation to substorms, VLF's, and magnetic impulses

Auroral zone electron precipitation occurs when trapped fluxes increase in the magnetosphere during magnetic bays. During precipitation, VLF wave activity is also enhaced. Figure 8 shows such a correlation for the event observed on August 17, 1967 (the VLF data here were recorded by the Stanford group at Siple Station, Antarctica; the data were kindly provided by Dr. Ho, Fig. 8 first appeared in WINCKLER, 1974). The X-rays here consisted of microbursts. That electron precipitation and VLF wave activity was correlated has been known for some time (see, for example, GURNETT and O'BRIEN, 1964).

Correlation between VLF and electron precipitation was sought at a finer time scale by MILTON and OLIVEN (1967) and by OLIVEN et al. (1968). The top panel of Fig. 9 shows microbusts detected simultaneously by a ballon-borne X-ray detector and direct electron microburst measurements made by Injun 3. This figure also shows VLF wave spectrum. Rising tone events were active during microburst observations but a one-to-one correlation was not found. It is interesting to note that the VLF risers here have durations similar to microbursts.

An extremely good correlation was observed between whistlers and X-ray bursts (microbursts) at Siple Station (ROSENBERG et al., 1971; FOSTER and ROSENBERG, 1976). The results shown in the middle panel of Fig. 9 show a one-to-one corre-

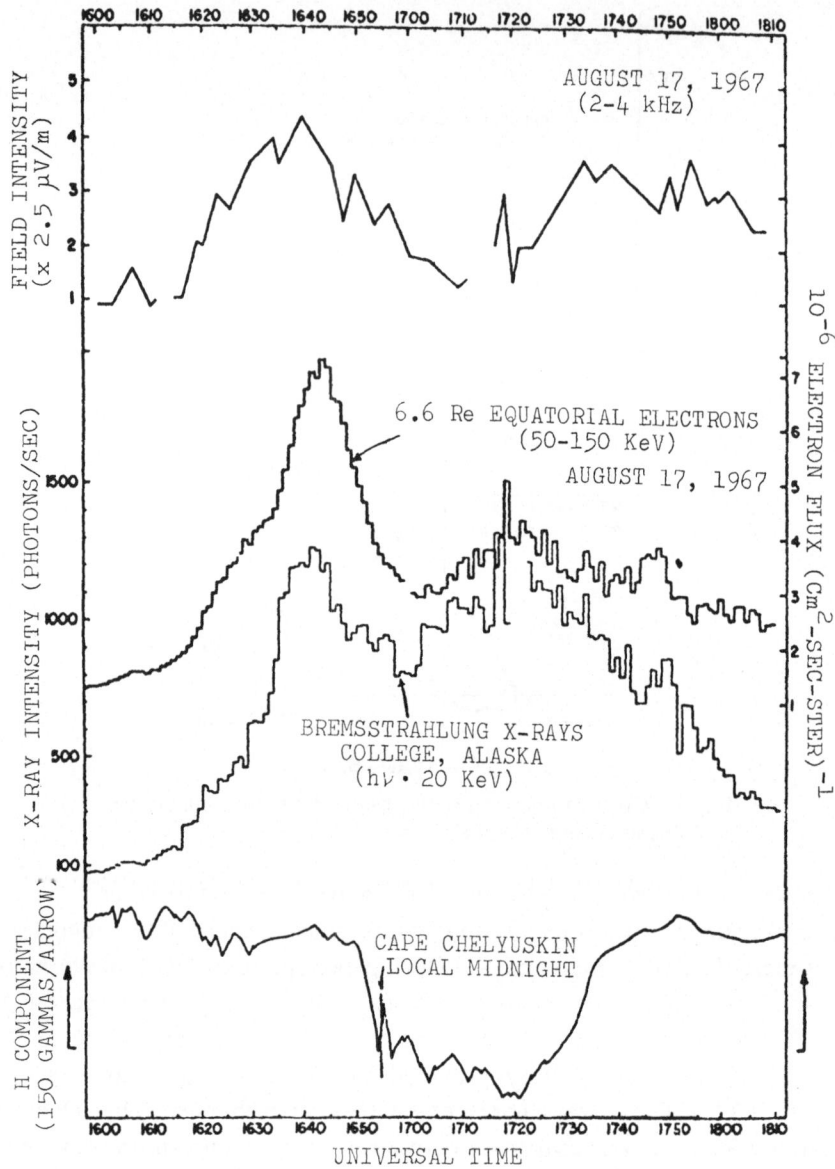

Fig. 8. Correlated events observed in trapped electron increases, precipitated fluxes that include microbursts, magnetic bays, and VLF wave activity.

spondence between the X-ray and VLF bursts. The rising tone VLF bursts detected here appear similar to those detected by Milton and Oliven (1967).

Micropulsation activity increases during substorms and therefore it is expected that micropulsation and electron precipitation will be correlated. Like the VLF wave activity, however, the micropulsations in general do not show a one-to-one correlation except in rare cases. The one instance when good correlation was found is shown in the bottom panel of Fig. 9. Here, one sees that X-ray microbursts and magnetic micropulsation impulses occurred together (McPherron et al., 1968).

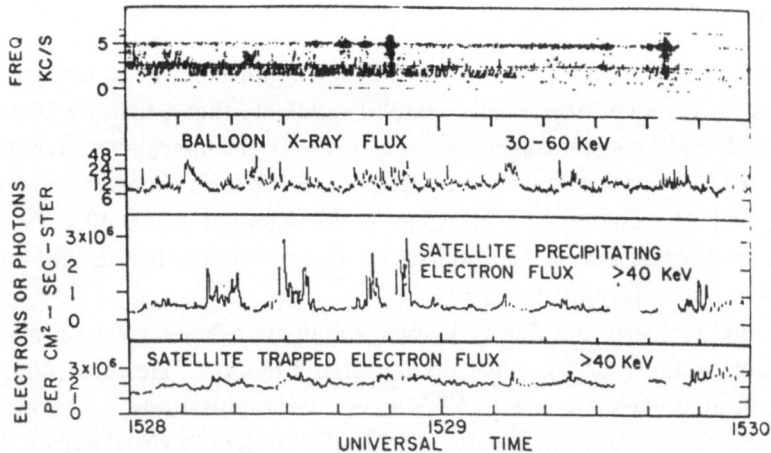

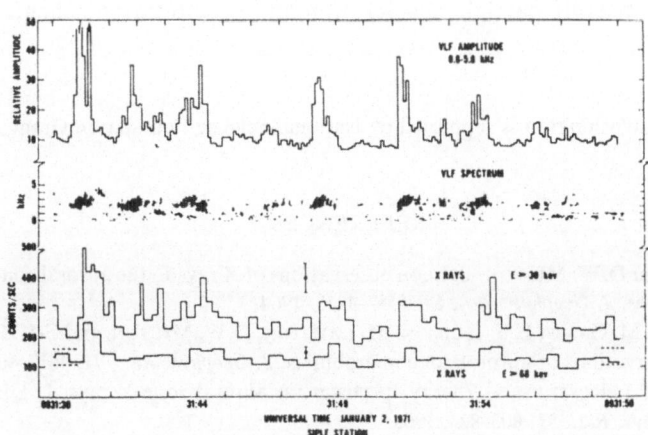

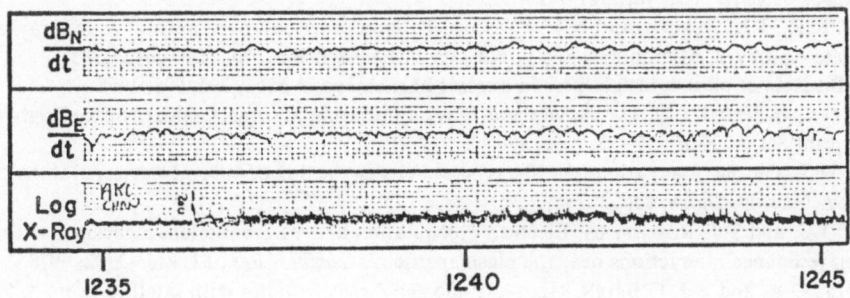

Fig. 9.   Fine-scale correlation of VLF and magnetic impulses with individual microburst events.

3. Conclusion

Characteristics of microbursts deduced from balloon-borne experiments and those revealed in a rocket and a polar satellite indicate that a major electron loss process on the dayside magnetosphere is associated with the microburst precipitation phenomena. All information that we have at the present time comes from electrons of energies ≥ 40 keV. There is no information for energies below 40 keV. Nor is there any direct information on microbursts for regions higher than several hundred kilometers from the surface of the earth.

At this time, virtually nothing is known about microburst mechanisms (there is only one short note by Coppi, 1965, on microburst theory). However, the observation of microburst correlation with VLF waves, micropulsations, and substorms, and the fact that microburst-like precipitation is also evident in visual auroral forms, indicate that a fundamental plasma process is at work here. Future observations of microburst phenomena should delineate more clearly the role of microburst precipitation in the overall particle acceleration and precipitation scheme in the magnetosphere.

The research in Washington is supported by National Science Foundation Grant ATM77-09080.

REFERENCES

ANDERSON, K.A. and D.W. MILTON, Balloon observations of X rays in the auroral zone, 3, High time resolution studies, *J. Geophys. Res.*, **69**, 4457–4470, 1964.

ANDERSON, K.A., L.M. CHASE, H.S. HUDSON, M. LAMPTON, D.W. MILTON, and G.K. PARKS, Balloon and rocket observations of auroral-zone microbursts, *J. Geophys. Res.*, **71**, 4617–4629, 1966.

BARCUS, J.R. and T.J. ROSENBERG, Energy spectrum for auroral-zone X rays, 1, Diurnal and type effects, *J. Geophys. Res.*, **71**, 803–823, 1966.

BARCUS, J.R., R.R. BROWN, and T.J. ROSENBERG, The spatial and temporal character of fast variations in auroral-zone X-rays, *J. Geophys. Res.*, **71**, 125–141, 1966.

BROWN, R.R., Observations of narrow microburst trains in the geomagnetic storm of August 4–6, 1972, *J. Geophys. Res.*, **78**, 1727–1729, 1973.

BROWN, R.R., J.R. BARCUS, and N.R. PARSONS, Balloon observations of auroral-zone X rays in conjugate regions, 2, Microbursts and pulsations, *J. Geophys. Res.*, **70**, 2599–2612, 1965.

COPPI, B., Role of plasma instabilities in auroral phenomena, *Nature*, **205**, 998, 1965.

CRESSWELL, G.R. and T.N. DAVIS, Observations on pulsating auroras, Goddard Space Flight Center Preprint Series, X-612-65-485, Nov. 1965.

EVANS, D.S., On the observation of a 10-cps periodicity in the flux of auroral electrons (abstr.), *Trans. Amer. Geophys. Union*, **48**, 72, 1967.

FOSTER, J.C. and T.J. ROSENBERG, Electron precipitation and VLF emissions associated with cyclotron resonance interactions near the plasmapause, *J. Geophys. Res.*, **81**, 2183–2192, 1976.

GURNETT, D.A. and B.J. O'BRIEN, High-latitude geophysical studies with satellite Injun 3, 5, Very-low-frequency electromagnetic radiation, *J. Geophys. Res.*, **69**, 65, 1964.

HUDSON, H.S., G.K. PARKS, D.W. MILTON, and K.A. ANDERSON, Determinations of the auroral-zone X-ray spectrum, *J. Geophys. Res.*, **70**, 4979–4982, 1965.

LAMPTON, H., Daytime observations of energetic auroral-zone electrons, *J. Geophys. Res.*, **72**, 5817, 1967.

MCPHERRON, R.L., G.K. PARKS, F.V. CORONITI, and S.H. WARD, Studies of the magnetospheric substorm, 2, Correlated magnetic micropulsations and electron precipitation occurring during auroral substorms, *J. Geophys. Res.*, **73**, 1697–1713, 1968.

MILTON, D.W. and M.N. OLIVEN, Simultaneous satellite and balloon observations of the same auroral zone precipitation event, *J. Geophys. Res.*, **72**, 5357, 1967.

O'BRIEN, G.J., High latitude geophysical studies with Injun 3, 3, *J. Geophys. Res.*, **69**, 13–43, 1964.

OLIVEN, M.N., D. VENKATESAN, and K.G. McCRACKEN, Microburst phenomena, 2, Auroral-zone electrons, *J. Geophys. Res.*, **73**, 2345–2353, 1968.

PARKS, G.K., Auroral zone microbursts, substructures, and a model for microburst precipitation, *Proc. Int. Conf. on X-rays in Space*, Calgary, Alberta, Canada, August 14–21, p. 849–874, 1974.

PARKS, G.K., Spatial characteristics of auroral-zone X-ray microbursts, *J. Geophys. Res.*, **72**, 215–225, 1967.

PARKS, G.K., D.W. MILTON, and K.A. ANDERSON, Auroral zone X-ray bursts of 5- to 25-millisecond duration, *J.Geophys. Res.*, **72**, 4587, 1967.

PARKS, G.K., F.V. CORONITI, R.L. McPHERRON, and K.A. ANDERSON, Studies of magnetospheric substorm, 1, Characteristics of modulated energetic electron precipitation occurring during auroral substorms, *J. Geophys. Res.*, **73**, 1685–1696, 1968.

PARKS, G.K., H.S. HUDSON, D.W. MILTON, and K.A. ANDERSON, Spatial asymmetry and periodic time variations of X-ray microbursts in the auroral time zone, *J. Geophys. Res.*, **70**, 4975–4978, 1965.

ROSENBERG, T.J., R.A. HELLIWELL, and J.P. KATSUFRAKIS, Electron precipitation associated with discrete very low frequency emissions, *J. Geophys. Res.*, **76**, 8445, 1971.

TREFALL, H.J., W.E. BJORDAL, S.L. ULLALAND, and J. STADSMA, On the extension of auroral-zone X-ray microbursts, *J. Atmos. Terr. Phys.*, **28**, 225–233, 1966.

TREILHOU, J.-P., A. SAINT-MARC, G. GASSET, A. BOUTONNET, I.A. JOULINE, and K.H.D. KANONIDI, Auroral and subauroral research by balloon-borne and ground-based measurememts, *in* European sounding-rocket and related research at high latitudes, *ESRO SP-97*, 87, 1973.

VENKATESAN, D., M.N. OLIVEN, P.J. EDWARDS, K.G. McCRACKEN, and M. STEINBOCK, Microburst phenomena, 1, Auroral-zone X-rays, *J. Geophys. Res.*, **73**, 2333–2343, 1968.

WINCKLER, J.R., Investigation of electron dynamics in the magnetosphere with electron beams injected from sounding rockets, Invited Paper presented at Int. Conf. on X-rays in Space, Calgary, Alberta, Canada, August 14–21, 1974.

WINCKLER, J.R., P.D. BHAVSAR, and K.A. ANDERSON, A study of the precipitation of energetic electrons from the geomagnetic field during magnetic storms, *J. Geophys. Res.*, **67**, 3717–3736, 1962.

ZUHLIN, I.A., J.-P. TREILHOU, and J. GASSET, Roentgen bremsstrahlung at sub-auroral latitudes, *Cosmic Res.*, **11**, 441, 1972.

MECHANISMS FOR THE FORMATION
OF AURORAL STRUCTURE

Observed Microstructure of Auroral Forms

T. Neil DAVIS

Geophysical Institute, Fairbanks, Alaska 99701, U.S.A.

(Received February 7, 1978)

Discrete auroras are primarily composed of linear elements, whereas diffuse auroras appear in a variety of shapes. Dynamical variations within discrete auroras are characterized by shear, evidently caused by field-aligned currents and charge excesses. In contrast, diffuse, pulsating aurora exhibits little shear, indicating lack of large currents and electric fields associated with this type of aurora.

1. Introduction

An inherent characteristic of most auroral forms is their highly structured nature. No general explanation for the structure exists, though two statements by ALFVÉN (1972) seem pertinent, "Space plasmas have often a complicated inhomogeneous structure" and "Currents produce filaments or flow in thin sheets."

Since the introduction of sensitive imaging systems for auroral observation about 15 years ago, it has been possible to record the often highly transient microstructure within auroras, though the great speed of the variation and weakness of some of the emitting structures presses to the limit the spatial (of order 1 mrad) and temporal (1/60 sec) resolution of devices now in use. This review deals mainly with observations with such imaging systems and also photographic cameras that have been made from the ground or aircraft by Canadian, Japanese, Soviet and American groups. In dealing with the dynamical morphology of auroral microstructure, these groups have emphasized relative measurements of intensity versus two spatial dimensions and time with only secondary importance being attached to observations of absolute intensity. Other imaging observations emphasizing measurement of absolute intensity have been made by S.J. Mende of Lockheed Palo Alto Research Laboratories.

A major difficulty in observation of auroral microstructure is caused by the alignment of the height extent of auroral forms along the direction of the local magnetic field. Exact determination of the shape of an auroral form from one observing point on the ground below can be accomplished only if the form is precisely in the magnetic zenith. Departure of the form by only a few degrees from the observer's magnetic zenith creates such a different perspective that errors in shape recognition and description are likely. This general difficulty makes it nearly impossible to accurately describe an entire auroral form, since virtually any form observed in the magnetic zenith extends, in at least one horizontal direction, well outside the zenith. Thus in describing the auroral microstructure, we are usually limited to those parts of the forms that lie near the zenith rather than entire forms.

We are concerned here with auroral shapes, developments from one shape to another, spatial motions and temporal variations. In considering the motions, it is natural to ask what is moving and what causes the apparent motion. Though it is now known that apparent auroral motions are caused by motions of the distant source region from which precipitating auroral primaries derive (i.e. 'successive dumping'), by motions of entire magnetic flux tubes feeding particles into the atmosphere ($E \times B$ motion) and by tipping of the flux tubes, it is impossible to tell from the observed apparent motions alone which type of motion is occurring. Therefore it is useful to bring into the interpretation of observed motions results obtained by other observations, particularly those of electric fields obtained by direct probe measurements or by the introduction of visible tracer ions above auroras.

2. Major Types: Discrete and Diffuse Auroras

Since the advent of topside imagery of auroras by the ISIS-2 (LUI and ANGER, 1973) and DMSP (PIKE and WHALEN, 1974) satellites it has become common to refer to 'diffuse' and 'discrete' aurora because the topside images show two regions of auroral precipitation aptly described by these terms. Least one become misled by these terms, it is necessary to recall that ground-based imaging observations show that both discrete and diffuse auroras can be highly structured. The primary distinction seems to be that diffuse auroras, even if structured, lack the high intensity and shear phenomena exhibited by 'discrete' auroras. Whereas 'diffuse' auroras have a variety of shapes ranging from linear to highly irregular, complex structures, the basic form of the 'discrete' aurora is linear. At times, the linear elements become highly distorted, but the basic linear characteristic remains.

3. Discrete Auroras

3.1 Shapes

Since the elemental shape of discrete aurora is linear it seems useful to retain the term arc for these structures and to give a new definition of an auroral arc based upon our current knowledge:

An auroral arc is the recognizable luminosity resulting from the impingement of a field-aligned sheet beam of electrons (or other charged particles) upon the atmosphere, the luminosity being approximately proportional to the energy deposited by particles in the energy range 100 eV to 100 keV. The lower border of the arc is sharp, in consequence of increasing atmospheric density with depth; it has an altitude, typically 80 to 300 km, defined by the maximum energy of the particles in the primary beam. The vertical extent to the diffuse upper border of the arc increases with the range of particle energy in the primary beam; the range varies from a few kilometers to more than 100 km. Arcs range in width from a few tens of meters to more than 1 km. They extend in length from of 100 km to thousands of kilometers. Their curvature ranges from slight to highly contorted.

This definition emphatically states that one can interpret the position, shape and intensity of an observed auroral arc in terms of the causative incoming particle beam. Also the definition is broadly enough stated to incorporate a number of forms listed in the Photographic Atlas of Auroral Forms (IUGG, 1963)—arcs, bands, draperies, partial corona and, as we shall see later, even rays. In light of current knowledge of causative processes there seems no reason to retain distinction between these different form types except for convenience of describing visual observations.

MAGGS and DAVIS (1968) measured the widths of 581 linear discrete and diffuse auroral forms as ranging from the instrumental cutoff at 70 to 4,440 m with the median width being 230 m. Experiments using rocket-borne electron accelerators (DAVIS *et al.*, 1971; HALLINAN *et al.*, 1977) indicate that minimum widths may be as small as 30 or 40 m, a few times the electron gyroradius. It is common to observe several narrow arcs spaced close together so as to appear as a single structure when viewed well off the direction of the local magnetic field. Though the individual arcs within such an ensemble may behave somewhat independently, discrete auroras never cross one another; hence, the ensemble may deform or drift as a unit.

3.2 Motions and deformations

The basic linear elements of discrete auroras are observed to undergo rotational

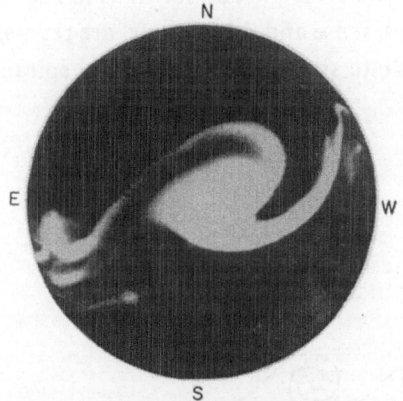

SPIRAL-200km DIA

CURLS-2.7km DIA

Fig. 1. At top, an all-sky photograph showing an auroral spiral which contains curls (auroral rays) along the arm leading to a second spiral in the east. At bottom, a TV image of several interconnected curls in the aurora.

developments which temporarily perturb the shape or cause major changes in shape. The rotational development having the largest scale size is the spiral, a vortex street structure ranging in size from 20 to 1,300 km (DAVIS and HALLINAN, 1976). HALLINAN (1976a, b) has shown the spiral to be an equilibrium configuration that develops when the current density exceeds approximately 10^{-5} amp m^{-2} in a field-aligned current sheet. Spirals in both the north and south hemispheres are clockwise (viewed antiparallel to the magnetic field direction **B**, a convention followed throughout this paper) indicating upward current; they both windup and unwind, i.e. they are reversible. One wishing to associate measurements made at two different altitudes on a magnetic field line threading discrete aurora should take special note of the fact that the development of a spiral shifts the orientation of the magnetic field enough to cause the feet of the field lines to be displaced up to half the spiral diameter (i.e. 10 to 650 km).

Similar to spirals in appearance are the much smaller vortex streets called curls, as described by HALLINAN (1969), HALLINAN and DAVIS (1970), and OGUTI (1974). Curls are of order 1 km in diameter, they are always counterclockwise and are irreversible. In contrast with the spiral equilibrium structure, curls are the result of instability in a charge sheet; the rotation sense indicates negative excess charge at the point about which the curl forms. Seen end-on curls are auroral rays. HALLINAN (1976b) notes that curls and spirals can occur in the same auroral form, as in Fig. 1, along with folds, which exhibit a clockwise rotation sense and, like spirals, are reversible. As shown in Fig. 2, folds are of intermediate size between curls and spirals.

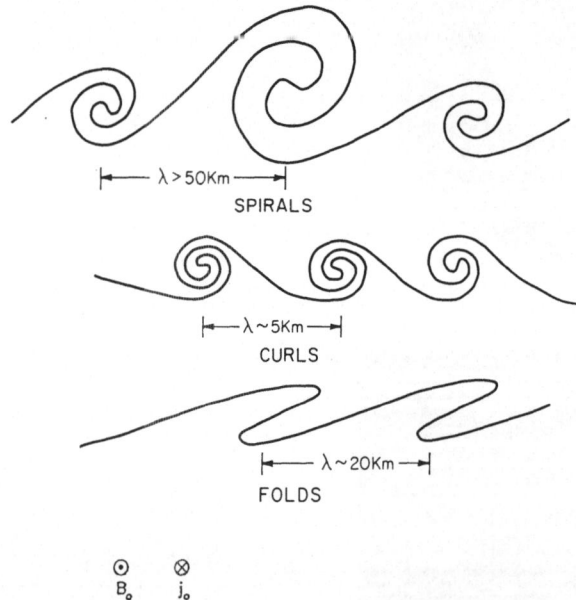

Fig. 2. Drawings of the cross sections of auroral arcs illustrating three major rotational distortions: spirals, curls and folds. The spirals and folds are clockwise, and the curls are counterclockwise. After HALLINAN (1976b).

HALLINAN (1976b) attributes folding to the current sheet mechanism and states that they occur at scale sizes where there is conflict between the counterclockwise shear of the charge sheet and clockwise shear of the current sheet such that the formation of spirals is suppressed.

It is not certain that all deformations to discrete auroral structures can be attributed to the charge sheet and current sheet shearing mechanisms discussed by Hallinan. One complex repetitive deformation occurring in the few minutes near the auroral breakups is the pulsing phenomenon called flickering (BEACH *et al.*, 1968). Sometimes visible by eye, flickering is a 10-Hz fluctuation in which field-aligned emitting elements seem to expand and contract and move erratically back and forth, but rotational motion appears to be involved. The variation has been detected by 42-MHz radar (SOFKO and KAVADAS, 1969).

Extensive observations of small-scale auroral deformations by OGUTI (1975) indicate the frequent occurrence of a deformation he has termed splitting (see Fig. 3). Figure 4 shows another example of such a development documented by DAVIS (1965). I have noticed that this deformation is likely to occur where there is a change in curvature of an arc, so it, too, may be the result of competing shear attributable to the charge sheet and current sheet mechanisms.

The observed rate of development of curls requires small-scale, transient horizontal electric fields sometimes as large as 1 V m^{-1}. Motions of rays along the arcs and relative motions of adjacent auroral arcs (Fig. 5), imply an inward-directed electric field in association with each arc, the motion being to the east on the north

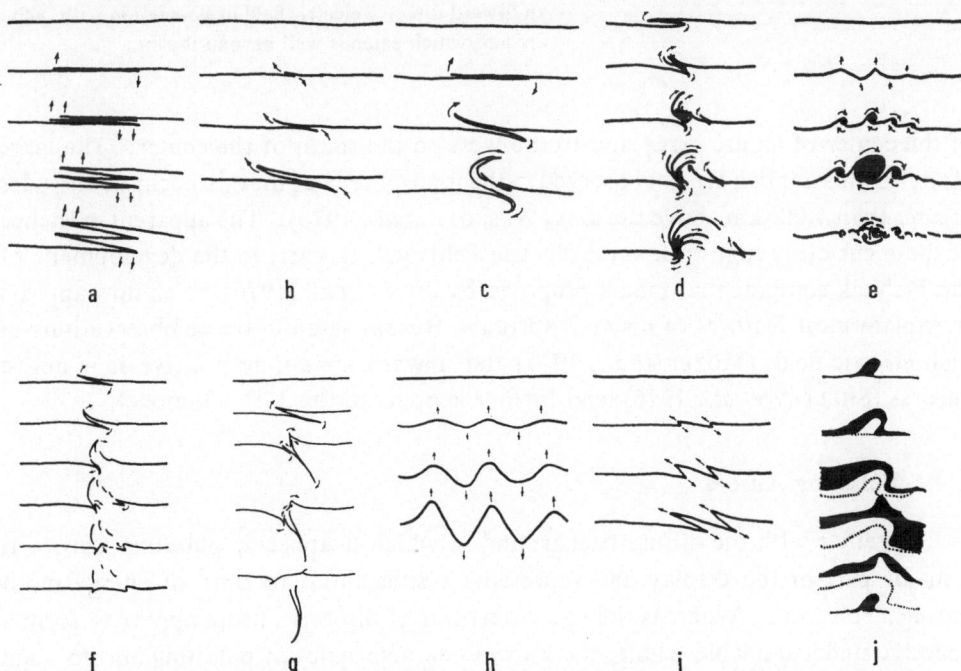

Fig. 3.   Schematic illustrations by OGUTI (1975) of various apparent deformations to auroral arcs. Parts A and B show the splitting deformation.

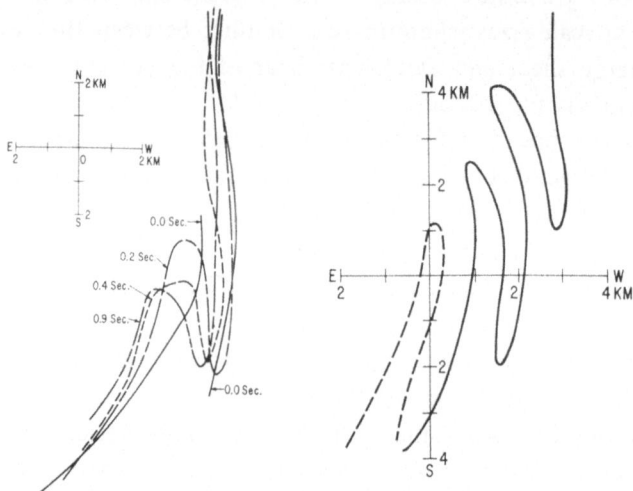

Fig. 4. An example of splitting observed near the magnetic zenith at Fort
     Churchill that led, within a few seconds, to the folded configuration at
     right. After DAVIS (1965).

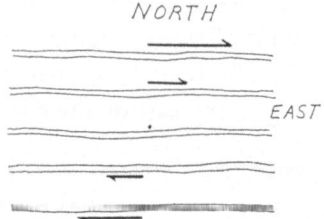

Fig. 5. The pattern of longitudinal motion observed in
     close-spaced, active auroral arcs. The pattern implies
     an inward-directed electric field in association with each
     arc and which extends well beyond the arc.

of the center of an arc array and to the west on the south of the center. The large
electric fields implied are not observed in the ionosphere but they do occur at altitudes
greater than 4,000 km above the arcs (WESCOTT et al., 1976). The apparent existence
of these curiously high transverse electric fields led, in part, to the development of
the V-shock equipotential model proposed by SWIFT et al. (1976) which now appears
to explain most features of discrete auroras. Recent satellite-borne observations of
high electric fields (MOZER et al., 1977) and upward streaming positive ions above
auroras (SHELLEY et al., 1976) lend further support to the V-shock model.

4. Pulsating Aurora

Together with the diffuse background in which it appears, pulsating aurora is
a major part of the display and represents a substantial fraction of energy input
during a substorm. Whereas the characteristics of discrete aurora appear to form a
cohesive understandable whole, the known characteristics of pulsating aurora seem
chaotic and without obvious interpretation. Another contrast is in the complete
lack of shearing phenomena within the pulsating aurora, whereas shearing develop-

ments are inherent in discrete aurora. The lack of shear in pulsating aurora does imply that substantial electric fields do not play a role in creating pulsating aurora.

Investigations, largely photometric, that have described the periodic variations in pulsating auroras are cited by OMHOLT (1971) and VALLENCE JONES (1974). Studies that have emphasized the shape, motions and detailed temporal changes in pulsating auroras using imaging data are those of CRESSWELL and DAVIS (1966), SCOURFIELD and PARSONS (1971), OGUTI (1975), ROYRVIK (1976) and ROYRVIK and DAVIS (1977).

A pulsating aurora is defined by ROYRVIK and DAVIS (1977) as one whose maximum intensity never exceeds $\sim 10\,\mathrm{KR}$ in N_2^+ 1NG 4278Å and which undergoes at

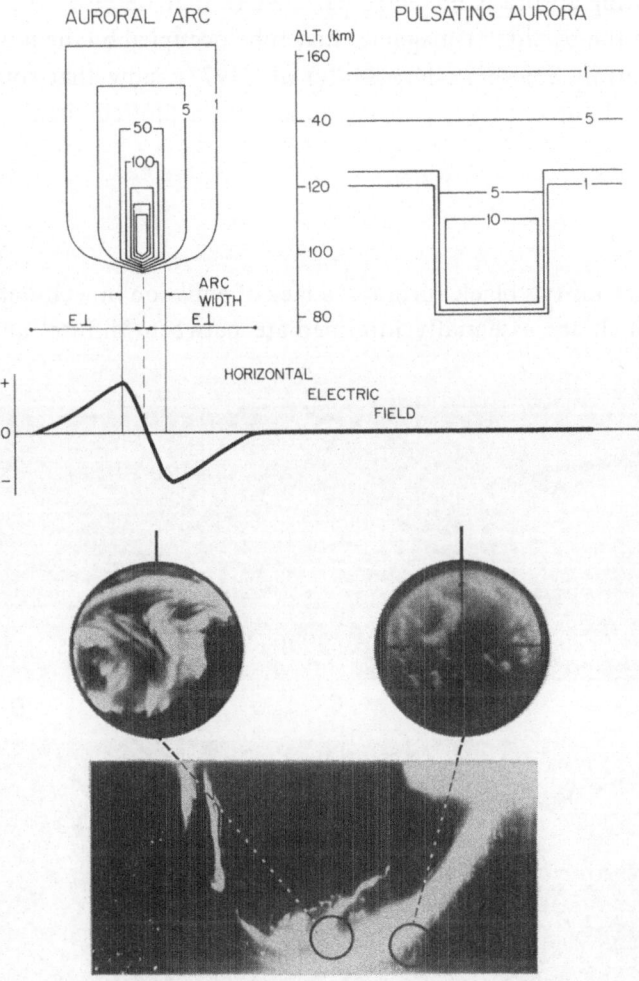

Fig. 6. Schematic models of the auroral arc, at left, and of pulsating aurora. Indicated below each drawing is the horizontal electric field associated with each structure. The pulsating form is shown to extend downward from a diffuse background emission. At bottom, all-sky photographs show the differing appearance of the two types of aurora, and the DMSP image indicates occurrence along the auroral oval.

least one full cycle wherein there is first a rapid increase in intensity and then a rapid decrease, an additional restriction being that the horizontal drift motion of the modulated form differs little from the drift of nearby forms. Pulsations occur in arc and arc-like forms and in patches of irregular shape, the size and shape remaining unchanged as the form pulses on and off or growing and contracting (SCOURFIELD and PARSONS, 1971; ROYRVIK and DAVIS, 1977). Quasiperiodic variations in the range 0.05 to 2 Hz also are complex, a given form can change from one type of temporal behavior to another, and adjacent forms can display disparate behavior. Pulse shapes vary from widely-spaced half-sine waves to closely-packed square waves. Clear-cut trends with latitude, local time or whatever time have not been found. This failure and the major difference in behaviors of proximate forms suggest that the cause of pulsating aurora is linked to parameters that describe the plasma and field quantities on the particular magnetic flux tube occupied by the pulsating form. Recent observations by STENBAEK-NIELSEN et al. (1977) show that some pulsating auroras have very limited height extents, much more limited than is implied by Fig. 6.

5. Black Aurora

More than a curiosity, 'black aurora' is a lack of emission in well-defined regions within auroras which are essentially intermediate between 'diffuse' and 'discrete.'

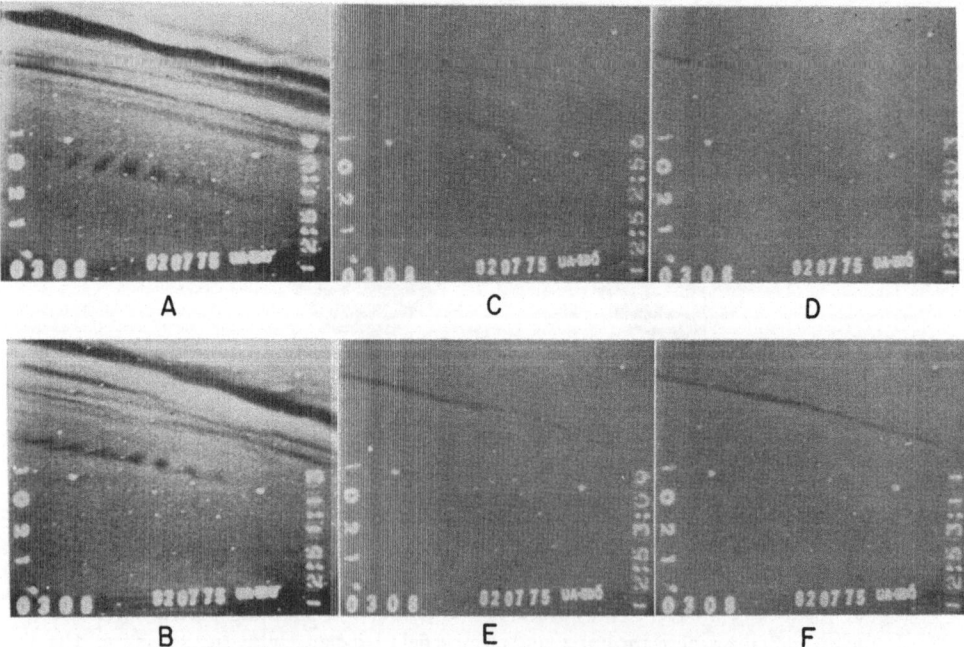

Fig. 7. Television images of 'black aurora,' void regions in a field of relatively uniform auroral emission. The vortex street in parts A and B is clockwise seen antiparallel to B, indicating an excess of positive charge in the black void region. After ROYRVIK (1976).

These auroras are intermediate both in the sense that they occur in the transition from one type to another (especially diffuse to discrete) and the auroras exhibit a degree of shear behavior intermediate between that of diffuse and discrete aurora. The black void areas, in the form of thin ribbons or elongated elliptical holes drift east after midnight and west before (NAZARCHUCK, 1975). Comparison of our observations with those of NAZARCHUCK (1975) and OGUTI (1975) suggests that the black auroras appearing in the transition from discrete to diffuse aurora at the end of the substorm expansion phase may differ from those observed near the end of the recovery phase. Figure 7, after ROYRVIK (1976) shows examples of the latter type. The row of holes is of particular interest—these clearly are a vortex street structure similar to curls, but they show clockwise rotation and thereby imply an excess of positive space charge in the holes, i.e. outside the normal electron aurora.

REFERENCES

ALFVÉN, H., Relations between cosmic and laboratory plasma physics, in *Cosmic Plasma Physics*, edited by K. Schindler, pp. 1–14, Plenum Press, New York, 1972.

BEACH, R., G.R. CRESSWELL, T.N. DAVIS, T.J. HALLINAN, and L.R. SWEET, Flickering, a 10-cps fluctuation within bright auroras, *Planet. Space Sci.*, **16**, 1525–1529, 1968.

CRESSWELL, G.R. and T.N. DAVIS, Observations on pulsating auroras, *J. Geophys. Res.*, **71**, 3155–3163, 1966.

DAVIS, T.N., Television cinema photography of auroras, in *Auroral Phenomena*, edited by M. Walt, pp. 15–19, Stanford University Press, Stanford, Ca., 1965.

DAVIS, T.N., T.J. HALLINAN, G.D. MEAD, J.M. MEAD, M.C. TRICHEL, and W.N. HESS, Artificial aurora experiment: Ground-based optical observations, *J. Geophys. Res.*, **76**, 6082–6092, 1971.

DAVIS, T.N. and T.J. HALLINAN, Auroral spirals, 1. Observations, *J. Geophys. Res.*, **81**, 3953–3958, 1976.

HALLINAN, T.J., The morphology of small-scale folds and curls in the aurora, M.S. Thesis, University of Alaska, Fairbanks, Alaska, 52 pp., May, 1969.

HALLINAN, T.J., Spiral-like auroral forms: Observations and a proposed theory, Ph.D. dissertation, University of Alaska, Fairbanks, Alaska, 90 pp., May, 1976a.

HALLINAN, T.J., Auroral spirals. 2. Theory, *J. Geophys. Res.*, **81**, 3959, 1976b.

HALLINAN, T.J. and T.N. DAVIS, Small-scale auroral arc distortions, *Planet. Space Sci.*, **18**, 1735–1744, 1970.

HALLINAN, T.J., H.C. STENBAEK-NIELSEN, and J.R. WINCKLER, The Echo IV electron beam experiment: Television observations of artificial auroral streaks indicating strong beam interactions in the high-latitude magnetosphere, *J. Geophys. Res.*, **83**, 3262–3272, 1978.

INTERNATIONAL UNION OF GEODESY AND GEOPHYSICS, *International Auroral Atlas*, Edinburgh University Press, Edinburgh, 1963.

LUI, A.T.Y. and C.D. ANGER, A uniform belt of diffuse auroral emission seen by the ISIS-2 scanning photometer, *Planet. Space Sci.*, **21**, 799–809, 1973.

MAGGS, J.E. and T.N. DAVIS, Measurements of the thickness of auroral structures, *Planet. Space Sci.*, **16**, 205–209, 1968.

MOZER, F.S., C.W. CARLSON, M.K. HUDSON, R.B. TORBERT, B. PARADY, J. YATTEAU, and M.C. KELLEY, Observations of paired electrostatic shocks in the polar magnetosphere, *Phys. Rev. Lett.*, **38**, 292–295, 1977.

NAZARCHUCK, G.K., A drift of inhomogeneities in the aurora striated band, submitted to Plasma Instabilities Symp., XVI, IUGG/IAGA General Assembly, Grenoble, France, 1975.

OGUTI, T., Rotational deformations and related drift motions of auroral arcs, *J. Geophys. Res.*, **79**, 3861–3865, 1974.

OGUTI, T., Metamorphoses of aurora, Memoirs of National Institute of Polar Research, Series A, No. 12, 101 pp., Tokyo, 1975.

OMHOLT, A., *The Optical Aurora*, 198 pp., Springer, New York, 1971.

PIKE, C.P. and J.A. WHALEN, Satellite observations of auroral substorms, *J. Geophys. Res.*, **79**, 985–1000, 1974.

ROYRVIK, O., Pulsating aurora: Local and global morphology, Ph.D. dissertation, 133 pp., University of Alaska, Fairbanks, Alaska, May, 1976.

ROYRVIK, O. and T.N. DAVIS, Pulsating aurora: Local and global morphology, *J. Geophys. Res.*, **82**, 4720–4740, 1977.

SCOURFIELD, M.W.J. and N.R. PARSONS, Pulsating auroral patches exhibiting sudden intensity-dependent spatial expansion, *J. Geophys. Res.*, **76**, 4518–4524, 1971.

SHELLEY, E.G., R.D. SHARP, and R.G. JOHNSON, Satellite observations of an ionospheric acceleration mechanism, *Geophys. Res. Lett.*, **3**, 654–656, 1976.

SOFKO, G.J. and A. KAVADAS, Periodic fading in 42-MHz auroral backscatter, *J. Geophys. Res.*, **74**, 3651–3658, 1969.

SWIFT, D.W., H.C. STENBAEK-NIELSEN, and T.J. HALLINAN, An equipotential model for auroral arcs, *J. Geophys. Res.*, **81**, 3931–3934, 1976.

WESCOTT, E.M.., H.C. STENBAEK-NIELSEN, T.J. HALLINAN, T.N. DAVIS, and H.M. PEEK, The Skylab barium plasma injection experiments. 2. Evidence for a double layer, *J. Geophys. Res.*, **81**, 4495–4502, 1976.

VALLANCE JONES, A., *Aurora*, 301 pp., D. Reidel, Hingham, Mass., 1974.

Birkeland Currents and Auroral Structure

Hugh R. ANDERSON

Rice University, Houston, Texas 77001, U.S.A.

(Received February 7, 1978)

A short history of studies of Birkeland currents and a summary of measurement techniques is provided. The observations made to date show that discrete auroral arcs and upward flowing Birkeland currents are usually associated. There is experimental uncertainty about whether these upward currents are strongest just above the arcs or are concentrated at the edges. A consistent pattern relating the location of the downward return current, the electrojet, and the electric field has not been found.

1. Introduction

The past few years have seen a rapidly increasing interest in magnetic field-aligned currents (Birkeland currents) of all scale sizes and in the phenomena that accompany them. The tables of contents in recent volumes of the Journal of Geophysical Research testifies to this, as does the number of papers on the currents that were presented at this symposium. My purpose is to describe observations of Birkeland currents that relate the currents to auroral forms. I will omit discussion of the larger scale circulation except to locate the auroral currents in it and to provide a brief historical background. For more extended general discussions see reviews by ARNOLDY (1974) and ANDERSON and VONDRAK (1975).

You will see that the experimentalists do not all deduce the same current morphology from their data. Although some of the differences must represent a variable phenomenon, it is my opinion that there are real experimental differences awaiting resolution.

2. Historical Sketch

The idea that field-aligned currents could feed the horizontal ionospheric currents that cause magnetic perturbations at high latitude was first suggested by BIRKELAND (1908, 1913). This idea was largely ignored for 50 years as attention was focussed on the horizontal currents, although ALFVÉN (1939–1940) discussed the relation between FAC and the auroral electrojet. Other analyses involving FAC were made by KERN (1962), FEJER (1963), and COLE (1963) with Kern suggesting that the FAC might accelerate auroral particles.

In 1964 BOSTRÖM focussed attention on the possible connection of FAC and aurora discussing the FAC that would flow associated with a uniform electric field

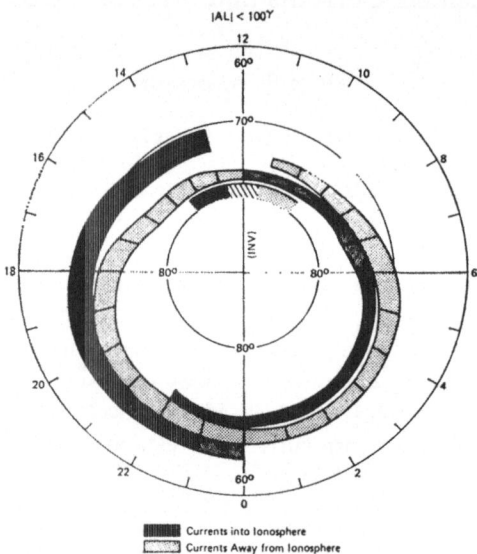

Fig. 1. Large scale Birkeland current pattern deduced by IIJIMA and POTEMRA (1976) from Triad data.

and a highly conducting strip, such as might occur in an auroral arc, imbedded in a less conducting ionosphere. SWIFT (1965) suggested that if the current exceeded 40 μA/m² the resulting instabilities would produce anomolously high parallel resistivity.

In the United States A.J. Dessler was responsible for much of the early interest in Birkeland currents, and the appreciation of their importance. The first in-situ measurement of transverse magnetic perturbations due to Birkeland currents was made by a vector magnetometer on satellite 1963-38C. Initially the magnetic fluctuations were interpreted as hydromagnetic waves (ZMUDA et al., 1966, 1977) but CUMMINGS and DESSLER (1967) pointed out that the latitudinal extent of the fluctuations was too small for waves of the appropriate wavelength and they were thereafter interpreted as being due to Birkeland currents (ZMUDA et al., 1970).

Later SCHIELD et al. (1969) proposed a complete current system connecting the polar ionosphere and magnetosphere. This contains a nearly incomprehensible figure which, however, correctly shows the pairs of Birkeland sheet currents on the morning and evening sides of the polar cap. This pattern has been identified by Zmuda, Armstrong, Potemra, Iijima, and others using satellite magnetometer data. Figure 1 from IIJIMA and POTEMRA (1976) shows the pattern. This current system appears in theoretical models of magnetospheric convection.

3. Measurement Techniques

Before describing the connection observed between Birkeland currents and the

Table 1. $J_{\parallel}$ techniques.

Advantages	Disadvantages
Measure ∇B along trajectory	
(1) Responds to total $J_{\parallel}$.	(1) Needs $\nabla \times B$ but uses ∇B.
(2) Gives some J_h.	$\therefore$ Not local.
	Model-dependent.
Count charge carriers in-situ	
(1) Local measurement.	(1) Hard to count *all* the particles.
(2) Easy for some particle energies.	
$J_{\parallel} = -\nabla_h \cdot J_h ,$ $J_h = \Sigma E$	
E from in-situ or radar.	
(1) Radar is ground-based.	(1) Highly accurate J_h required to differentiate.
(2) Both J_h, $J_{\parallel}$ are obtained.	(2) Radar requires integration time.
	(3) In situ gets Σ from particle flux.

All methods have the space vs. time variation problem.

aurora we will note the principal measurement techniques. Birkeland currents may be measured in three ways, which are summarized in Table 1.

(1)   In-situ vector magnetic measurement.

Advantage: The total current is measured, regardless of what carries it if curl B can be determined.

Disadvantages: Since curl B has not yet been measured, the currents inferred from measurements of ∇B along a track are not unique but are model-dependent.

(2)   In-situ measurement of charged particle fluxes.

Advantage: The measurement is truly local.

Disadvantage: Nobody has yet counted particles of all energies down to zero because electron thermal speed exceeds greatly the probable drift speed. Total current is difficult to assess.

(3)   Calculation of $J_{//}$ from $-\nabla_h \cdot J_h$. Here J_h is obtained by measuring E_h, and calculating Σ from measured values of n_e, or from values deduced from the precipitating flux or the auroral luminosity. The field E_h can be measured in-situ by probes or by incoherent scatter radar.

Advantage: The radar measurement is done from earth. With either method both J_h and $J_{//}$ are obtained.

Disadvantages: Calculation of $\nabla_h \cdot J_h$ introduces large uncertainties due to fluctuations in J_h. The radar technique also requires a certain homogeneity in space and time so that suitable integrations can be made. In addition some assumptions about neutral winds must be made.

There are also three methods of observing and locating the aurora:

(1)   Measure the precipitating particles in-situ.

(2)   Image the auroral emission.

(3)   Observe enhanced electron densities resulting from particle precipitation with radar.

4. Observations

A relationship between the large scale current system and the nightside aurora has been presented by KAMIDE and ROSTOKER (1977) using Triad magnetic data and particle fluxes and images obtained by the DMSP satellites. As Fig. 2 shows, the upward currents are associated with the bright equatorward aurora in the morning and with the region of discrete aurora in the evening. There is evidence that peak upward current density generally coincides with the discrete auroral forms in the evening. The downward current coincides with the diffuse aurora and eastward electrojet in evening, and in the morning with the poleward aurora, although the intensities are not correlated. The authors conclude that the downward current is carried by upward drifting thermal electrons.

A similar description of the evening sector is given by TSUNODA et al. (1976) based on Triad magnetometer data and the observation of diffuse radar aurora by the Homer radar. However, KLUMPAR (1976) and KLUMPAR et al. (1976) report slightly different results using particle and magnetic field measurements made by ISIS-2. The differences are first that on the premidnight side the equatorward downward current extends 3° southward of the plasma sheet precipitation commonly identified as the diffuse aurora. Secondly in the morning sector and the poleward region of upflowing current, variations in the electron flux generally coincide with strong $J_{//}$. As Fig. 3 shows, some downward $J_{//}$ is carried by upward electron fluxes of 5 eV to 15 keV, but the correspondence is not exact.

We will now examine data obtained by rocket and incoherent scatter radar on distinct, selected auroral forms. These data show one of two patterns. The first has upward current sheets at the edge of or just outside the region of energy precipitation,

AURORA, FIELD-ALIGNED CURRENT AND AURORAL ELECTROJET

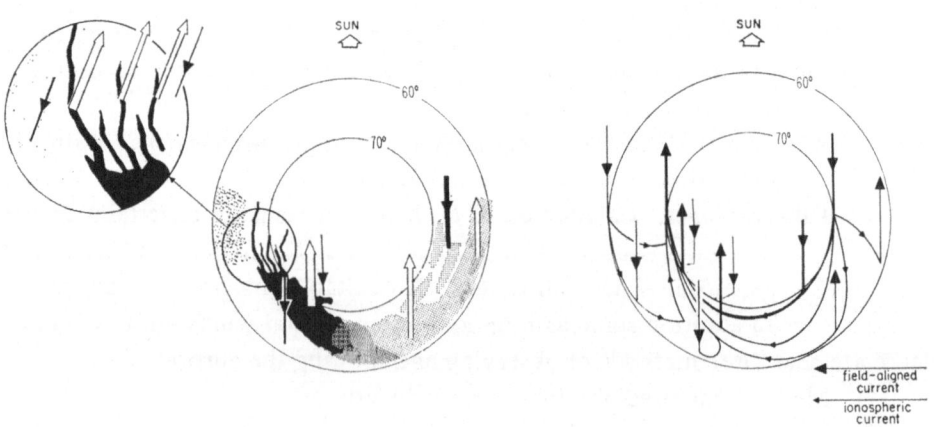

Fig. 2. The relationship between auroral regions and currents deduced by KAMIDE and ROSTOKER (1977).

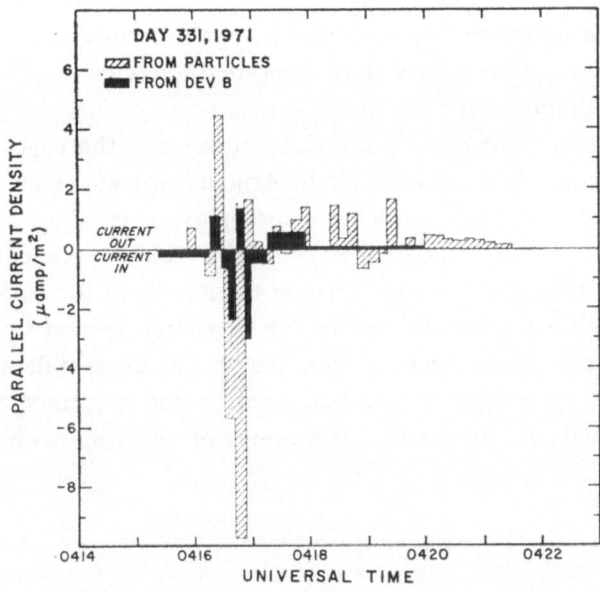

Fig. 3. Observation of aurora particle and total currents measured in
the post-midnight sector by particle detectors and a magnetometer
on ISIS-2 (KLUMPAR *et al.*, 1976).

Table 2.   Results summary, $J_{//}$.

Method and No.	Conditions and time	Electrojet	Author
Upward sheets at edges of arcs			
Particles			
3 (1 N edge only)	Pre-breakup		Arnoldy *et al.*
	Breakup		
	Post breakup		
2	Early quiet		Maehlum and Mostue
$\nabla_h \cdot J_h$	Early quiet	East	Evans *et al.*
Upward sheets over arcs			
∇B			
2 (return S)	Early quiet	W	
	Pre-breakup	E	Rice
1 (return N)	Breakup	E+W	
1 (return not obs)	Breakup	W	U. of Minn.
Particles			
1 (return S)	Late breakup	W	Berkeley
$\nabla \cdot J$			
1 (return N)	Early quiet	E	SRI
Opposite sheets at edges of arcs			
∇B			
1 (return S)	Early	E	Rice
Opposite sheets, dynamic aurora			
∇B			
1 (return N)	Breakup	W	Rice

that is of the auroral arc. The second pattern has the upward sheet coincident with the aurora, and a downward return current, if any is measured, to the north or south. A third rare pattern does show oppositely directed sheets at edges of the aurora. Table 2 summarizes these observations.

Most of the reports of upward Birkeland currents at the edges of auroral arcs rely on particle data. Arnoldy *et al.* (R.L. Arnoldy, private communication, 1977) describe three rocket flights, one each into prebreakup, breakup and post breakup aurora. On each of these, the flux of 100–500 eV electrons carried a net upward current and was stronger at the edges than at the middle of the aurora. The aurora was defined by light emission and by the flux of energy carried by more energetic electrons. Figure 4 shows some of these data. The current densities were a few $\mu A/m^2$. No downward current was measured by the detectors, which were not sensitive below 100 eV. Similar concentrations of electrons with $E \lesssim 500$ eV into

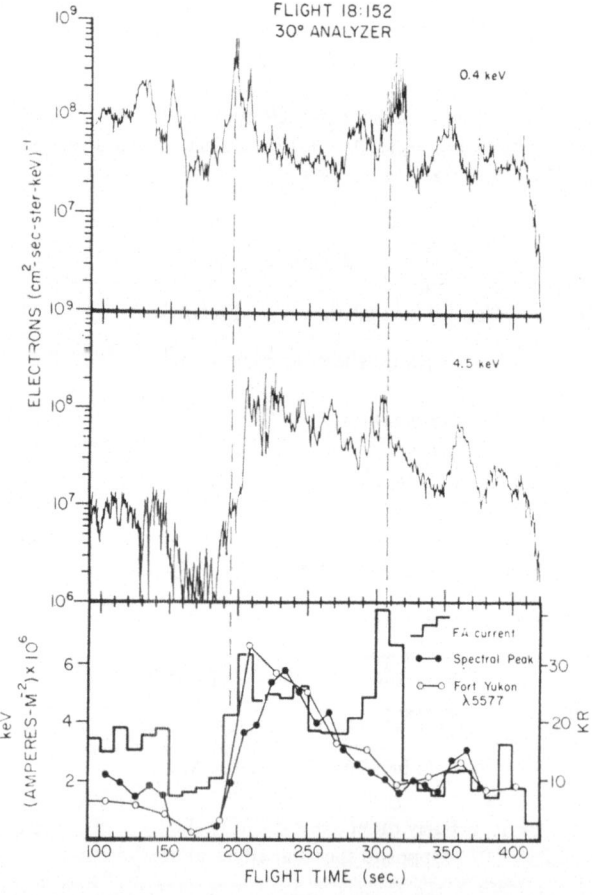

Fig. 4. Top panels: Particle fluxes measured by rocket-borne instrumentation. Bottom panel shows Birkeland currents carried by electrons and the location of the aurora deduced from particle data and ground based photometers (R.L. Arnoldy, private communication, 1977).

field-aligned fluxes at the edges of auroral arcs have been reported by WHALEN and McDIARMID (1972) and by MAEHLUM and MOSTUE (1974). The former group flew a rocket into an early evening substorm over several auroral bands. A sheet of upward current 10 km thick of density $2 \times 10^{-4} A/m^2$ lay at the poleward edge of the bands and was carried by electrons below 550 eV. There was an eastward electrojet. The latter group made two measurements over stable premidnight auroral arcs using mother-daughter payloads to separate spatial from temporal variations. In both cases 500 eV electron fluxes were field-aligned or concentrated at the edges of the auroral arcs. More energetic electrons were isotropic and produced the aurora.

A different technique was used by EVANS *et al.* (1977). A rocket was flown over a stable early evening arc, crossing an eastward electrojet and dropping into the Harang discontinuity. The electric field and flux of electrons with <0.1 to 16 keV were measured. Horizontal height-integrated conductivity was calculated from the electron flux and horizontal current from $J_h = \Sigma E_h$. The Birkeland current was then calculated from the horizontal divergence of the horizontal current under the assumption of uniformity along the arc. Upward current is seen to flow over the aurora but somewhat concentrated towards the edges as shown in Figs. 5 and 6. The upward current density was 25 $\mu A/m^2$ at the edges. The upward current carried by the measured energetic electrons was about 15 $\mu A/m^2$.

It seems evident that in some auroral arcs the flux of lower energy electrons is greatest at the edges of the region of energy precipitation, although this is not always the case (see for example, PAZICH and ANDERSON, 1975). It is much less certain that the net upward Birkeland current is strongest at the edges. Associated downward current has not been observed.

The other major group of observations shows an upward Birkeland sheet current centered over the auroral arc; an adjacent downward return current is frequently

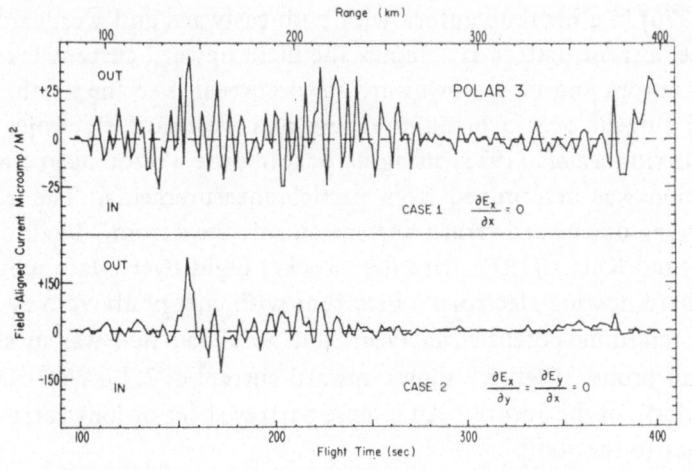

Fig. 5. Birkeland currents calculated by EVANS *et al.* (1977) from the horizontal divergence of horizontal current. Two different assumptions about the horizontal electric field were used.

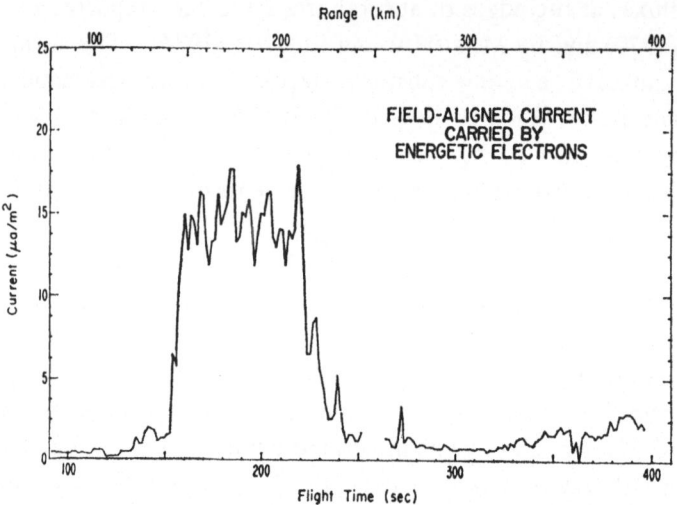

Fig. 6.   Birkeland current carried by measured energetic electrons over the same
          aurora as in Fig. 5.

seen as well.   Our group at Rice calculates the currents from vector changes in
magnetic field measured along the rockets' trajectories.   The position of the aurora
is determined by on-board particle energy spectrometers and by photometric and
all-sky camera observations from the ground beneath.   As shown in Fig. 7, upward
current was observed over an auroral arc by rockets Sq-4 and N/T-3, with an equal
and opposite sheet current just to the south.   In one case the aurora was in the early
evening but the electrojet flowed eastward.   The second was a pre-breakup aurora
with an eastward flowing electrojet.   The westward electrojet was still to the east
(see PARK and CLOUTIER, 1971; CLOUTIER et al., 1970; VONDRAK et al., 1971; and
CASSERLY and CLOUTIER, 1975).   Figure 8 shows data obtained by SESIANO and
CLOUTIER (1976) in a breakup aurora with both eastward and westward electrojets.
Although the current pattern is complex the main upward current is centered over
the brightest aurora and main downward sheet current is to the north.

Upward current over a breakup aurora and westward electrojet was also observed by KINTNER et al. (1973) using a rocket-borne vector magnetometer.   The
auroral position was determined from particle measurements.   The trajectory pre-
cluded observing downward return current outside the aurora.   BERING et al. (1973)
and CARLSON and KELLY (1977) describe a rocket flight over a late substorm aurora
with a westward flowing electrojet.   Electrons with energy above 5 eV were measured with a retarding potential analyzer, and total ion flux was measured with a
split Langmuir probe.   Figure 9 shows upward current of 7.2 $\mu A/m^2$ carried by electrons in and north of the aurora.   An intense narrow sheet of ions carried 170 $\mu A/m^2$
downward just to the south.

Figure 10 shows the current over a early evening quiet arc observed by de la
Beaujardiere et al. (de la Beaujardiere, private communication, 1977) using the
Chatanika incoherent scatter radar.   Here the electric field is measured by the radar

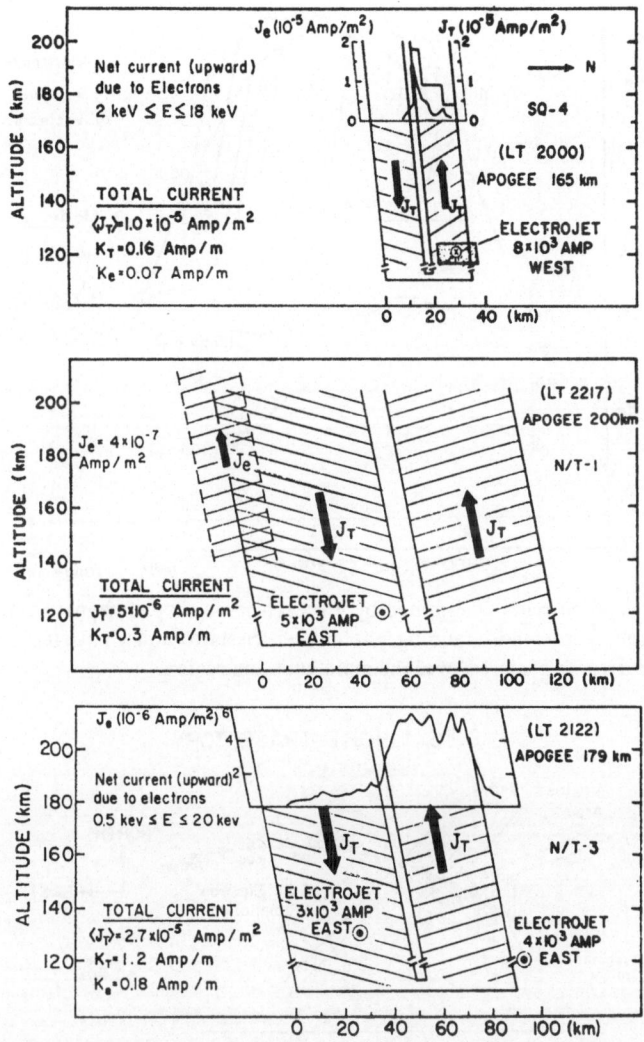

Fig. 7. Birkeland current data obtained by Rice University group using sounding rockets instrumented with particle detectors and a vector magnetometer. Upper panel: Upward current is centered on the auroral arc (PARK and CLOUTIER, 1971). Middle panel: Fig. 11 shows this data in more detail. Few particle data were obtained (CLOUTIER et al., 1973). Bottom panel: Upward current is centered on the aurora (CASSERLY and CLOUTIER, 1975).

and the conductivity is calculated from electron density measured by the radar. Thence horizontal current is computed and the vertical Birkeland current set equal to the divergence of the horizontal current. The upward current is centered over the aurora with a downward current to the north.

It seems certain that in reasonably stable electron arcs an upward Birkeland current is usually centered over the arc. The fraction of it carried by energetic electrons varies. An equal and opposite return current flows adjacent to the arc but at this time it is not known what determines whether it is to the north or south.

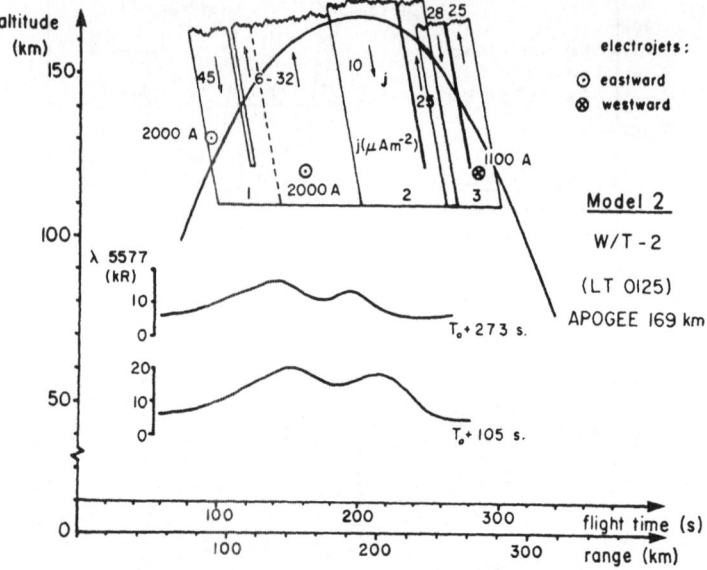

Fig. 8. Birkeland currents determined by a rocket-borne vector magnetometer, and the
aurora from a meridian scanning photometer (SESIANO and CLOUTIER, 1976).   The
upward current is centered over the main emission region.

8:56 FLIGHT TRAJECTORY

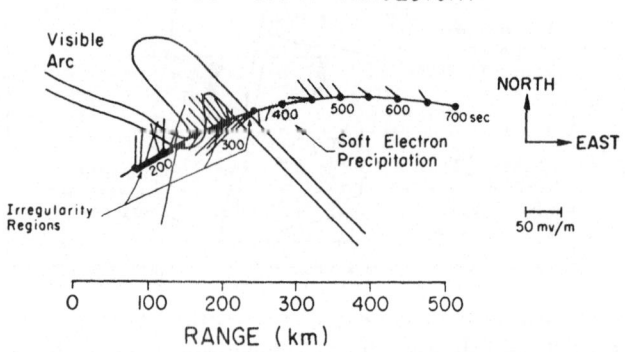

Fig. 9. Upward current carried by soft electrons in and to the north of an auroral
arc. An intense downward flux of ions was observed just south of the aurora
at 2205 (BERING et al., 1973; CARLSON and KELLEY, 1977).

Perhaps it is worth noting that in the cases where the upward current is concentrated
at the edges of the aurora, the current is still centered on the arc. In several of
those cases experimental technique did not permit detection of downward currents,
which are surely carried by drifting thermal electrons or ions.

Two more observations by the Rice group do not fit into either major category.
The observations of rocket N/T-1, shown in Fig. 7 and again in Fig. 11 were made
over an early evening arc with eastward electrojet (CLOUTIER et al., 1973). The arc
position was deduced entirely from optical observations made from the ground.
The arc split early in the rocket flight and then coalesced. Here the Birkeland sheet

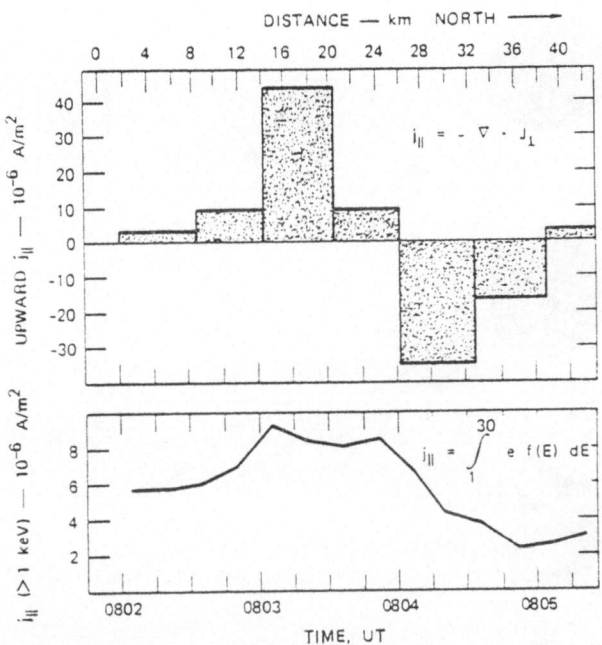

Fig. 10. Birkeland current (top panel) and flux of energetic electrons (bottom panel) deduced from Chatanika radar data (de la Beaujardiere, personal communication, 1977).

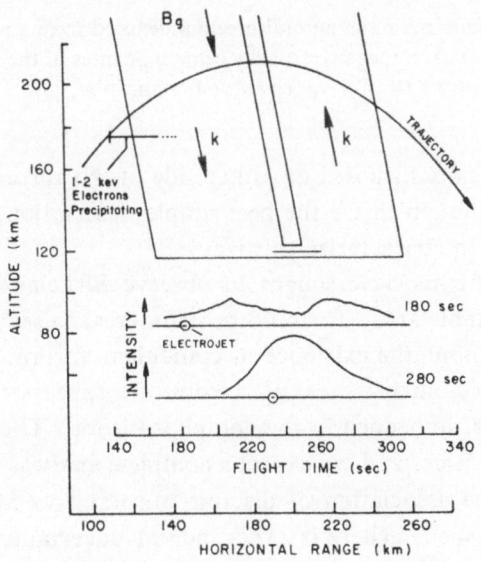

Fig. 11. Birkeland current deduced from rocket-borne vector magnetometer and meridian scanning photometer observations of the aurora. Currents flow at the edges of the aurora (CLOUTIER et al., 1973).

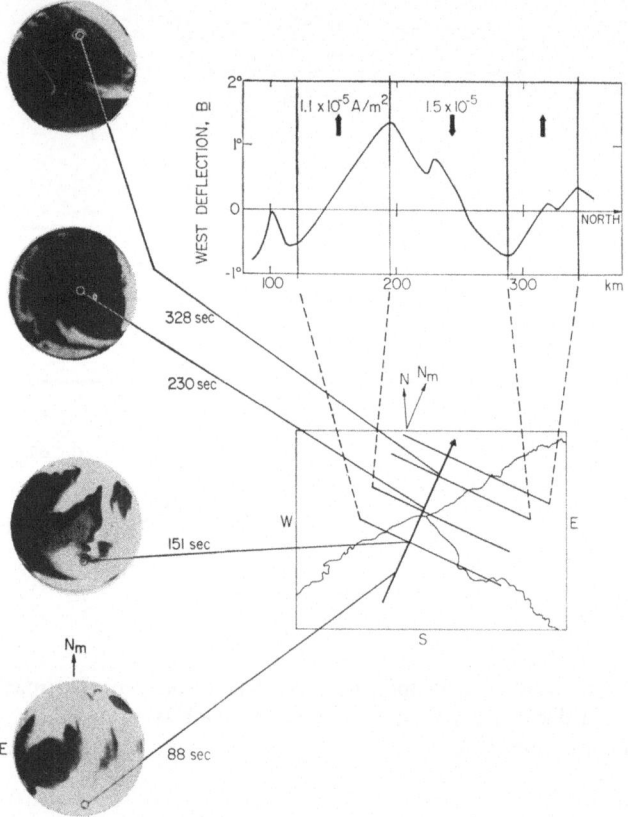

Fig. 12. Birkeland currents in an auroral breakup deduced from a rocket-borne magnetometer, and Ft. Yukon (negative) all sky camera pictures of the aurora as the rocket passed over the aurora (R. Coley, private communication, 1977).

currents are symmetrically disposed on either side of the aurora, the downward current to the south. Although this is the best simple description, it may be that the aurora was too active to characterize this way.

Our most recent flights have sought to observe Birkeland currents in auroral situations other than stable arcs. It would be of interest to see whether paired sheet currents can flow without the existence of coincident aurora. Figure 12 shows results from a flight into a strong breakup aurora. It appears that the simple sheet currents are still there, imbedded in the complex aurora. These data are being analyzed by R. Coley at Rice, and we expect a complete analysis within a few months.

In my opinion the association of discrete auroral arcs and upward Birkeland currents is well established. There is experimental uncertainty about whether the currents are concentrated at the edges of the arcs. A consistent pattern relating the location of the downward return current, the electrojet, and the electric field, has not yet been found. The possibility that simple paired sheet currents can exist in the midst of complex aurora is suggested but not experimentally proven.

This work was supported by Grant ATM75-22871 from NSF and NGL44-006-012 from NASA.

REFERENCES

ALFVÉN, H., A theory of magnetic storms and of the aurorae (1), *Kgl. Sv. Vetenskapsakad. Handl., Ser. 3*, **8**, 1–39, 1939–1940 (Partial reprint in *EOS, Trans. Am. Geophys. Union*, **51**, 181–193, 1970).

ANDERSON, H.R. and R.R. VONDRAK, Observations of Birkeland currents at auroral latitudes, *Rev. Geophys. Space Phys.*, **13**, 243–262, 1975.

ARNOLDY, R.L., Auroral particle precipitation and Birkeland currents, *Rev. Geophys. Space Phys.*, **12**, 217–231, 1974.

BERING, E.A., M.C. KELLEY, and F.S. MOZER, Split Langmuir probe measurements of current density and electric fields in an aurora, *J. Geophys. Res.*, **78**, 2201–2213, 1973.

BIRKELAND, K., *The Norwegian Aurora Polaris Expedition 1902–1903*, Vol. 1, *On the Cause of Magnetic Storms and the Origin of Terrestrial Magnetism*, sect. 1 and 2, H. Aschehoug, Christiania, Norway, 1908 and 1913.

BOSTRÖM, R., A model of the auroral electrojets, *J. Geophys. Res.*, **69**, 4983–5000, 1964.

CARLSON, C.W. and M.C. KELLEY, Observations and interpretation of particle and electric field measurements inside and adjacent to an active auroral arc, *J. Geophys. Res.*, **82**, 2349–2360, 1977.

CASSERLY, R.T., Jr. and P.A. CLOUTIER, Rocket-based magnetic observations of auroral Birkeland currents in association with a structured auroral arc, *J. Geophys. Res.*, **80**, 2165, 1975.

CLOUTIER, P.A., H.R. ANDERSON, R.J. PARK, R.R. VONDRAK, R.J. SPIGER, and B.R. SANDEL, Detection of geomagnetically aligned currents associated with an auroral arc, *J. Geophys. Res.*, **75**, 2595–2600, 1970.

CLOUTIER, P.A., B.R. SANDEL, H.R. ANDERSON, P.M. PAZICH, and R.J. SPIGER, Measurement of auroral Birkeland currents and energetic particle fluxes, *J. Geophys. Res.*, **78**, 640–647, 1973.

COLE, K.D., Motions of the aurora and radio aurora and their relationship to ionospheric currents, *Planet. Space Sci.*, **10**, 129–163, 1963.

CUMMINGS, W.D. and A.J. DESSLER, Field-aligned currents in the magnetosphere, *J. Geophys. Res.*, **72**, 1007–1014, 1967.

EVANS, D.S., N.C. MAYNARD, J. TRØM, T. JACOBSEN, and A. EGELAND, Auroral vector electric field and particle comparisons. 2. Electrodynamics of an arc, *J. Geophys. Res.*, **82**, 2235–2249, 1977.

FEJER, J.A., Theory of auroral electrojets, *J. Geophys. Res.*, **68**, 2147–2158, 1963.

IIJIMA, T. and T.A. POTEMRA, Field-aligned currents in the dayside cusp observed by Triad, *J. Geophys. Res.*, **81**, 5971–5979, 1976.

KAMIDE, Y. and G. ROSTOKER, The spatial relationships of field-aligned currents and auroral electrojets to the distribution of nightside auroras, submitted to *J. Geophys. Res.*, 1977.

KERN, J.W., A charge separation mechanism for the production of polar auroras and electrojets, *J. Geophys. Res.*, **67**, 2649–2666, 1962.

KINTNER, P.M., L.J. CAHILL, Jr., and R.L. ARNOLDY, Current system in an auroral substorm, report, School of Phys., Univ. of Minn., Minneapolis, 1973.

KLUMPAR, D.M., Field-aligned currents in the evening MLT-sector and their association with primary and secondary auroral particle fluxes, *EOS, Trans. Am. Geophys. Union*, **57**, 988, 1976 (Abstr.).

KLUMPAR, D.M., J.R. BURROWS, and M.O. WILSON, Simultaneous observation of field-aligned currents and particle fluxes in the post-midnight sector, *Geophys. Res. Lett.*, **3**, 395, 1976.

MAEHLUM, B.N. and H. MOESTUE, High temporal and spatial resolution observations of low energy electrons by a mother-daughter rocket in the vicinity of two quiescent arcs, *Planet. Space Sci.*, **21**, 1957, 1974.

PARK, R.J. and P.A. CLOUTIER, Rocket-based measurement of Birkeland currents related to an auroral arc and electrojet, *J. Geophys. Res.*, **76**, 7714–7733, 1971.

PAZICH, P.M. and H.R. ANDERSON, Rocket measurement of auroral electron fluxes associated with field-aligned currents, *J. Geophys. Res.*, **80**, 2152, 1975.

SCHIELD, M.A., J.W. FREEMAN, and A.J. DESSLER, A source for field-aligned currents at auroral latitudes, *J. Geophys. Res.*, **74**, 247–256, 1969.

SESIANO J. and P.A. CLOUTIER, Measurements of field aligned currents in a multiple auroral arc system, *J. Geophys. Res.*, **81**, 116–122, 1976.

SWIFT, D.W., A mechanism for energizing electrons in the magnetosphere, *J. Geophys. Res.*, **70**, 3061–3074, 1965.

TSUNODA, R.T., R.I. PRESNELL, and T.A. POTEMRA, The spatial relationship between the evening radar aurora and field-aligned currents, *J. Geophys. Res.*, **81**, 3791–3802, 1976.

VONDRAK, R.R., H.R. ANDERSON, and R.J. SPIGER, Rocket-based measurement of particle fluxes and currents in an auroral arc, *J. Geophys. Res.*, **76**, 7701–7713, 1971.

WHALEN, B.A. and I.B. MCDIARMID, Observations of magnetic-field-aligned auroral-electron precipitation, *J. Geophys. Res.*, **77**, 191, 1972.

ZMUDA, A.J., J.H. MARTIN, and F.T. HEURING, Transverse magnetic disturbance at 1,100 kilometers in the auroral region, *J. Geophys. Res.*, **71**, 5033–5046, 1966.

ZMUDA, A.J., F.T. HEURING, and J.H. MARTIN, Dayside magnetic disturbances at 1,100 kilometers in the auroral oval, *J. Geophys. Res.*, **72**, 1115–1117, 1967.

Note Added in Proof

An additional observation of Birkeland currents has been made by WESCOTT *et al.* (1975) using a shaped charge barium injection to trace magnetic field lines. By comparing the barium's location with a model field and observing displacement of the barium streak when an auroral arc moved through it, they deduced that a Birkeland sheet current of $\sim 3 \times 10^{-2}$ A/m flowed upward over the north side of the auroral arc.

WESCOTT, E.M. *et al.*, *J. Geophys. Res.*, **80**, 951–967, 1975.

Relationships between Particle Precipitation and Auroral Forms

J.L. BURCH[*,**] and J.D. WINNINGHAM[***]

*Southwest Research Institute, San Antonio, Texas 78284, U.S.A.
**University of Texas at San Antonio, San Antonio, Texas 78285, U.S.A.
***University of Texas at Dallas, Richardson, Texas 75080, U.S.A.

(Received February 7, 1978)

The present state of knowledge on the relationships between high-latitude particle precipitation and the aurora is reviewed. Attention is focused on the large-scale relationships between auroral forms and magnetospheric particle populations, on the relationships between satellite and sounding-rocket measurements, and on the interaction of auroral electrons with the atmosphere. While significant progress is being made in relating the large-scale features of the aurora to magnetospheric plasma domains, and in understanding the way in which auroral electrons deposit their energy in the atmosphere, only slight progress has been made in relating satellite data to the small-scale phenomena associated with auroral arcs.

1. Introduction

The purpose of this paper is to review the present state of knowledge on the relationship between high-latitude particle precipitation and the aurora. Due to length limitations the discussion covers only electron observations, and is somewhat restricted in scope, focusing chiefly on very recent measurements. The paper is divided into three main topics: (1) the large-scale relationships between auroral forms and the particle populations of the magnetosphere as determined from satellite measurements; (2) the relationship between satellite and sounding-rocket observations, particularly field-aligned pitch-angle distributions and upward field-aligned currents measured in the vicinity of auroral forms; and (3) recent results on the interaction of auroral electrons with the atmosphere.

2. Satellite Observations of Particle Precipitation and the Aurora

2.1 Nightside observations

The initial results of ACKERSON and FRANK (1972) and WINNINGHAM and HEIKKILA (1973), using data from ISIS 1 and 2 Injun 5, combined with all-sky camera and aircraft optical observations, have indicated a close relationship between inverted-'V' type electron precipitation and discrete auroral forms in the evening-side auroral oval, the afternoon-quadrant cleft and the polar cap. The availability of global auroral images from ISIS and DMSP has subsequently allowed significant extensions of this work to be made.

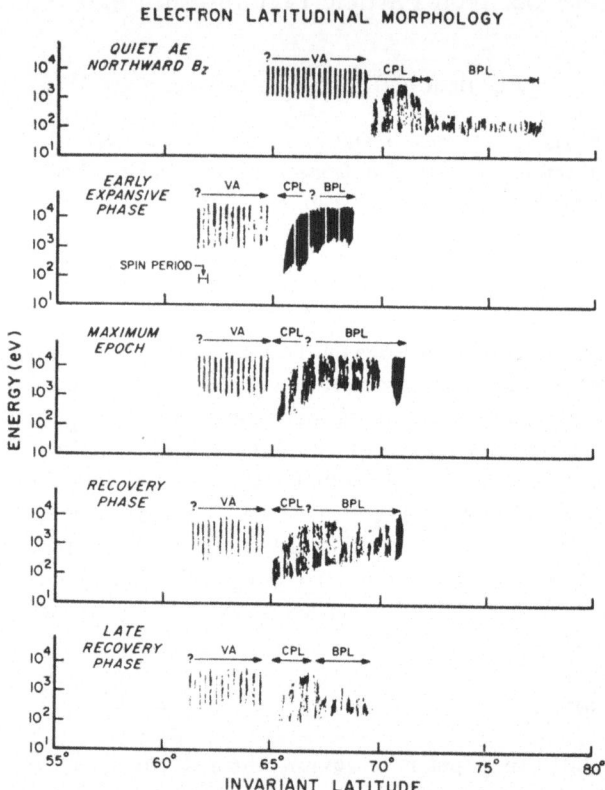

Fig. 1.   Schematic representation of the latitudinal morphology of
electron precipitation in the 21-03 MLT sector during substorms
as deduced by Winningham *et al.* (1975).

The nightside patterns of electron precipitation and their evolution during substorms has been described extensively by Hoffman and Burch (1973) and Winningham *et al.* (1975).   Figure 1 shows schematic ISIS spectrograms (not real data) of nightside electron precipitation as presented in the Winningham *et al.* paper. The patterns of precipitating electrons are generally classifiable into diffuse and discrete zones attributed by Winningham *et al.* to sources in the central plasma sheet (CPS) and boundary plasma layer (BPL) respectively.   As seen in Fig. 1, the occurrence of substorms is signaled by energization of the BPL electrons along with a general spreading in latitude of the entire precipitation zone and a distinct separation between the CPS and BPL.

The recent work of Lui *et al.* (1977) has established the relationship between these precipitation zones and auroral forms derived from ISIS global images.   Figures 2 and 3 (taken from the Lui *et al.* paper) show electron data and auroral intensity profiles for both a substorm period (Fig. 2) and a quiet period (Fig. 3).   In the Fig. 2 example, the auroral intensity profile indicates a discrete form located along the satellite path between 70.5 and 72.5° invariant latitude.   This form and the diffuse

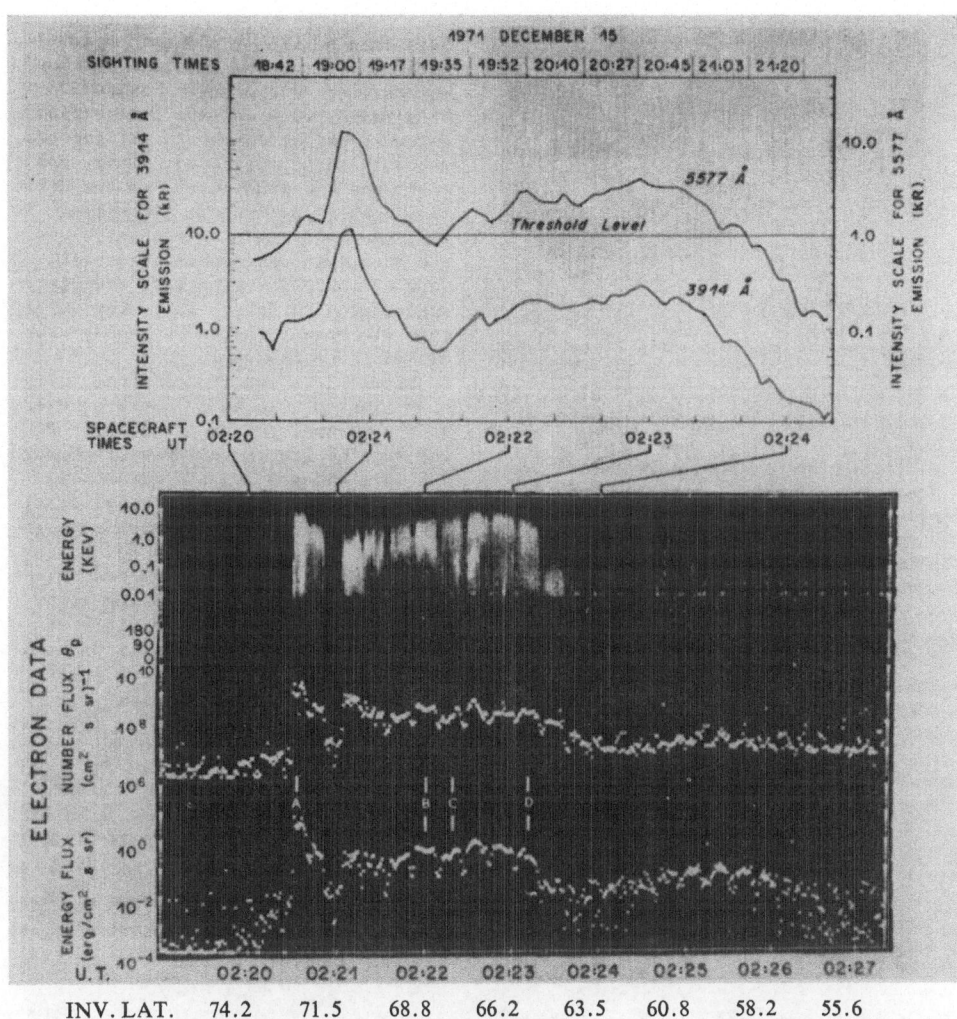

Fig. 2.   Auroral intensity profiles along the magnetic projection of the ISIS 2 satellite orbit, and electron and proton data on December 15, 1971.   (From Lui *et al.*, 1977)

aurora (located between 64 and 70°) were separated by a region of weak emission. The electron data in Fig. 2 show a narrow region of structured and intense precipitation (the BPL) near $\lambda = 72.5°$, corresponding well with the discrete auroral emissions.   As is the case for the auroral emissions, a region of very weak precipitation exists between the BPL and the CPS.   The broad and uniform region of electron fluxes in the CPS are consistent with spatial coincidence with the diffuse auroral emissions.

This same correspondence between discrete auroral forms and the BPL and between diffuse auroral emissions and the CPS is maintained in the quiet cases examined by Lui *et al.* in which the BPL appeared to be attached to the poleward part of the CPS, as shown in Fig. 3.

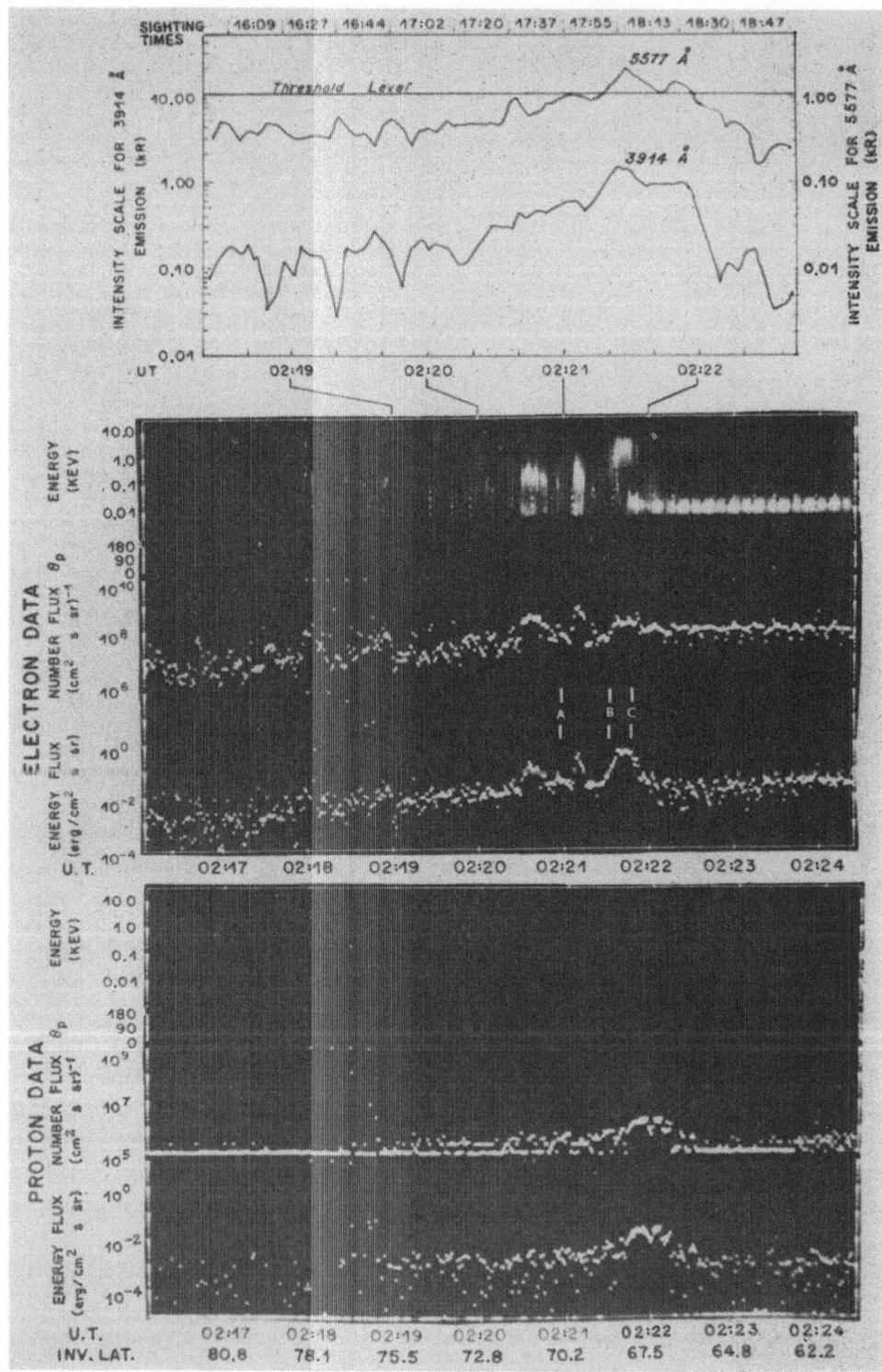

Fig. 3.   Same as Fig. 2, except for December 21, 1971.

As noted by LUI *et al.* (1977) and earlier by BURCH *et al.* (1976) and others, the distribution functions of electrons precipitating into the diffuse aurora region are generally Maxwellian in character, while those observed within the structured precipitation region are well-described by Maxwellian distributions that have been accelerated by a field-aligned electrostatic potential drop.

2.2   Dayside observations

Figure 4 is a DMSP image of dayside aurora taken from SNYDER and AKASOFU (1976). This picture shows a newly-observed feature—the V-shaped auroras fanning out from a central region near noon. A recent paper by REIFF *et al.* (1977) suggests that the dark region corresponds to the dayside convection 'throat' described by

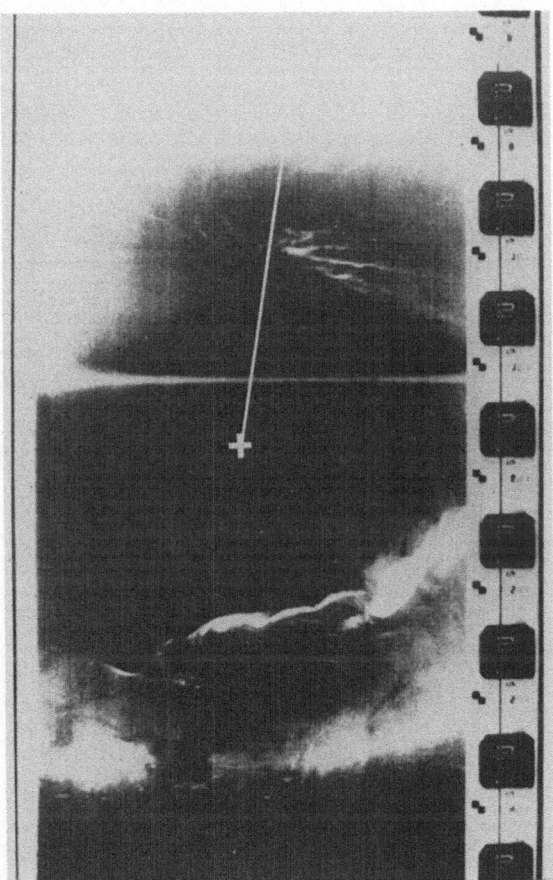

Fig. 4.   An example of a DMSP photograph which was taken approximately along the noon-midnight meridian. The location of the invariant pole is shown by a cross, together with the isochronal line for magnetic noon at 1332 UT. The location of the south pole is shown by a dot; satellite DMSP 8531; orbit 5960; time, 1332 UT; date, May 10, 1975. (From SNYDER and AKASOFU, 1976)

Heelis *et al.* (1976), while the auroral forms are associated with the convection shear reversals that occur at local times remote from the throat. This suggestion is based on the expected relationship between the dayside aurora and intense particle beams associated with the field-aligned currents required to support the convection shear reversals. Although the afternoon-quadrant auroral forms are probably associated with the inverted-'V' electron precipitation that is often observed there (McDiarmid and Burrows, 1975; Burch *et al.*, 1976; Winningham *et al.*, 1977) it

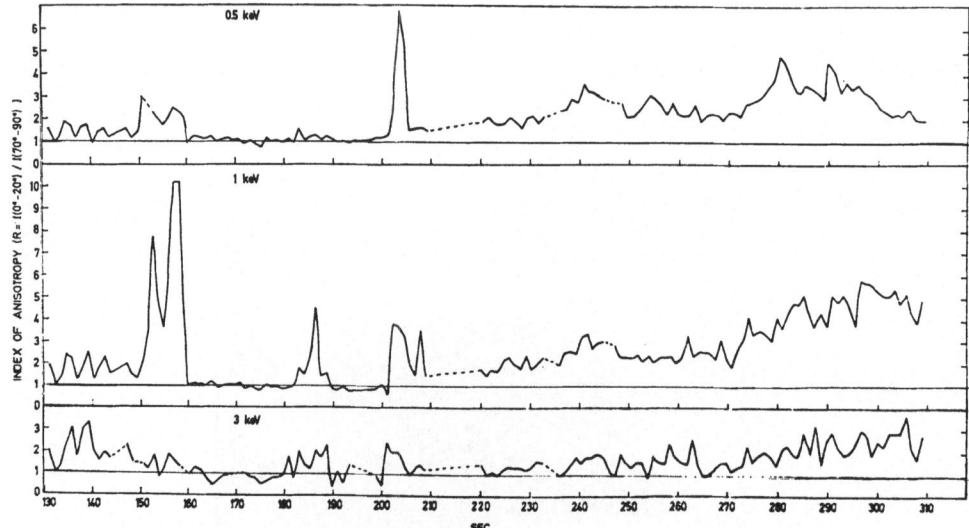

Fig. 5. Ratio between electron fluxes observed along and normal to the geomagnetic field lines. Auroral arcs were observed between 160 and 180 sec and between 190 and 200 sec. (Taken from Maehlum and Mostue, 1973)

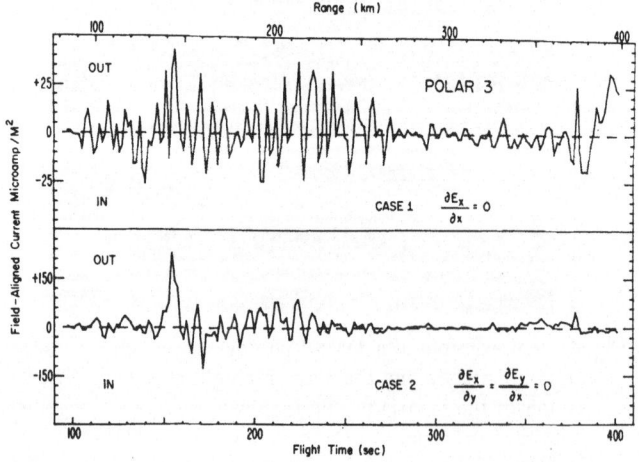

Fig. 6. Field-aligned current pattern as obtained by examining the divergence of the horizontal current system. (From Evans *et al.*, 1977)

is not clear what particle population produces the morning-quadrant auroras, since very little electron energization is observed in that sector of the cleft.

3.   Satellite versus Rocket Measurements

A long-standing unanswered question in the field of auroral physics concerns relationship between the large-scale (100 to 300 km north-south) electron inverted-'V' structures and the much narrower (.1 to 10 km) auroral arcs.   The lack of structure

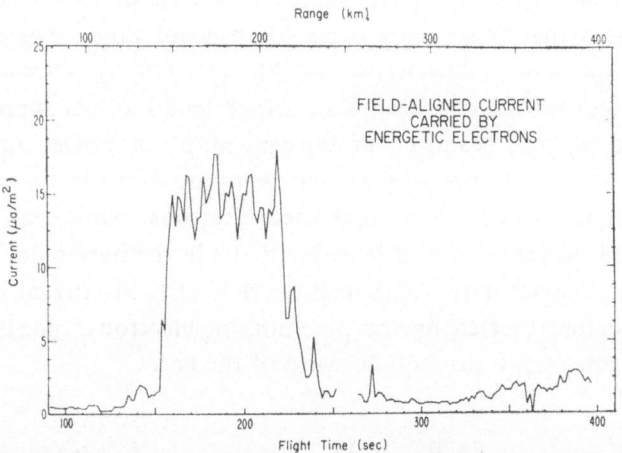

Fig. 7.   Net current out of the ionosphere carried by the energetic electrons measured by the particle detectors on board the sounding rocket Polar 3, as reported by EVANS *et al.* (1977).

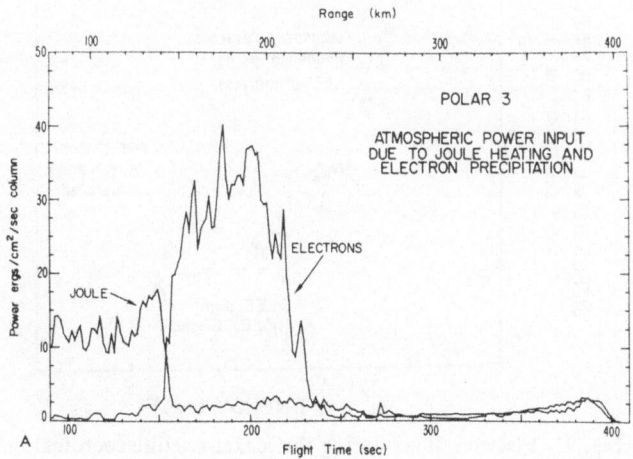

Fig. 8.   Energy dissipated into the auroral atmosphere through Joule heating and by precipitating auroral electrons.   That contribution provided by the particles defines the visual auroral arc. (From EVANS *et al.*, 1977)

evident in many satellite measurements of inverted-'V' electron beams results almost surely from the sparsity of measurements near zero pitch angle. An example of the structure observed near zero pitch angle has been presented by LIN and HOFFMAN (1977). Their Atmosphere Explorer data exhibit a high degree of spatial structure at very small pitch angles and only at energies near and just below the spectral peak as this peak moves to higher and then to lower energies during the satellite's traversal of the inverted 'V.' One might suppose, then, that narrow field-aligned beams embedded in an inverted-'V' are responsible for the narrow regions of enhanced light output.

Rocket measurements of particle fluxes in the vicinity of auroral arcs have, however, indicated that this simple explanation is not valid. For example, Fig. 5 displays the rocket data of MAEHLUM and MOSTUE (1975), showing field-aligned beams at the edges of auroral arcs. Data which lend indirect support to this idea that field-aligned electron beams occur adjacent to, but not over, auroral arcs have been presented by EVANS *et al.* (1977), and reproduced in Figs. 6 and 7. Figure 7 displays the current carried by the measured electrons, while Fig. 6 displays the total field-aligned current deduced from gradients in the horizontal currents as derived from magnetometer data. Although the field-aligned current over the arc is roughly equal to that carried by the precipitating electrons, the strongest current appears in a narrow region just equatorward of the arc.

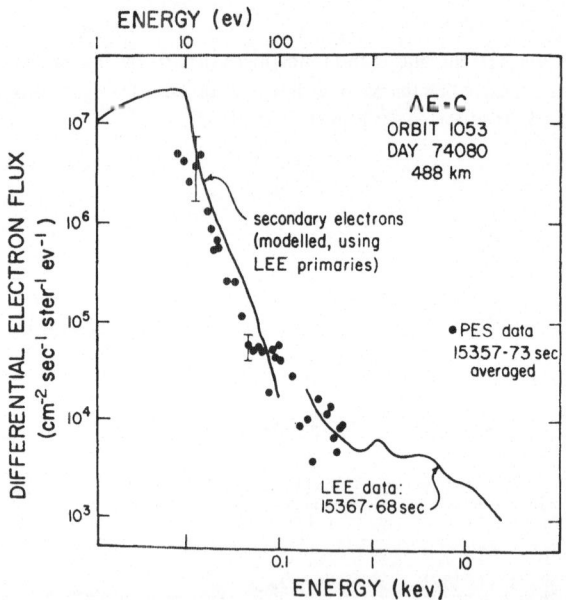

Fig. 9.   Electron fluxes during the rocket-satellite coordination experiment of REES *et al.* (1977). The LEE (Low Energy Electron Experiment) measurements represent a 1-s spectral scan, while the PES (Photoelectron Spectrometer) data are averaged over a 15-s interval. The theoretical secondary electron spectrum is arbitrarily terminated at 100 eV.

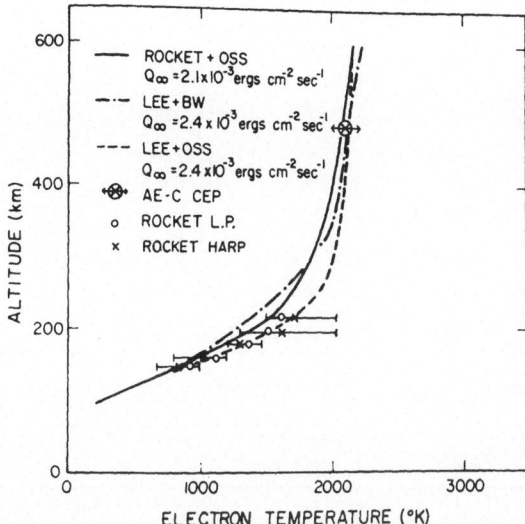

Fig. 10. Temperature measurements obtained with the rocket-borne
Langmuir probe (LP), the hyperbolic electrostatic analyzer (HARP),
and the satellite-borne cylindrical electrostatic probe (CEP) in the
experiment of REES *et al.* (1977). The solid curves are altitude profiles of electron temperature deduced from model computations by
using various input parameters indicated in the figure.

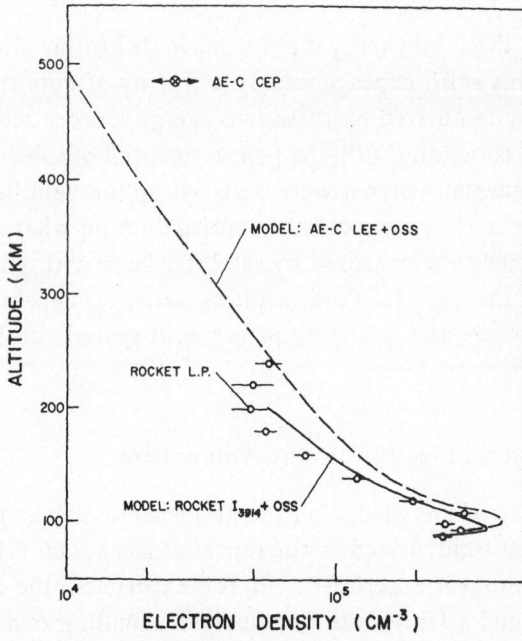

Fig. 11. Altitude profiles of electron density in the experiment of REES
et al. (1977). Total ion measurements made by the satellite-borne
cylindrical electrostatic probe (CEP) at 491 km and electron measurements by the Langmuir probe (LP) on the rocket between 90 and 240
km are shown together with model computations using the AE-C
input data and rocket input data.

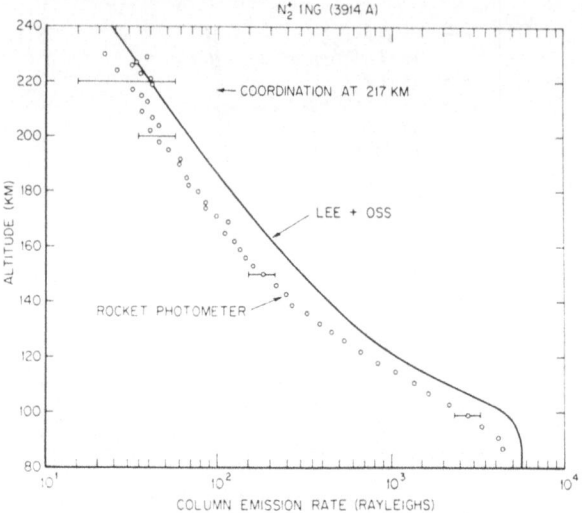

Fig. 12. Column emission rate of $N_2^+ 1NG(0, 0)$ (3914 Å) measured by
the upward looking rocket-borne photometer and computed from
the particle flux observed at the satellite altitude. LEE is the AE-C
Low Energy Electron Experiment, while OSS is the AE-C Open
Source Spectrometer which measures neutral gas densities. (From
Rees *et al.*, 1977)

Figure 8 shows the total energy input from Joule heating and from precipitating
electrons in the Evans *et al.* experiment. It is worthy of note that a smooth transition between regions dominated by these two energy sources occurs, with the strong
field-aligned current coinciding with the poleward cutoff of substantial Joule heating,
as expected if the Pederson current were diverted up the field line at that point.

We still, however, have no definite information on what relationship, if any,
the large-scale inverted-'V's measured by satellites have with phenomena related to
the data of Figs. 5 through 8. Polar-orbiting satellites capable of comprehensive
measurements of particle and field parameters with spatial resolutions significantly
below 1 km will first be required.

4.   Effects of Auroral Electrons in the Atmosphere

The present state of knowledge on the atmospheric effects produced by auroral
electron beams is well-summarized by the report of Rees *et al.* (1977) on their recent
coordinated satellite/rocket experiment. In this experiment the satellite Atmosphere
Explorer C (AE-C) and a University of Michigan sounding rocket crossed the same
L-shell with an east-west separation of 40 km. Data on the input electron beam were
obtained from AE-C, while the atmosphere was represented by a model atmosphere
normalized to the simultaneous AE-C neutral gas data. Electron temperatures and
densities were obtained both by satellite- and rocket-borne instruments. The output
secondary electrons were measured at AE-C, while the output 3914 Å nitrogen emis-

sions were monitored by rocket-borne photometers. Figures 9 through 12 show the remarkable agreement between model calculations and actual data for this unique input-output experiment, with the 3914 Å emission exhibiting the only significant disagreement.

5. Summary and Conclusions

Significant advances have occurred in our knowledge of the large-scale relationships between auroral morphology and magnetospheric particle populations and in the interaction of auroral electron beams with the atmosphere. By contrast, our ability to relate the large-scale phenomena measured from satellites to the smaller scale phenomena related to auroral arcs, and measured adequately only by rockets, is progressing only very slowly. It is crucial that coordinated sounding-rocket and satellite measurements be made with comparable spatial resolution if significant progress in this latter area is to be made.

This work was supported in part by National Aeronautics and Space Administration Contract NAS8-32584.

REFERENCES

ACKERSON, K.L. and L.A. FRANK, Correlated satellite measurements of low-energy electron precipitation and ground-based observations of a visible auroral arc, *J. Geophys. Res.*, **77**, 1128-1136, 1972.

BURCH, J.L., S.A. FIELDS, W.B. HANSON, R.A. HEELIS, R.A. HOFFMAN, and R.W. JANETZKE, Characteristics of auroral electron acceleration regions observed by Atmosphere Explorer C, *J. Geophys. Res.*, **81**, 2223-2230, 1976.

EVANS, D.S., N.C. MAYNARD, J. TROIM, T. JACOBSEN, and A. EGELAND, Auroral vector electric field and particle comparisons, 2, Electrodynamics of an arc, *J. Geophys. Res.*, **82**, 2235-2249, 1977.

HEELIS, R.A., W.B. HANSON, and J.L. BURCH, Ion convection velocity reversals in the dayside cleft, *J. Geophys. Res.*, **81**, 3803-3809, 1976.

HOFFMAN, R.A. and J.L. BURCH, Electron precipitation patterns and substorm morphology, *J. Geophys. Res.*, **78**, 2867-2884, 1973.

LIN, C.S. and R.A. HOFFMAN, Rapid fluctuations of inverted V electron fluxes, *EOS, Trans. AGU*, **58**, 1210, 1977.

LUI, A.T.Y., D. VENKATESAN, C.D. ANGER, S.-I. AKASOFU, W.J. HEIKKILA, J.D. WINNINGHAM, and J.R. BURROWS, Simultaneous observations of particle precipitations and auroral emissions by the ISIS 2 satellite in the 19-24 MLT sector, *J. Geophys. Res.*, **82**, 2210-2226, 1977.

MAEHLUM, B.N. and H. MOSTUE, High temporal and spatial resolution observations of low energy electrons by a mother-daughter rocket in the vicinity of two quiescent auroral arcs, *Planet. Space Sci.*, **21**, 1957-1967, 1973.

McDIARMID, I.B., J.R. BURROWS, and E.E. BUDZINSKI, Average characteristics of magnetospheric electrons (150 eV to 200 keV) at 1,400 km, *J. Geophys. Res.*, **80**, 73-79, 1975.

REES, M.H., A.I. STEWART, W.E. SHARP, P.B. HAYS, R.A. HOFFMAN, L.H. BRACE, J.P. DOERING, and W.K. PETERSON, Coordinated rocket and satellite measurements of an auroral event, 1, Satellite observations and analysis, *J. Geophys. Res.*, **82**, 2250-2258, 1977.

REIFF, P.H., J.L. BURCH, and R.A. HEELIS, Flow-aligned dayside auroral arcs, *Geophys. Res. Lett.*, 1978 (in press).

SNYDER, A.L., Jr. and S.-I. AKASOFU, Auroral oval photographs from the DMSP 8531 and 10533 satellites, *J. Geophys. Res.*, **81**, 1799–1804, 1976.

WINNINGHAM, J.D., S.-I. AKASOFU, F. YASUHARA, and W.J. HEIKKILA, Simultaneous observations of auroras from the South Pole station and of precipitating electrons by ISIS 1, *J. Geophys. Res.*, **78**, 6579–6594, 1973.

WINNINGHAM, J.D., F. YASUHARA, S.-I. AKASOFU, and W.J. HEIKKILA, The latitudinal morphology of 10-eV to 10-keV electron fluxes during magnetically quiet and disturbed times in the 2100-0300 MLT sector, *J. Geophys. Res.*, **80**, 3148–3171, 1975.

WINNINGHAM, J.D., T.W. SPEISER, E.W. HONES, Jr., R.A. JEFFRIES, W.H. ROACH, D.S. EVANS, and H.C. STENBAEK-NIELSEN, Rocket-borne measurements of the dayside cleft plasma: The Tordo experiments, *J. Geophys. Res.*, **82**, 1876–1888, 1977.

Photometric Investigation of Precipitating Particle Dynamics

S.B. MENDE

Space Science Laboratory, Lockheed Palo Alto Research Laboratories,
Palo Alto, California, U.S.A.

(Received February 7, 1978)

The theoretical basis and the experimental evidence for the interpretation of photometric data in terms of precipitating particle morphology will be reviewed. Particular emphasis will be laid on the interpretation of auroral morphology in terms of the spectroscopic signature of the precipitating particles. The meridian scanning photometric data will be discussed briefly to emphasize the auroral morphology related findings. The new type of monochromatic TV system which was specially developed for two-dimensional spectrophotometric auroral imaging will be described. The auroral morphology related findings will be discussed according to the local time sector of the observations. The early evening auroras are characterized by soft electron precipitation on the equatorward edge of the auroral oval. This precipitation is not a steady diffuse region but it is structured and the structures show coherent convective motion often in the westward direction. Harder precipitating features superimpose on this soft precipitation. These harder features (electrons 2 keV or larger) are observable by all-sky cameras and they participate in the substorm auroral dynamics. Following substorm onsets and northward expansions the electron auroras harden considerably and the protons expand poleward. During early morning the diffuse soft regions occur poleward of the hard auroral features. These early morning hard auroral features often show very hard pulsating patches.

1. Introduction

Auroras which can be observed by the naked eye are most dynamic and thus probably the most difficult to interpret in terms of simple physical principles. In fact, they occur relatively infrequently consisting of a small class of auroras, which are so to speak, 'the tip of the iceberg.' With the advance of high sensitivity photoelectric detection techniques, the existence of auroras were observed which occur far more frequently and which are below the visual threshold.

Photoelectric techniques permit the study of the structure of the subvisual aurora. In this paper we will be able to present pictures which represent the morphology of these subvisual structures.

Photoelectric techniques permit the quantitative study of auroral structures. In fact, it permits the interpretation of auroral morphology in terms of precipitating particle flux morphology. This is an important feature if one is to bear in mind that rockets and satellites are essentially single-point observations and thus are subject to space-time uncertainties. The observation of the optical signatures readily provides the space-time morphology of the precipitating particles.

Photometric observations permit the separation of different types of particles such a protons and electrons. In addition, they allow the discrimination between higher energy (greater than 1 keV) and lower electron (lower than 1 keV). Thus, one can observe auroral morphology in the dimension of particle energetics.

Using aircraft photometer data EATHER and MENDE (1972) has shown the existence of a parmanent subvisual optical auroral oval generated by soft electrons precipitation. They have also shown that there is good correlation between the mean energy of the auroral particles and the intensity of auroras. Thus the more intense visible auroras are generally produced by more energetic auroral particles ranging from 3 up to 30 keV.

The purpose of this paper is to summarize the findings of recent photometric and TV studies which classify auroral structures according to the energy and type of the precipitating particles.

There is general agreement that the auroral structures and auroral morphology corresponds closely to the precipitating particle dynamics. The type of work of EATHER and MENDE (1972) based on the calculations of REES and LUCKEY (1974) permit the separation of the optical data into the properties of the precipitating particles according to type, flux and energy and thus get one step closer to the morphology of the precipitating particle generating mechanisms.

For the study of the dynamics a set of systematic measurements and a convenient means of data presentation had to be developed. In this paper we will review of the auroral morphology related findings of recent spectrophotometric and monochromatic TV experiments.

2. Instrumentation and Data Presentation

Meridian scanning photometer was used for example by ROMICK and BELON (1967), FUKUNISHI (1975) and EATHER et al. (1976). The reader's familiarity with scanning photometer technique is assumed. EATHER et al. (1976) developed a new data presentation technique which permits the easy interpretation of the latitude-time morphology of auroras using scanning photometer data. The presentation is called the KEOGRAM (derived from the Eskimo-word Keoeeit for aurora). This presentation essentially consists of representing a photometer channel as a computer generated grey-scale presentation, where they grey scale represents auroral intensity. The ordinate is the latitude and the abscissa is time (local and/or UT). The KEOGRAMS of EATHER et al. (1976) use a latitude value which is obtained from the zenith angle of the photometer and the altitude of the emission as derived from the spectral ratios. EATHER et al. (1976) also uses airglow and Van Rhijn correction. A typical KEOGRAM is illustrated on Fig. 1.

The scanning photometer measurements of EATHER et al. (1976) are restricted because they monitored only the latitude-time intensity morphology of the aurora.

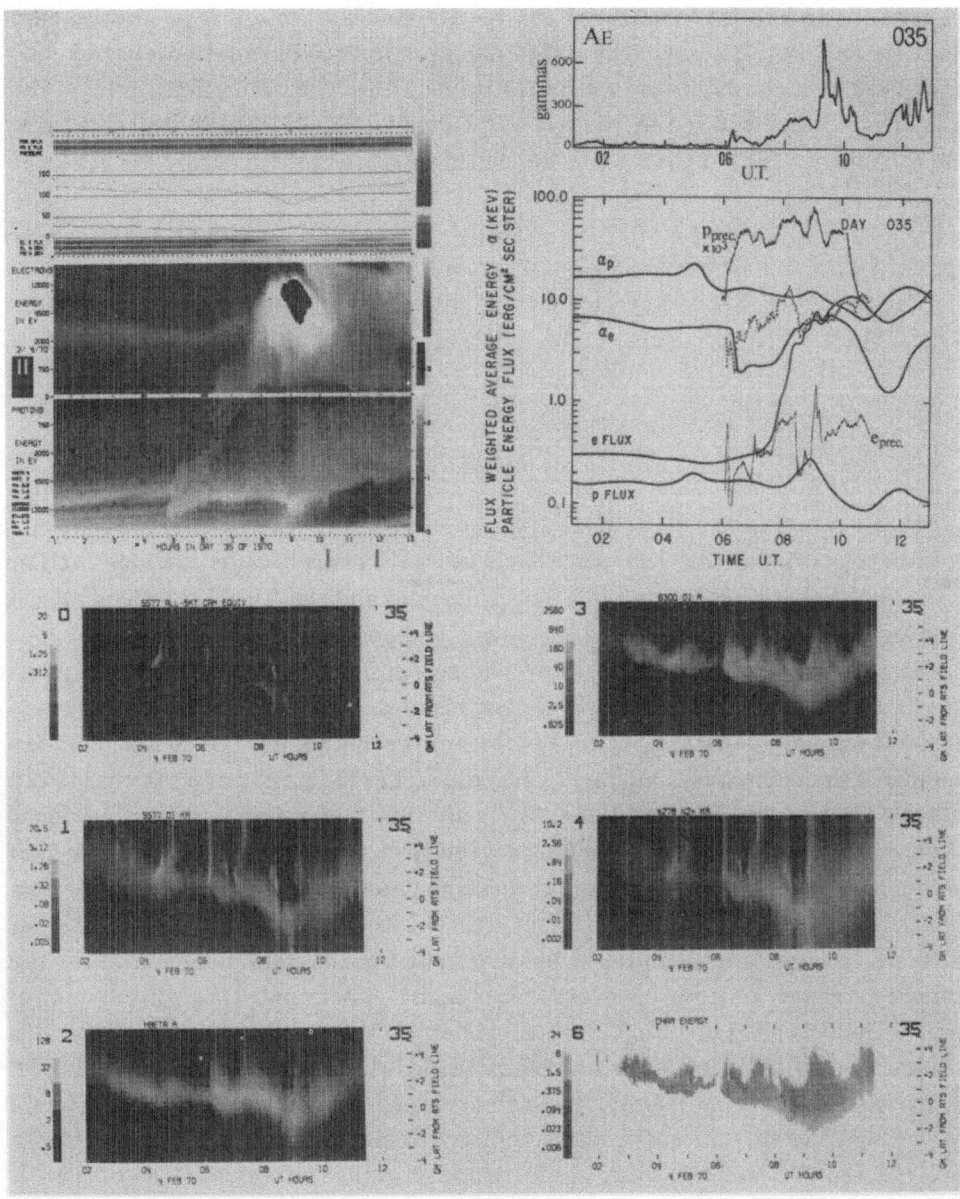

Fig. 1. Composite of KEOGRAMS and ATS-5 spectrograms. ATS-5 particle spectrograms
(parallel detectors), KEOGRAMS, ATS-5 energy fluxes and average energies, precipitated energy fluxes and average energies, and AE indices for day 032, 1970. Left column,
from top to bottom: (a) Top line plot is the *H* component of the magnetic field at ATS-5
(in gammas), and bottom line plot is detector pitch angle (in degrees). (b) Electron spectrogram. (c) Proton spectrogram. (d) All-sky camera equivalent KEOGRAM. (e) 5577
OI KEOGRAM. (f) H$_\beta$ KEOGRAM. Right column, from top to bottom: (g) AE plot.
(h) ATS electron and proton energy fluxes and electron and proton average energies (flux
weighted). Also shown are precipitated proton and electron fluxes derived from photometric data at the predicted ATS-5 field line position, and the average energy of precipitated electrons (plotted as crosses). (i) 6300 OI KEOGRAM. (j) 4278 N$_2^+$ KEOGRAM.
(k) Characteristic energy KEOGRAM.

These observations do not reveal the two-dimensional morphology and the corresponding changes in precipitation. In order to monitor the low level precipitation
accessible only with photometers and to observe the two-dimensional auroral morphology simultaneously, a new type of instrument was developed. Some preliminary
results have been discussed by MENDE and EATHER (1976) which were generated by
this instrument. The instrument is described in some more detail by MENDE et al.
(1977).

The all-sky imaging photometer (ASIP, see Fig. 1 of MENDE and EATHER (1976))
is essentially a monochromatic all-sky camera with a low light level TV system as
its prime detector. The system images the entire sky through narrow band interference filters which are sequentially selected. One of the filters is centered on
4278 N_2^+ band because of the proportionality between the N_2^+ radiation and the energy
of the incoming precipitating particles. The resulting TV monitor photograph represents the spatial distribution of the deposited energy of the particle precipitation.
Another filter is centered on the λ 6300 Å OI forbidden transition and the resulting
picture represents approximately the fraction of electron precipitation which is produced by electrons in the less than 2 keV range. This is because the long lifetime
λ 6300 Å OI radiation is generated at high altitudes and the λ 6300 Å OI emission is
enhanced only when the electrons of lower energy stop at higher altitudes. The
comparison of the 2 television monitor photographs taken through the two filters
gives us a good description of both the total electron precipitation and the hardness-
softness mean-energy-parameter of the electron precipitation. For detailed understanding of the technique, one might refer the reader to EATHER and MENDE (1972),
REES and LUCKEY (1974). Additional filters are used to obtain the proton produced
H-beta emission intensity. These filters are the H-beta (4861) filter and another filter
which is used to provide the continuum electron precipitation adjacent to the H-beta
wavelength, at 4820 Å. The two filters, the H-beta (4861) and this control filter
(4820) are closely matched in the width and height of the transmission profile. The
subtraction of the emission detected through the H-beta and through the control
filter respectively gives us a 2-dimensional picture of the pure H-beta emission. This
technique works quite satisfactorily until the electron auroras get very intense and
the electron induced contamination becomes much larger than the pure H-beta
emission. When the aurora reaches these brightness levels, the subtraction of the
two large signals from each other can produce spurious features.

The presentation of the electron auroral hardness parameter was in terms of a
color representation where the 4278 emission was represented in blue and the 6300
in red color on the TV monitor (MENDE and EATHER, 1976). The resulting color
display indicates at a glance the relative intensity of the soft and hard precipitation.
Because of the relatively large cost involved in the reproduction and printing of
such color images, a new display technique is applied in the present paper. Diagonally
cross-hatched grids were superimposed on the TV monitor. Cross-hatch in the direction from top left to bottom right was superimposed on the TV frame representing
the 6300 picture and cross-hatch in the direction from bottom left to top right was

2-22, 1976

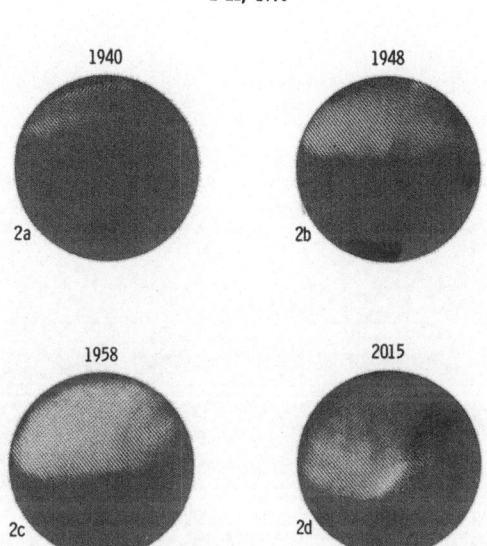

Fig. 2. All-sky imaging photometer (ASIP) TV monitor photograph
samples taken at Kiruna. 4278 frame was photographed through
cross-hatch bars running from bottom-left to top-right. 6300 frame
was superimposed by double exposure through bars running from
top-left to bottom-right. In soft auroras these latter bars (6300 emission) dominate. North is top and west is right.

superimposed on the TV frame representing the 4278 picture. The two TV frames
were double-exposed on the same photographs. The dominant direction of the cross-
hatching indicates the dominant emission intensity. In Fig. 2 examples of these are
shown. For example Fig. 2a shows a 4278 arc on the northern horizon (very top,
cross hedge bottom-left to top right) and a 6300 arc south from it (cross hedge top
left to bottom right). The dotted regions for example some parts of 2c represent
strong emissions in both wavelength.

The ATS-6 data was provided by courtesy of C.E. McIlwain and it is presented
in the usual form of spectrograms.

3.   Soft Particles Dynamics

On the family of KEOGRAMS (Fig. 1) is the frame labeled 5577 all-sky camera
equivalent. This shows how rarely an all-sky camera or a visual observer would
detect auroras during a typical night such as Feb. 4, 1970. Comparing this to the
frame labeled 6300 it is evident that the soft electron precipitation which is primarily
responsible for this feature (6300) was present as a single continuous belt all the
time. The presence of the continuous omni-present low intensity optical aurora
was detected by EATHER and MENDE (1972) from aircraft observation. It was photographed as a continuous belt by ISIS-II (LUI and ANGER, 1973) and was also identified
as the 'diffuse aurora' by LUI et al. (1973). The fact that this region is caused by

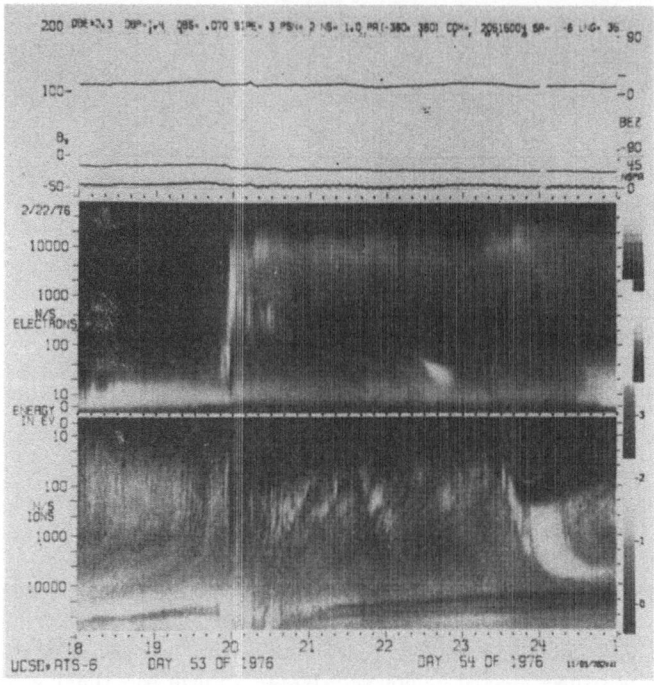

Fig. 3.   ATS-6 spectrogram for the day Feb. 22, 1976.

low energy electrons and that it is associated with the inner boundary of the low
energy electrons at synchronous altitude was shown by EATHER *et al.* (1976). Note
that the zero latitude on the KEOGRAMS of Fig. 1 is the calculated position of
the ATS-5 field line conjugate and also note that the appearance of the soft electrons
at 0620 UT (see the ATS-5 spectrogram presented on Fig. 1) coincides with the time
of crossing the zero latitude by the equatorward boundary of the 6300 emission.
Thus, it appears that ATS-5 and its field-line conjugate are simultaneously enveloped
by the soft electrons. EATHER *et al.* (1976) showed this relationship to hold true
for several different nights.

The soft electrons observed by synchronous satellite are often referred to as
plasma-sheet electrons and thus we have the association of the plasma sheet with
the soft diffuse auroral zone. The question arises whether this region is really diffuse
and whether the soft electron precipitation happens to be a steady drizzle. Such a
situation would be predicted by a type of steady pitch-angle diffusion instability of
plasma-sheet electrons.

During a quiet day, Feb. 22, 1976, observations were made at Kiruna and the
ASIP (TV) data are presented on Fig. 2. The corresponding ATS-6 spectrogram
between 1800–0100 is shown on Fig. 3. Figure 2 was taken at 1940. This picture
is somewhat typical of the very quiet times. The region of 6300 emission is signified
by the poleward (north) side of the picture. The striping shows significant dominance
of the 6300 with the exception over very faint auroral arc right on the northern
horizon showing where the harder electrons occur. The regions of soft electrons

are very active and the intensity fluctuations ($\pm 30\%$ in amplitude) show a dominant westward motion. In the next picture (Fig. 2b) at 1948, we have seen that the soft electron precipitation has significantly advanced equatorwards. It has strongly intensified and consists exclusively of 6300 emission with virtually no 4278 present. At 1948, a significant increase in the electrons occur at ATS-6. Note ATS-6 is parked so that its detectors closely look towards the 0 deg pitch angle looking at particles downcoming on the northern hemisphere. At 1948, we observe the arrival of few hundred electron volts of electrons also a gradual increase of 2–5 kV electrons is observed. Figure 2c shows that the electron precipitation region has expanded equatorward and it has suddenly hardened considerably, cross-hatching in the 4278 direction has appeared. The appearance of the 4278 emission precisely coincides with the arrival of the electrons in the 5–10,000 keV range at ATS-6. The hard electrons last only a few minutes and the soft electrons return as observed both at ATS-6 and from the ground. In the next frame, 2d, taken at 2015, we note the presence of some 4278 representative cross-hatching. This is once again coincident with the electrons turning on at around 2015 with energy of around 10 keV.

Thus, during quiet times, the ASIP instrument observes a continuous zone of soft electron precipitation characterized by large intensity fluctuations drifting dominantly in the westward direction. As the activity increases, this region moves inward (equatorwards). At quiet times, 1 to 1 correlation can be made with the boundary of the soft precipitation by assuming that ATS-6 is at its calculated field line conjugate. Hard electron precipitation is superimposed on the soft region. The predominant motion observed in the soft electron precipitation agrees with the expected convection direction of the magnetospheric plasma in the lower to sub-auroral regions of the magnetosphere.

4.   Harder Electron Morphology, Early Evening Hours

As the electron energy increases the intensity increases in a similar manner (EATHER *et al.*, 1972). Energetic precipitation morphology is therefore more readily observable and often can be documented by all-sky cameras.

Observing the KEOGRAM set, Fig. 1, we observe sporadic intensification especially in the 4278 frame. The lack of corresponding increase in the 6300 shows that electron hardening accompanies the intensifications. These intensifications correspond to plasma injections as observed generally by the dispersion traces of the ATS-5 protons. These intensifications can be uniquely associated with plasma injection events and thus possibly substorms. What is the morphological characteristic of these intensifications? We already know from the KEOGRAMS that they occur in the evening hours and that they are superimposed on the diffuse soft zone. An ASIP (TV) color composite and corresponding KEOGRAM was published by MENDE and EATHER (1976) and on plate 2 of that paper they have given several examples of this. The broad red 6300 region and the superimposed blue-white 4278 intense hard arc can be clearly identified.

Hence, the 4278 intensifications of the KEOGRAMS are the early evening auroral arcs. The picture agrees with earlier interpretation (Akasofu, 1968) in which the substorm onset is signified by the intensification of the arc/arcs in the evening side. The arcs in the early evening side do not as a rule break up and participate in the morphological display of the auroral activity typical of midnight local time. Nevertheless, these arcs respond to injections at synchronous altitude by brightening up and retaining their arc-like morphological character. Since the appearance (brightening) of these features are simultaneous with injections they cannot be regarded as precursors to the injections. An attempt was made by Eather et al. (1976) to associate these arcs to McIlwain's (McIlwain, 1974) injection boundary concepts.

5.   Harder Electron Morphology, Midnight Region

For the observation of harder precipitation observations were carried out at Kiruna (65.2° geomagnetic). For a more disturbed day, March 30, 1976, the ASIP data is presented in Fig. 4. Figures 4a and 4b are taken greater and lesser sensitivity (exposure time) respectively.

Figure 4a was taken at 1957 and it shows a reduced sensitivity exposure picture enabling us to resolve the softness of the main arcs which are saturated in the long exposure picture. From this picture we can see that the northern arc is very much softer than the center and southern arc.

Figure 4c and 4d represent the auroral situation at 2007 UT. This can be regarded as the beginning of the expansive phase as defined by Akasofu (1968) when the equatorward arc brightens up prior to the break-up of the auroral arc structures. On the KEOGRAMS of Eather et al. (1976), this is shown as a brightening of the KEOGRAM trace which is superimposed on the diffuse soft auroral regions. The shorter exposure frame 4d, shows the brighter arc. Approximately, this is all what shows up on conventional all-sky camera pictures. Close examination of this picture, one can distinctly see the separation of the high altitude 6300 region from the low altitude 4278 region on this picture. Note that because of perspective/distortion of the all-sky lens of the ASIP (TV), the lower altitude regions are shown more equatorwards than the high altitude regions. The typical development of substorm behavior is shown on Fig. 4e and f, the disturbed striated region is propagating inward from the east. As it has been observed in several events, the auroral electrons at this stage are still fairly soft and one can see mixed patches of soft and hard precipitation. Although some caution has to be exercised in making such observations because the altitude perspective can confuse the situation. Another problem comes from the fact that the 4278 and 6300 frames are taken 4 sec apart and during intense breakup the auroral forms can move large distances during this time. The photographs taken at 2020 UT (4g) and thereafter are the short exposure type. The complex aurora structure shows both soft and hard features. Note at this time ATS-6 (Fig. 5) still does not see any real hard electrons.

The ATS-6 synchronous satellite is enveloped by hard electrons just after 2030.

3-30, 1976

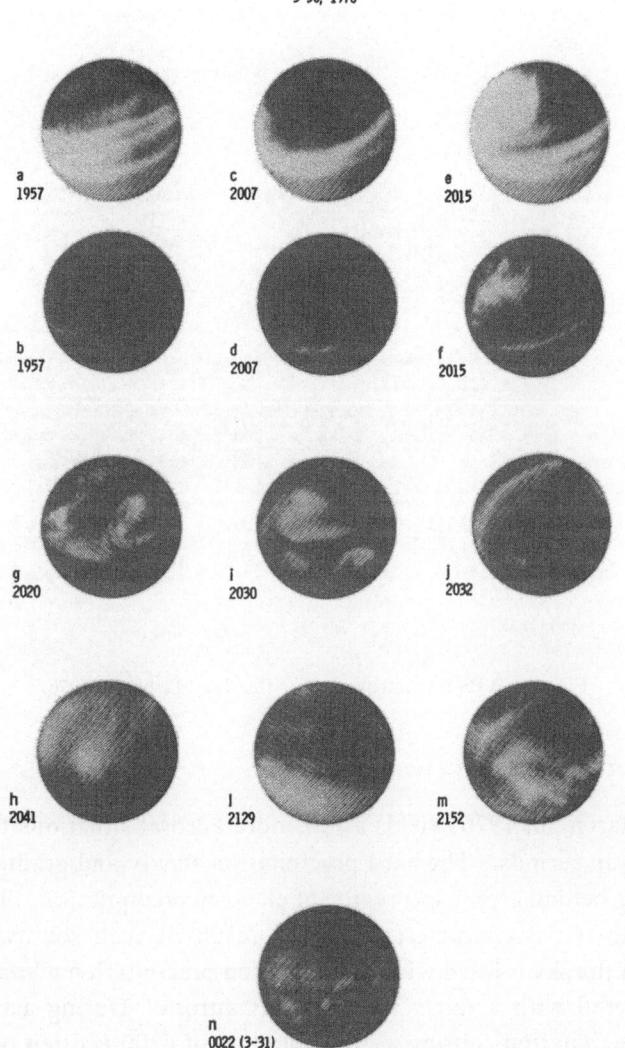

Fig. 4. ASIP TV monitor photographs prepared similarly to Fig. 2 for day 3, 30, 1976. The pairs a and b, c and d and e and f were taken one stortly after the other with different exposures to cover the large range of auroral intensities. b, d and f were taken with 1/8th and 1/4th of the 4278 and 6300 exposures respectively.

The ASIP picture taken at 2030 shows a considerable hardening of the electron precipitation when compared to 2020. The process described here is somewhat typical of the northward expansion of the aurora as has been described by Akasofu. The northward expansion is almost complete by 2032, Fig. 4j.

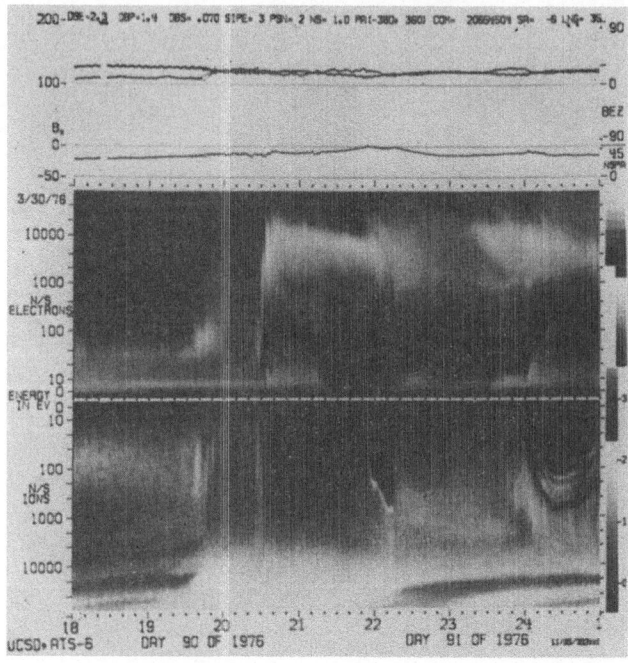

Fig. 5.   ATS-6 spectrogram for the day March 30, 1976.

6.   Early Morning Precipitation Features

The day March 30, 1976, yields a few more auroral situations typical of post-substorm break-up periods. The hard precipitation slowly and gradually drift equatorwards leaving behind a very soft region of electron precipitation. The frame taken at 2129 UT, Fig. 4l, is characteristic of the result of such southward drift. The southern half of the sky is filled with hard electron precipitation while the northward regions are covered with a fairly uniform soft aurora. During active nights, the soft electron precipitation causing a uniform glow of 6300 is often observed. Note that one cannot rule out a priori the possibility of some airglow mechanism in the wake of prior activity causing the enhanced 6300 emission.

The configuration of Fig. 4l is changed by a new substorm which occurs at 2152, Fig. 4m. Typically during substorm auroral breakup events, the fast moving auroral forms are a complex mixture of soft and hard electron precipitation.

A typical early morning configuration is shown in picture 4n taken on 31 March 0022 UT, this frame shows patches of auroral pulsations. These pulsations are very hard and they are superimposed on a hard diffused background.

7.   Proton Morphology

In Fig. 6a we are presenting proton pictures taken on March 30, 1976, at 1957. This frame shows very clearly that the proton observing technique works very well

3-30, 1976

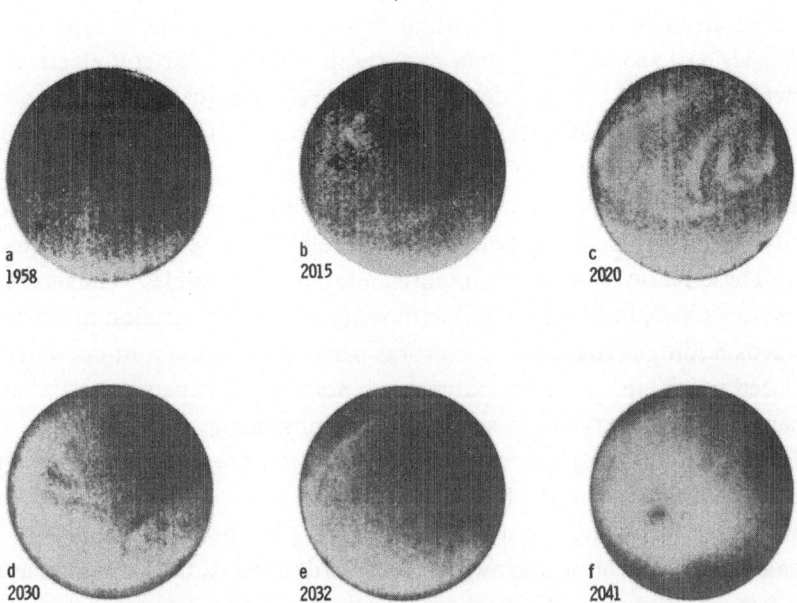

Fig. 6.   ASIP TV monitor photographs at corrected background subtracted H_β (4861) all-sky emission.   The white areas represent regions of proton precipitation.

and the intense arcs (Fig. 4a, b) disappear completely leaving the proton arc showing up near on the southern horizon.   Since the proton arc in generally co-located with the soft precipitation region (EATHER *et al.*, 1976), this region is likely to have the appropriate soft electron and proton spectra which correspond to the region containing the ATS-6 satellite.   The protons intensity at 2015 just at the onset of the break-up of the arc (5e, f).

The protons in Fig. 6c at 2020 UT show a southern bright arc which is certainly a proton arc.   The very complex structures which are shown on the picture are probably residues from the contamination subtraction.   Note that they are coincident with the bright auroral forms shown on Fig. 6g.   At 2032 at the end of the northward (poleward expansion) protons are still dominantly in the southern part of the sky (6e).   At 2040 frame, 5h, the sky is filled with hard diffuse precipitation.   The corresponding proton frame, Fig. 6f, shows that in the ten minute period the protons have expanded northward following the electron expansion and the southern part of the sky, the previous location of the protons is void.   Such northward expansion of protons during substorm expansion phase has been identified by many authors, e.g. EATHER and MENDE (1971) and FUKUNISHI (1975).

8.   Summary Discussion

Spectro-photometric techniques permit the classification of auroral structures according to the properties of the precipitating particles.   Recent developments in observing techniques permit the direct recording of the all-sky auroral situation and the separation of auroras according to energy and particle type (protons or electrons). Using a special TV system pictures in 6300 Å of the soft electron induced auroras were obtained. These dominate the pre-midnight sector and occur as a permanent feature on the equatorward side of the auroral oval. It is, however, not a steady diffuse region but it is structured and the structures show a coherent convective motion. On very quiet days on the equatorward side this motion is westward. On this dynamic soft precipitation harder precipitating features are often superimposed. These features are more intense and therefore visible. The intensification of these features have been associated with synchronous altitude ion injection events.

Towards midnight these harder auroras participate in the well-documented substorm-related break-up, surge formation and northward expansion. In several instances which were observed the spectral hardening of the electrons from few keV to 10–20 keV takes place a few minutes after auroral break-up and is in agreement with the appearance of hard electrons at synchronous altitude at the field lines corresponding a northward expansion following the auroral morphology changes after break-up. The proton northward (poleward) expansion is often delayed a few minutes after the electrons. Early morning configurations show an extensive diffuse soft electron region polewards and a hard electrons region equatorwards. The hard electron region is very patchy and this is the region where pulsating auroras are observed.

Much of the work covered in this paper and the time spent in preparing the manuscript is supported by the National Science Fundation, Atmospheric Science Section under grant ATM-76-00926 and by a Lockheed independent research program.

REFERENCES

Akasofu, S.-I., *Polar and Magnetospheric Substorms*, Springer, New York, 1968.

Eather, R.H. and S.B. Mende, Systematics in auroral energy spectra, *J. Geophys. Res.*, 77, 660–673, 1972.

Eather, R.H., S.B. Mende, and R.J.R. Judge, Plasma injection at synchronous orbit and spatial and temporal auroral morphology, *J. Geophys. Res.*, 81, 2805, 1976.

Fukunishi, H., Dynamic relationship between proton and electron auroral substorms, *J. Geophys. Res.*, 80, 553, 1975.

Lui, A.T.Y. and C.D. Anger, A uniform belt of diffuse auroral emission seen by the ISIS-2 scanning photometer, *Planet. Space Sci.*, 21, 799–809, 1973.

Lui, A.T.Y., P. Perrault, S.-I. Akasofu, and C.D. Anger, The diffuse aurora, *Planet. Space Sci.*, 21, 857, 1973.

Mende, S.B. and R.H. Eather, Monochromatic all-sky observation and auroral precipitation patterns, *J. Geophys. Res.*, 81, 3771, 1976.

Mende, S.B., R.H. Eather, and E.K. Aamodt, Instrument for monochromatic observation of all-sky auroral images, *Appl. Opt.*, 16, 1691, 1977.

McIlwain, C.E., Substorm injection boundaries, in *Magnetospheric Physics*, edited by B.M. McCormac, p. 143, D. Reidel, Dordrecht, Netherlands, 1974.

Rees, M.H. and D. Luckey, Auroral electron energy derived from ratio of spectroscopic emissions. 1. Model computations, *J. Geophys. Res.*, 79, 5181, 1974.

Romick, G.J. and A.E. Belon, The spatial variation of auroral luminosity (I), *Planet. Space Sci.*, 15, 475, 1967.

Generation Mechanisms for Magnetic-Field-Aligned
Electric Fields in the Magnetosphere

Carl-Gunne Fälthammar

Department of Plasma Physics, Royal Institute of Technology,
100 44 Stockholm, Sweden

(Received February 7, 1978)

Magnetic-field-aligned electric fields in the magnetosphere can be generated in several different ways.

Current driven wave instabilities can lead to either anomalous *resistivity* or *electric double layers*. Both can support potential drops of many kilovolts. In the former case this requires wave turbulence extended over large distances, because observed wave amplitudes allow only moderate field strength. In the latter case the voltage drop is concentrated to one or more (perhaps numerous) thin regions, sometimes referred to as electrostatic shocks, each of the order of several tens of Debye lengths thick. Once established, such layers may exist with or without substantial wave turbulence. It has also been proposed that wave-particle interaction may establish a *collisionless* version of the *thermoelectric effect*. In the hot magnetospheric plasma potential drops can be created as a consequence of the *magnetic mirror force*. In a magnetically trapped plasma this, in principle, requires no current, although some leakage current is expected. The mirror effect has its most dramatic impact in regions of upward Birkeland currents, where it limits the current density to moderate levels unless large voltages are applied (many times the voltage-equivalent of the thermal energy of the magnetospheric plasma). The distribution of this voltage depends critically on the distribution functions of the plasma particles involved, and it may well degenerate into electric double layers.

Observational data now available indicate that more than one of the mechanisms mentioned are operative in the magnetosphere but it is not yet possible to evaluate their relative importance.

1. Introduction

Recent literature in the area of magnetospheric physics reflects a considerable interest in magnetic-field-aligned electric fields. Such electric fields can have important consequences for magnetospheric theory (Fälthammar, 1977).

The previous lack of interest in this important problem area was due to the prevalence of overidealized models of the magnetospheric plasma. In these models there was no place for mechanisms that could support magnetic-field-aligned electric fields of any significance. For this reason, early suggestions of their importance (Alfvén, 1958) went unheeded.

The reconsideration has come as a consequence of refined observations, especially of particle distribution functions, that have led numerous authors to postulate magnetic-field-aligned electric fields in explanation of their findings. It is at this time impossible to say to what extent the various postulates are correct, because direct measurements of parallel electric fields are still very scarce. However, these direct measurements (Section 2) although few, are exceedingly important because they (1) make it possible to say with certainty that magnetic-field-aligned electric fields do occur in the magnetosphere, and (2) give us some guidance to the magnitude and distribution of these fields.

With realistic plasma models one can identify several possible mechanisms capable of supporting significant electric field strengths along the magnetic field (BLOCK and FÄLTHAMMAR, 1976). Here we will systematically consider the existing possibilities (Section 2). Specific mechanisms will be discussed in the light of recent experimental evidence (Sections 3–7). The electric fields supported by different mechanisms will have certain distinguishing characteristics, especially in terms of strength and spatial distribution. It is concluded (Section 8) that probably more than one of the possible mechanisms is operative in the magnetosphere, but much more experimental evidence is required.

2. General Comments and Classification of Mechanisms

Electric fields in the magnetosphere arise both from charge separation and from induction due to time-varying magnetic fields. In either case the integral $\int E \cdot ds$ between two spatially separated points generally differs from zero. If, as was previously believed, the magnetospheric plasma had a very large conductivity, its charge carriers would move to rearrange themselves so as to make each magnetic field line essentially an equipotential. This follows from the assumption of (a generalized) Ohms law, $i = \sigma \cdot E$ being valid, and $\sigma_{//}$ (the conductivity along the magnetic field) being large. If, in any local region, $E \cdot B$ were non-zero, a current $\sigma_{//}(E \cdot B)/B$ would flow to readjust the charge distribution and suppress $E \cdot B$ to (essentially) zero.

Two comments are appropriate here. First, even when the primary cause of the electric field is induction by a time-varying magnetic field, charge separation, too, is essential in determining the spatial distribution of the electric field. Even if $\partial B/\partial t$ were known exactly, this would not be enough to calculate the actual electric field in the magnetosphere. As a vast number of charge carriers are present in the plasma, minute rearrangements are enough to thoroughly change the electric field distribution from the 'pure' induction field, that would apply in a vacuum. Secondly, it is interesting to note that (as emphasized by PELLINEN et al., 1978) an electric induction field can only be locally extinguished by charge rearrangement. If charges are redistributed so that $E_{//}$ vanishes, $E_{\perp}$ will typically be strengthened and certainly be non-vanishing.

If a finite $E_{//}$ is to exist other than as a brief transient, its force on the charge carriers has to be balanced in some way. In a collision-dominated plasma, the balancing force is collisional friction due to the motion caused by the electric field.

As we will be dealing with essentially collisionless plasmas, the collisional friction is insignificant and will be disregareded. The problem of finite $E_{//}$ being possible then centers around the existence of other mechanisms by which the force from the d.c. $E_{//}$ on the charge carriers can be balanced. Locally this problem is exactly the same whether the electric field as a whole is inductive or potential. We therefore need not make any further distinction between these two cases.

In a collisionless plasma there exist the following four possibilities of balancing the force from the d.c. $E_{//}$.

2.1   Forces from a.c. electric fields

In addition to the d.c. electric field, $E_{//}$, there may be an a.c. electric field $\tilde{E}$. For a steady state to exist, an average charge carrier (runaways excluded) must essentially lose as much momentum to the a.c. field as it gains from the d.c. field. This puts certain requirements on the intensity of $\tilde{E}$, which will be discussed later.

a)   It is known theoretically as well as experimentally that the force from a.c. electric fields of wave turbulence can cause substantial friction to the current-carrying electrons and establish a situation of *anomalous resistivity* (Section 4).

b)   The a.c. electric force on the charged particles is likely to vary much with the particle energy and pitch angle. It has been proposed that as a consequence a *collisionless* kind of *thermoelectric effect* and a corresponding non-vanishing $E_{//}$ may occur in the interface between plasmas with widely differing distribution functions (Section 5).

2.2   Forces from the d.c. magnetic field

A converging d.c. magnetic field exerts a *magnetic mirror force* $-\mu \cdot (\text{grad } B)_{//}$ on each charge carrier. On high latitude magnetic field lines this force is sufficient to support magnetic-field-aligned potential drops of many kilovolts (Section 6).

2.3   Forces from a.c. magnetic fields

Depending on the nature of the wave turbulence, it may contain magnetic a.c, fields as well. Such fields may therefore accompany anomalous resistivity and colli-sionless thermoelectric effect, but in the magnetosphere, their role in the force balance is negligible.

2.4   Inertia forces

A final possibility is that the force from the electric field is, locally, balanced only by the inertia of the charge carriers, all of which run away and carry their momentum to other parts of the current path. The relation between $E_{//}$ and $i_{//}$ is then not local but depends on the entire circuit. Such is the situation in the case of *electric double layers* (Section 7).

3.   Observed Voltages and Field Strengths

Even before direct measurements were available, it was clear that the magnetic-field-aligned electric fields ought to be large and/or extended enough to give voltage differences of the order of several kilovolts. Otherwise they could not play the role that they appear to play in particle acceleration. Such voltages are also enough to make $E_{//}$ interesting in affecting magnetospheric plasma dynamics by partial 'unfreezing' (electrical decoupling) of magnetic field lines. The recent direct observations have confirmed the expected voltages and added key information on the strength and spatial distribution of the fields.

Artificial ion-cloud experiments using shaped charges (WESCOTT et al., 1976; HAERENDEL et al., 1976) have given direct measurements of magnetic-field-aligned potential drops. In one case Haerendel et al. reported a potential drop of 7.4 kV, in good agreement with the magnitudes envisaged on other grounds. Limited spatial resolution precludes establishing the field strength in this case, and HAERENDEL et al. (1976) only conclude that the voltage drop occurs over 1,000 km or less. This gives a lower limit of 7.4 mV/m to the field strength, but the actual field strength may well be much higher.

Direct probe measurements of d.c. electric fields at high altitude were for the first time performed by MOZER et al. (1977). These measurements revealed transverse as well as parallel electric fields of the order of hundreds of mV/m in thin regions, of the order of 3 km in transverse extent, and usually occurring in pairs.

The paired regions of transverse field are expected as a consequence of magnetic-field-aligned potential drops located on field lines between them (SHAWHAN et al., 1978). MOZER et al. (1977) have found one clear case of very large magnetic-field-aligned electric field, hundreds of mV/m. As the large field strength implies a small vertical extent, it is consistent that such case should occur only in a small fraction of the satellite passages. The remarkably large field strength observed puts strong requirements on the responsible generation mechanisms.

4. Anomalous Resistivity

Anomalous resistivity was the first mechanism to be invoked as a means to support substantial electric fields in plasmas with negligible classical resistivity. An extensive literature exists on the subject. For a recent review of the theory, with a view to space applications, see PAPADOPOULOS (1977) and references therein.

In the present paper the emphasis will be on a physics-oriented discussion of anomalous resistivity as a candidate mechanism for supporting the kind of magnetic-field-aligned electric fields now known to occur in the magnetosphere.

By definition anomalous resistivity in a strict sense implies that as a consequence of wave-particle interaction, rather than collisional friction, there exists a local relation of the form

$$E_{//} = \eta_{eff} i_{//} \tag{1}$$

between current density $i_{//}$ and electric field $E_{//}$. (The subscript $//$ indicates that we are considering the components along the geomagnetic field.)

The wave turbulence required for this is usually assumed to be the end result of an instability driven by the current itself. However, in principle it may also be sustained by external means, such as an injected high-energy electron beam that is unstable in the local plasma environment (PAPADOPOULOS and COFFEY, 1974a, b; PAPADOPOULOS, 1975).

The wave turbulence will act differently on particles in different parts of velocity space. In particular, particles of sufficient initial velocity will become runaways. Those particles will become decoupled from the main population of charge carriers. The current they carry will not be determined by local conditions but depend on the whole electric circuit involved.

It is therefore interesting to note that the auroral precipitation often carries currents that are substantial fractions of the total Birkeland current. If the precipitating electrons are to be considered runaways from a region of anomalous resistivity, care must be taken in applying the anomalous resistivity concept. The electric field-electric current relation is then not controlled by the anomalous resistivity alone. Other effects, such as magnetic mirror or inertia forces must be considered, and the whole circuit must be taken into account.

The resistivity of a plasma, anomalous as well as classical, can be written in the form

$$\eta_{eff} = \frac{m_e}{e^2 n_e} \, \nu_{eff} \equiv \frac{1}{\varepsilon_0 \omega_{pe}} \, \frac{\nu_{eff}}{\omega_{pe}} \tag{2}$$

where ω_{pe} is the electron plasma frequency and ν_{eff} the effective collision frequency *for momentum loss*. In the case of anomalous resistivity ν_{eff} results from wave particle interaction. The value of ν_{eff} then depends on the actual distribution of the wave fields in k-ω-space and the distribution of charged particles in velocity space in the final turbulent state. Whereas it is straightforward to calculate onset criteria and growth rates of candidate instabilities, theoretical calculation of the nonlinear development of wave fields and particle distributions up to the final state is very difficult.

In calculating ν_{eff} it is therefore often assumed that the end result of instability growth is a turbulent state with certain simple properties. That this is not always an entirely safe method is clear from the experimentally established fact that sometimes the instability leads to double layer formation (Section 7) instead of anomalous resistivity.

Table 1.   Estimates for ν_{eff}/ω_{pe} (from SCHRIJVER, 1973b).

Author	Method	Value of ν_{eff}/ω_{pe}
Kalinin *et al.*	Experiment (linear)	1.25–$2.5 \cdot 10^{-2}$
Wharton *et al.*	Experiment (linear)	0.83–$1.25 \cdot 10^{-2}$
Schrijver	Experiment (linear)	$0.53 \cdot 10^{-2}$
Zavioski *et al.*	Experiment (toroidal)	$0.42 \cdot 10^{-2}$
Hamberger *et al.*	Experiment (toroidal)	0.33–$1.33 \cdot 10^{-2}$
Biskamp	Computer simulation	0.33–$0.67 \cdot 10^{-2}$

On the other hand there is no doubt of the existence of anomalous resistivity as a phenomenon, as it has been observed experimentally in the laboratory (see e.g. SCHRIJVER (1973a, b) and references therein). In the present discussion on the potential role of anomalous resistivity in space plasma we will largely build on results of laboratory experiments and numerical simulations rather than purely theoretical results. Values of the ratio ν_{eff}/ω_{pe} have been compiled by SCHRIJVER (1973b). They are reproduced in Table 1.

In the present context we are especially interested in how large values of $E_{//}$ can be achieved. If the resistivity is sustained by current-driven instability, the electron drift velocity has to have a certain minimum value. We may write the corresponding current density as

$$i_{//} = \beta e n_e \sqrt{\frac{2kT_e}{m_e}} . \tag{3}$$

The value of β depends on the instability involved, being about unity for two-stream and Buneman instabilities, less than unity for the ion acoustic instability (typically 0.01–0.3). Using Eqs. (1)–(3) one can now rewrite the d.c. electric field strength as

$$E_{//} = \frac{kT_e}{e\lambda_D} \cdot \sqrt{2} \, \beta \cdot \frac{\nu_{eff}}{\omega_{pe}} . \tag{4}$$

Using the values of Table 1 we find that $E_{//}$ is, at most, of the order of 1% of the thermal voltage equivalent divided by the Debye length. Alternatively it may be written

$$E_{//} = \sqrt{n_e kT_e/\varepsilon_0} \cdot \sqrt{2} \, \beta \cdot \frac{\nu_{eff}}{\omega_{pe}} \simeq 10^{-8} \beta \sqrt{n_e T_e} \quad \text{(MKS units)} . \tag{5}$$

The latter equation relates the achievable value of $E_{//}$ to the electron-gas pressure. For example with typical plasma-sheet parameters ($n_e = 10^6 \, \text{m}^{-3}$, $T_e = 10^6 \, \text{K}$) we find $E_{//} = 10 \, \text{mV/m}$.

A particularly interesting question is how a d.c. electric field supported by anomalous resistivity is related to the a.c. electric field needed to support it. From consideration of the balance of momentum in the direction of the d.c. magnetic field it follows (FÄLTHAMMAR, 1977; SHAWHAN et al., 1978) that if the d.c. electric field $E_{//}$ is sustained by anomalous resistivity, then the rms value, E_{rms}, of the a.c. field has to fulfill the condition

$$E_{rms} \gg E_{//} . \tag{6}$$

This is essentially because the electrons gain momentum at the rate $e n_e E_{//}$ from the d.c. field. The momentum lost to the a.c. field depends on the actual distribution of waves and particles, but a high upper limit to its value is $e n_e E_{rms}$.

The charge carriers' momentum component along the direction of the d.c. magnetic field can in principle also be changed by a.c. magnetic fields, $\tilde{b}$, of turbulent waves. The instantaneous rate of momentum change for an individual particle is $e \boldsymbol{v}_\perp \times \boldsymbol{b}$. The vector $\boldsymbol{v}_\perp$ rotates at the electron gyro frequency around the total magnetic field, vector $\boldsymbol{B} = \boldsymbol{B} + \boldsymbol{b}$, which is utterly dominated by the d.c. geomagnetic field. Therefore waves of any other frequency than the electron gyro frequency have no net average effect. And even at that frequency only particles in an appropriate gyration phase lose momentum, those with opposite phase gain momentum at the same rate. The a.c. magnetic fields are therefore in practice not important in supporting an $E_{//}$. Furthermore they would by necessity be accompanied by a.c. electric fields such that Eq. (6) would still hold.

The result (6) is quite general and applies for any anomalous resistivity whether driven by the plasma current itself or an externally imposed agent, and regardless of what instabilities are responsible for the turbulence.

For current-driven resistivity we may use the results of computer simulations to obtain a less general but more precise relation between $E_{//}$ and E_{rms}. According to BISKAMP et al. (1972) the effective collision frequency in the state of saturated turbulence satisfies the relation

$$\frac{\nu_{eff}}{\omega_{pe}} = \alpha \frac{\varepsilon_0 E_{rms}^2}{2 n_e k T_e} \tag{7}$$

with $\alpha = 0.2$–0.3. It then follows from Eqs. (1)–(3) that

$$E_{rms} = \frac{1}{\beta} \left(\frac{\omega_{pe}}{\alpha \nu_{eff}} \right)^{1/2} \cdot E_{//} . \tag{8}$$

With the values of ν_{eff}/ω_{pe} given in Table 1, this means that E_{rms} exceeds $E_{//}$ by a factor of 12 to 39.

One important problem encountered in applying anomalous resistivity to space plasma is that of heat balance. Unlike the mirror effect (Section 6) and electric double layers (Section 7), the anomalous resistivity leads to local deposition of the dissipated power in the form of heat given to the local plasma (except for a certain amount that may be radiated away). The magnitude of this power per unit volume is, according to Eqs. (2) and (7)

$$P = \frac{E_{//}^2}{\eta_{eff}} = \frac{E_{//}^2}{E_{rms}^2} 2\alpha^{-1} n_e k T_e \omega_{pe} . \tag{9}$$

P can be quite large. For example, with $\beta = 1$ and the values of $E_{rms}/E_{//}$ found above,

$$P \gtrsim \frac{n_e k T_e}{40 \tau_{pe}} \tag{10}$$

where τ_{pe} is the electron plasma period.

If this power were taken up by the local plasma, it would be very rapidly heated, essentially with a time constant of only tens of electron plasma periods, or less.

Comparing with observational data we may now draw two main conclusions:

1) Extensive surveys by Hawkeye (GURNETT and FRANK, 1976) showed that the rms electric fields were rarely larger than 10 mV/m, although observations by

FREDRICKS et al. (1973) show that rms fields as strong as 90 mV/m exist occasionally. In view of Eqs. (6) and (7) such wave fields can support d.c. fields of the order of mV/m or less. Although unspectacular as a local phenomenon, such fields may well add up to significant potentials of several kilovolts if extended over a few Earth radii.

2) The recent discovery by MOZER et al. (1977) of $E_{//}$ with very high local field strengths—hundreds of mV/m—would require a.c. fields of several V/m if they were sustained by anomalous resistivity. Thus anomalous resistivity does not appear adequate to explain these observations but may well be operating *in addition* to the main mechanisms (which in this case is probably double layers, Section 7).

5.   Collisionless Thermoelectric Effect

The average force that a charge carrier experiences from the turbulent a.c. fields will be different for particles of different energy. In a collision-dominated plasma, where the friction force decreases rapidly with energy the result is the ordinary thermoelectric effect, which is capable of supporting a d.c. electric potential between plasmas of different temperatures, without requiring a net electric current. In the classical case we can write the magnetic-field-aligned component of the electric field

$$E_{//} = -\frac{k}{n_e e} T_e^{-\gamma/2} \frac{d}{ds}(n_e T_e^{1+\gamma/2}) \tag{11}$$

where γ is the thermal diffusion coefficient which has the value 1.4 for singly charged particles in a fully ionized plasma and d/ds is the derivative along the magnetic field. The thermoelectric field vanishes only if the density n_e and temperature T_e vary in a very special way, such as to keep $n_e T_e^{1+\gamma/2}$ constant.

It has been proposed (HULTQVIST, 1971, 1972) that in principle the same phenomenon can occur in a collisionless plasma with energy-dependent wave-particle interaction replacing the Coulomb collisions. Of course the value of the coefficient γ would depend on how the prevailing wave particle interaction varies with energy.

To evaluate whether the thermoelectric effect plays a role in the magnetosphere will be possible only when we have much better knowledge of the height variations of the total particle distribution function (or at least of density and temperature) and, in particular, of the distribution and properties of particle scattering wave fields.

Finally we note that the arguments leading to Eq. (6) apply equally well in the case of the collisionless thermoelectric effect as it does for anomalous resistivity.

6.   Magnetic Mirror Effect

The magnetic mirror effect as a means to support magnetic-field-aligned electric fields in the magnetosphere was proposed by ALFVÉN and FÄLTHAMMAR (1963).

Alfvén and Fälthammar considered the case of magnetically trapped particles.

What is required in this case is a differential anisotropy between ions and electrons but no current is needed. Further theoretical studies of this case have been made by PERSSON (1963, 1966) and WHIPPLE (1976). The existence of the mechanism in laboratory plasma has been established most clearly by GELLER *et al.* (1974).

The potential that can be supported depends on the particle energies involved and the amount of differential anisotropy. A simple illustrative example has been treated quantitatively by BLOCK and FÄLTHAMMAR (1976) who also quantitatively discussed more realistic cases.

In the special case (delta function distributions in energy and pitch angle) treated by ALFVÉN and FÄLTHAMMAR (1963), the electric field strength is

$$E_{//} = -K \frac{dB}{ds} \tag{12}$$

where

$$K = \frac{1}{eB} \cdot \frac{W_{e\perp} W_{i//} - W_{i\perp} W_{e//}}{W_{i//} + W_{e//}} \tag{13}$$

and $W_{i//}$ $(W_{i\perp})$, $W_{e//}$ $(W_{e\perp})$ are the kinetic energy of ions and electrons respectively, due to the parallel (transverse) motion. Whipple has generalized the theory to arbitrary distributions and showed that in the general case Eq. (12) holds with

$$K = \frac{1}{eB} \cdot \frac{\iint W_{e\perp} g_e \, dW d\mu - \iint W_{i\perp} g_i \, dW d\mu}{\iint g_e \, dW d\mu + \iint g_i \, dW d\mu} \tag{14}$$

where

$$g = m^{-3/2} \left(\frac{df^+}{dW} + \frac{df_e^-}{dW} \right)_\mu (W - eV - \mu B)^{-1/2} \tag{15}$$

(in customary notation). Applying this to the magnetosphere, WHIPPLE (1976) develops two methods for analyzing observational particle data in search of signatures of magnetic-field-aligned electric fields.

The magnetic mirror effect may be even more important in the presence of large Birkeland currents. In high-latitude flux tubes, the repulsive magnetic force on downgoing charge carriers has interesting consequences (RASSBACH, 1973; KNIGHT, 1973; LEMAIRE and SCHERER, 1974; LENNARTSSON, 1976, 1977; FÄLTHAMMAR, 1977).

For upward Birkeland currents the charge carriers available are upgoing ions from the ionosphere and downgoing electrons from the thin hot external plasma. Although the ions are unimpeded by the mirrors and therefore may be important (RASSBACH, 1973) their capacity for carrying current is rather limited. The downgoing electrons, on the other hand suffer from magnetic mirroring, and it can be shown (FÄLTHAMMAR, 1977) that for moderate electric fields this current-carrying capability can be written as a *conductance per unit area* (at ionospheric level), given by

$$\frac{di}{dV} = \frac{1}{2\pi} \varepsilon_0 \omega_{pe2} / \lambda_{D2} \tag{16}$$

where the subscript 2 refers to the external plasma.

Table 2. Conductance per unit area, referred to ionospheric level.

Source plasma	n_{e2}[1] (cm^{-3})	T_{e2}[1] (K)	di/dV (μA(kV)$^{-1}$m^{-2})
Plasma sheet	1	10^6	3
Magnetosheath (front)	30	$2 \cdot 10^6$	60
Solar wind	8	$1.5 \cdot 10^5$	60

[1] Siscoe (1973).

A corresponding *conductivity* in the sense of a local relation between current density and electric field strength does not exist in this case. Typical values of this conductance for different types of source plasma are given in Table 2. We notice that the conductance is remarkably limited so that commonly occurring current densities will require voltage drops of several kilovolts.

Unlike the total potential drop, V, the *distribution* of the electric fields along the magnetic flux tube is very difficult to calculate, as it depends on the distribution functions of both ions and electrons and their variations along the flux tube.

The above expression (16) and corresponding numerical values in Table 2 apply provided the loss of the source plasma is kept filled by continual replenishment. If not, the current will choke to still lower values.

The full current-voltage characteristic in the case of a Maxwellian source plasma has been derived by KNIGHT (1973) and LEMAIRE and SCHERER (1974). If we neglect the contribution from outgoing ionospheric electrons, which is cut off at a very small voltage drop, we can write the current density

$$i_{//} = en_{e2} \sqrt{\frac{kT_{e2}}{2\pi m_e}} \cdot \frac{B_1}{B_2} \left[1 - \left(1 - \frac{B_2}{B_1} \right) \exp \left\{ - \frac{eV}{kT_{e2}(B_1/B_2 - 1)} \right\} \right] \qquad (17)$$

the subscripts 1 and 2, referring, respectively, to the ionospheric level and the magnetosheath (KNIGHT, 1973). A quantitative diagram of $i_{//}$ versus V is given in Fig. 1.

So far we have discussed only the completely a diabatic situation and disregarded the effects of scattering. In the case discussed above, where the source plasma is continually replenished, the downgoing (although not the upgoing) plasma is isotropic and has a filled loss cone. Then mere scattering, without energy degradation, will not increase the current unless the wave distribution is such as to systematically decrease the particles' magnetic moment even in such a distribution. On the other

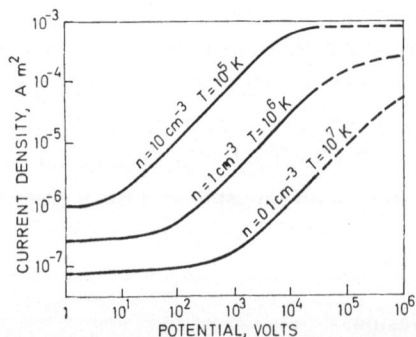

Fig. 1. Logarithmic plot of the current-voltage characteristic for three combinations of source plasma parameters typical for the plasma sheet.

hand, if the replenishment is incomplete so that the current is partly choked by a deficiency of charge carriers with small pitch angles then pitch-angle scattering will alleviate the choking and tend to restore the conductance calculated above. If the scattering becomes intense enough, it may itself impede the current and establish an anomalous resistivity.

In this context it is important to notice that the mirror effect can operate even when $i_{//}$ is far below the random thermal current, so that conditions may not be favorable for intense turbulence.

Although the mirror force can support large potentials along a flux tube, it can locally balance only rather moderate electric forces. As a generous upper limit to the electric field strength that it can balance locally we may take

$$E_{\max} = (V + kT_{e2}/e)|(\text{grad } B)_{//}|/B \qquad (18)$$

where V is the potential drop to the source plasma and T_{e2} is the temperature of the latter. With $(V + kT_{e2})/e$ of the order of 10 kV, it means that in the altitude range 2,000 to 8,000 km a high upper limit is 4–15 mVm^{-1}. Thus the mirror force alone cannot support the large *field strength* observed by MOZER et al. (1977) of parallel fields as large as hundreds of mV/m. However, it may facilitate formation of double layers or electrostatic shocks (cf. KAN, 1976; LENNARTSSON, 1977) and also influence the structure of such layers due to the (remote) mirroring of electrons trapped below the layer (LENNARTSSON, 1977).

To determine experimentally to what extent the mirror effect does play a role for maintaining parallel electric fields in the magnetosphere, comprehensive measurements of particle and wave spectra are needed together with measurements of the parallel electric field itself.

7.   Electric Double Layers

Plasma instabilities sometimes develop into a final state that is very different from the turbulent state assumed in anomalous resistivity theory, namely into a double layer. The existence of such states has been known experimentally for a long time (LANGMUIR, 1929). Recent experimental studies have been made by TORVÉN and BABIĆ (1975, 1976) and QUON and WONG (1976). Important theoretical results have been obtained from simplified models (LANGMUIR, 1929; BLOCK, 1972; KNORR and GOERTZ, 1974), and the BGK-solutions of the Vlasov equations (BERNSTEIN et al., 1957) provide a valuable tool for theoretical analysis of steady-state double layers. However, important problems remain unanswered, for example what the conditions are for instabilities to lead to double layer formation rather than to homogeneous turbulence.

Laboratory experiments as well as theoretical analysis are indispensable for improving the understanding of the double layer phenomenon. Numerical simulation is a valuable additional tool (GOERTZ and JOYCE, 1975).

Thus the double layer is primarily a phenomenon that is known with certainty to exist in a laboratory plasma, and which may also exist in space plasma. Its main characteristics, as summarized in a recent review by BLOCK (1977) are:

1) The potential drop through the layer is at least equal to the thermal voltage equivalent (kT_e/e) of the coldest plasma bordering the layer.

2) The electric field is much stronger inside the double layer than outside, i.e. the positive and negative charges very nearly cancel each other.

3) Quasi-neutrality is locally violated. (Related to this fact is the thinness of the double layer—typically some tens of Debye lengths.)

For space applications, it is also interesting to note that typically, but not necessarily, the mean free path is much larger than the thickness of the double layer.

Within the double layer each charge carrier is subject to a strong electric force which is, essentially, balanced only by the charge carrier's own inertia. Thus in a double layer all charge carriers 'run away.' This is a very important characteristic of the double layer as a means to support a voltage drop, because the power released is *not* deposited locally but can be carried away by the accelerated high-energy electrons down to low altitude where the atmosphere is an efficient energy sink. (Some dissipation may of course occur on the way down due to instability of the particle beam.) Thus the problem of energy deposition, which is serious in the case of anomalous resistivity, is essentially absent in the case of double layers.

It can be shown that the thickness L and the average electric field $\bar{E}$ of a double layer with voltage drop V can be written

$$L = \frac{1}{\gamma}\left(\frac{eV}{kT_e}\right)^{1/2}\lambda_D \tag{19}$$

$$\bar{E} = \frac{V}{L} = \frac{\gamma e n_e}{\varepsilon_0}\left(\frac{eV}{kT_e}\right)^{1/3}\lambda_D = \gamma\left(\frac{eV}{kT_e}\right)^{1/2}\frac{kT_e}{e\lambda_D} . \tag{20}$$

The factor γ depends on the shape of the space charge distribution, which in turn may depend on the potential drop V or the current density i, as discussed further by SHAWHAN et al. (1977). Here we will only note that γ is typically in the range 0.01 to 0.1. Recently several authors have discussed the possible role of double layers and related phenomena (electrostatic shocks) to the space plasma (BLOCK, 1972, 1976; KAN, 1976; SWIFT, 1975; SWIFT et al., 1976; KAN and AKASOFU, 1976; SHAWHAN et al., 1977), for a recent review, see BLOCK (1977).

As shown by SHAWHAN et al. (1978) the model studied by KNORR and GOERTZ (1974) leads to such a value of γ that Eq. (20) becomes

$$\bar{E} = 3\cdot10^{-5}n_e^{1/2}V^{1/2} \quad \text{(MKS units)} \tag{21}$$

while the model of LANGMUIR (1929) implies an average field that depends on the current density i and is given by

$$\bar{E} = 500i^{1/2}V^{1/4} \quad \text{(MKS units)} . \tag{22}$$

The experiment by QUON and WONG (1976) showed a sufficiently wide and steady double layer to allow its internal electric field distribution to be resolved. The parameters were: $n_e = 10^{14} \, \text{m}^{-3}$, $T_e = 4 \cdot 10^4 \, \text{K}$, $i = 10 \, \text{A m}^{-2}$. The potential drop was 15 V and occurred over a few centimeters $(20-30\lambda_D)$ with a maximum field strength of 500 V/m and an average value, $\bar{E}$, of about 300 V/m.

As the theoretical models are based on special simplifying assumptions, their direct application to the space plasma is questionable. Slightly safer may be to take the experimental data just quoted and scale them to space conditions using the theories only as a guide to the functional dependence on the parameters.

Table 3, from SHAWHAN et al. (1978) shows the numerical values of $\bar{E}$ for typical space parameters. Values calculated directly from the theoretical models are higher than any fields measured to date. The ones scaled from experiment fall very well within the range of values actually measured by MOZER et al. (1977). Furthermore, as emphasized by BLOCK (1977) the electric field strength, of the order of 100 mV/m and density, 1-50 cm^{-3}, observed by MOZER et al. (1977), does mean that condition 3) above is satisfied. (As the observed spatial extent is only of the order of kilometers, div $E \approx 0.1/10^3$, which means a net charge density ε_0 div E that corresponds to 0.01 particles cm^{-3}, and thus a relative charge imbalance of $2 \cdot 10^{-4}-10^{-2}$. This is a quasineutrality violation of a magnitude typical for double layers.

Table 3. Estimates of parallel electric fields in double layers
(from SHAWHAN et al., 1978).

Model	Potential drop (V/m)	
	1 kV	10 kV
Goertz and Joyce for $n=1$ cm^{-3}	1	4
Goertz and Joyce for $n=50$ cm^{-3}	10	30
Langmuir for $i=10^{-6}$A m^{-2}	3	5
Quon and Wong exp. scaled by Eq. (21) for $n_e=1$ cm^{-3}	0.25	0.8
Quon and Wong exp. scaled by Eq. (22) for $i=10^{-6}$A m^{-2}	0.25	0.4

8. Concluding Remarks

Of the mechanisms discussed, three have been demonstrated in the laboratory, the remaining two are based on theoretical considerations. The characteristics of each process and the expected magnitude of field strengths that it should be able to support under space conditions have been summarized by SHAWHAN et al. (1978) and are given in Table 4.

Strictly speaking none of the mechanisms has been proved to operate in space. The one closest to being confirmed is the double layer. As Table 4 shows, it is the only mechanism that can explain the strong $E_{//}$ recently measured by MOZER et al. (1977). Furthermore, the deviation from quasineutrality, which occurs only in the case of this mechanism does indeed take place in the region of those strong electric fields. In addition, conditions in the regions of high-latitude Birkeland currents are such that the magnetic mirror effect and anomalous resistivity should come into play. Thus it appears likely that at least three different mechanisms contribute to supporting parallel electric fields in the high-latitude magnetospheric plasma. How-ever, it still remains to verify this experimentally and to assess their relative im-portance.

Table 4. Mechanisms supporting parallel electric fields (from SHAWHAN et al., 1978).

Mechanism	Physical process	Field at $1R_E$
1. Anomalous resistivity	Electron current retarded by waves, some electrons accelerated (runaway)	$<3\,\text{mV m}^{-1}$
2. Thermo-electric effect	Potential to retard diffusion of hot electrons in pres-ence of wave turbulence	$<3\,\text{mV m}^{-1}$
3. Magnetic mirror	Hot electrons and ions mirror at different points on field line causing change separation	$<15\,\text{mV m}^{-1}$
4. Double layer	Localized charge separation maintained by $E \cdot J$ power input transformed into ordered linear kinetic energy	$\sim 1,000\,\text{mV m}^{-1}$

REFERENCES

ALFVÉN, H., On the theory of magnetic storms and aurorae, *Tellus*, **10**, 104, 1958.

ALFVÉN, H. and C.-G. FÄLTHAMMAR, *Cosmical Electrodynamics, Fundamental Principles*, pp. 162–167, Clarendon Press, Oxford, 1963.

BERNSTEIN, I.B., GREENE, J.M., and KRUSKAL, M.D., Exact nonlinear plasma oscillations, *Phys. Rev.*, **108**, 546, 1957.

BISKAMP, D., K.U. VON HAGENOW, and H. WELTER, Computer studies of current-driven ion-sound turbulence in free dimensions, *Phys. Lett.*, **39A**, 351, 1972.

BLOCK, L.P., Potential double layers in the ionosphere, *Cosm. Electrodyn.*, **3**, 349, 1972.

BLOCK, L.P., Double layers, in *Physics of the Hot Plasma in the Magnetosphere*, edited by B. Hultqvist and L. Stenflo, p. 229, Plenum Press, New York and London, 1976.

BLOCK, L.P. and C.-G. FÄLTHAMMAR, Mechanisms that may support magnetic-field-aligned electric fields in the magnetosphere, *Ann. Geophys.*, **32**, 161, 1976.

BLOCK, L.P., A double layer review, TRITA-EPP-77-16, Royal Institute of Technology, Stockholm, 1977.

FREDRICKS, R.W., F.L. SCARF, and C.T. RUSSELL, Field-aligned currents, plasma waves, and anoma-lous resistivity in the disturbed polar cusp, *J. Geophys. Res.*, **78**, 2133, 1973.

FÄLTHAMMAR, C.-G., Problems related to macroscopic electric fields in the magnetosphere, *Rev. Geophys. Space Phys.*, **15**, 457, 1977.

GELLER, R., N. HOPFGARTEN, B. JACQUOT, and C. JACQUOT, Electric fields parallel to the magnetic field in laboratory plasma in a magnetic mirror field, *J. Plasma Phys.*, **12**, 467, 1974.

GOERTZ, C.K. and G. JOYCE, Numerical simulation of the plasma double layer, *Astrophys. Space Sci.*, **32**, 165, 1975.

GURNETT, P.A. and L.A. FRANK, A region of intense plasma wave turbulence on auroral field lines, University of Iowa Report 76-12, 1976.

HAERENDEL, G., E. RIEGER, A. VALENZUELA, H. FÖPPL, H.C. STENBACK-NIELSEN, and E.M. WESCOTT, First observation of electrostatic acceleration of barium ions into the magnetosphere, in *European Programmes on Sounding-Rocket and Balloon Research in the Auroral Zone*, European Space Agency Report, ESA SP 115, p. 203, 1976.

HULTQVIST, B., On the production of a magnetic-field-aligned electric field by the interaction between the hot magnetospheric plasma and cold ionosphere, *Planet. Space Sci.*, **19**, 749, 1971.

HULTQVIST, B., On the interaction between the magnetosphere and the ionosphere, in *Solar-Terrestrial Physics*/Proc. of the International Symp., Leningrad, USSR, May 11–19, 1970, E.R. Dyer (gen. ed.), pp. 176–198, D. Reidel, Dordrecht, 1972.

KAN, J.R., Energization of auroral electrons by electrostatic shock waves, *J. Geophys. Res.*, **80**, 2089, 1976.

KAN, J.R. and S.-I. AKASOFU, Energy source and mechanisms for accelerating the electrons and driving the field-aligned current of the discrete auroral arc, *J. Geophys. Res.*, **81**, 5123, 1976.

KNIGHT, S., Parallel electric fields, *Planet. Space Sci.*, **21**, 741, 1973.

KNORR, G. and C.K. GOERTZ, Existence and stability of strong potential double layers, *Astrophys. Space Sci.*, **31**, 209, 1974.

LANGMUIR, I., The interaction of electron and positive ion space charges in cathode sheaths, *Phys. Rev.*, **33**, 954, 1929.

LEMAIRE, J. and M. SCHERER, Ionosphere-plasma sheet field-aligned currents and parallel electric fields, *Planet. Space Sci.*, **22**, 1485, 1974.

LENNARTSSON, W., On the magnetic mirroring as the basic cause of parallel electric fields, *J. Geophys. Res.*, **81**, 5583, 1976.

LENNARTSSON, W., On the role of magnetic mirroring in the auroral phenomena, *Astrophys. Space Sci.*, **51**, 461, 1977.

MOZER, F.S., N.K. HUDSON, R.B. TORBERT, B. PARADY, and J. YATTEAU, Observations of paired electrostatic shocks in the polar magnetosphere, *Phys. Rev. Lett.*, **38**, 292, 1977.

PAPADOPOULOS, K. and T. COFFEY, Nonthermal features of the auroral plasma due to precipitating electrons, *J. Geophys. Res.*, **79**, 674, 1974a.

PAPADOPOULOS, K. and T. COFFEY, Anomalous resistivity of the auroral plasma, *J. Geophys. Res.*, **79**, 1558, 1974b.

PAPADOPOULOS, K., Non-linear stabilization of beam plasma interactions by parametric effects, *Phys. Fluids*, **18**, 1769, 1975.

PAPADOPOULOS, K., A review of anomalous resistivity for the ionosphere, *Rev. Geophys. Space Phys.*, **15**, 113, 1977.

PELLINEN, R. and W. HEIKKILA, Energization of charged particles to high energies by an induced substorm electric field within the magnetotail, *J. Geophys. Res.*, **83**, 1544, 1978.

PERSSON, H., Electric field along a magnetic line of force in a low-density plasma, *Phys. Fluids*, **6**, 1756, 1963.

PERSSON, H., Electric field parallel to the magnetic field in a low-density plasma, *Phys. Fluids*, **9**, 1090, 1966.

QUON, B.M. and A.Y. WONG, Formation of potential double layers in plasmas, *Phys. Rev. Lett.*, **37**, 1393, 1976.

RASSBACH, M.E., Upward Birkeland currents, *J. Geophys. Res.*, **78**, 7553, 1973.

SCHRIJVER, H., The time dependence of turbulent resistivity in linear plasma column, *Physica*, **70**, 339, 1973a.

SCHRIJVER, H., Turbulent resistivity in hydrogen and noble-gas plasmas, *Physica*, **70**, 358, 1973b.

SHAWHAN, S.D., C.-G. FÄLTHAMMAR, and L.P. BLOCK, On the nature of large auroral zone electric fields at one R_E altitude, *J. Geophys. Res.*, **83**, 1049, 1978.

SISCOE, G.L., The particle environment in space, in *Photon and Particle Interaction with Surfaces in Space*, edited by R.J.L. Grard, p. 23, D. Reidel, Hingham, Mass., 1973.

SWIFT, D.W., On the formation of auroral arcs and acceleration of auroral electrons, *J. Geophys. Res.*, **80**, 2096, 1975.

SWIFT, D.W., H.C. STENBAEK-NIELSEN, and T.J. HALLINAN, An equipotential model for auroral arcs, *J. Geophys. Res.*, **81**, 3931, 1976.

TORVÉN, S. and M. BABIČ, Current chopping space charge layers in a low pressure arc plasma, Proc 12th International Conf. on Phenomena in Ionized Gases, Eindhoven, Netherlands, August 18–22, 1975, Part 1, p. 124, North-Holland Publ. Co., 1975.

TORVÉN, S. and M. BABIČ, Current limitation in low pressure mercury arcs, Royal Institute of Technology, TRITA-EPP-76-09, 1976.

WESCOTT, E.M., H.C. STENBAEK-NIELSEN, T.J. HALLINAN, and T.N. DAVIS, The Skylab plasma injection experiments, 2. Evidence for a double layer, *J. Geophys. Res.*, **81**, 4495, 1976.

WHIPPLE, E.C., The signature of parallel electric fields in a collisionless plasma, *J. Geophys. Res.*, **82**, 1525, 1976. Cf. also LENNARTSSON, W., *J. Geophys. Res.*, **83**, 2228, 1978 and WHIPPLE, E.C., *J. Geophys. Res.*, **83**, 2229, 1978.

Review of Auroral Currents and Auroral Arcs

G. ATKINSON

*Department of Communications, Communications Research
Centre, Ottawa, Ontario, Canada*

(Received February 7, 1978)

Discrete auroral arcs involve both magnetic-field-aligned electric fields and currents. Theories agree that the currents close below the arcs through the E region ionosphere. Postulated closures above the arcs include: (1) currents due to pressure gradients; (2) polarization currents; (3) currents in the conjugate ionosphere; (4) currents on the magnetopause. The energization of particles to several keV by parallel electric fields requires an energy input and hence a force doing work in the outer magnetosphere. For the above, this force is identified as: (1) diamagnetic repulsion of the plasma by the earth's field; (2) inertial forces as a flux tube gains and loses momentum due to interaction with the ionosphere; (3) magnetic stresses due to conjugate points not being those required for a minimum energy magnetosphere; (4) discontinuities in the magnetic stresses at the magnetopause arising from either the different histories of flux tubes or topological features such as the cleft. Any of the above result in a shear in the magnetic stresses from one side of the current sheet to the other consistent with the existence of the current sheet. Thus, if the magnetospheric convective flow is decoupled from the ionosphere by a non-zero resistivity, rapid convection occurs in thin sheets on each side of the arc producing: (1) a potential drop along field lines; (2) a net downward component of the Poynting vector to the region of particle acceleration. It is concluded that the process is one of energy transfer from magnetic fields to plasma and the possibility arises that this could occur at any tangential discontinuity in space where the electric current is subject to a non-zero resistivity. The process is compared with merging and possible flows are presented.

1. Introduction

The high-latitude auroras are a complicated and varied phenomenon and are rich in structures and substructures. While all seem to be caused by precipitating particles, it is becoming increasingly clear that the sources of these particles and the precipitation mechanisms are diverse. Thus diffuse aurora, discrete aurora, polar-cap aurora, proton aurora, pulsating aurora and multiple parallel arcs all require separate investigation.

The discussion in this review will be limited to discrete arcs. In addition to the commonest of these, namely those found near the northern edge of the pre-midnight auroral oval (AKASOFU, 1976), this may include those found elsewhere in the oval and in the polar cap and also multiple parallel arcs. The aim is to examine how existing theoretical models relate to the large-scale picture of the magnetosphere.

A further limitation will be to discuss only those discrete arcs in which the primary electrons are accelerated by electric fields parallel to magnetic field lines and further which are the site of upward magnetic-field-aligned electric currents. There is now substantial evidence that many (although possibly not all) discrete arcs have these properties. Some of this evidence will be documented in the next few paragraphs.

The evidence for the existence of field-aligned (Birkeland) currents comes primarily from magnetometer measurements on satellites and rockets. Representative satellite results are reported in ZMUDA and ARMSTRONG (1974), SUGIURA (1975), ARMSTRONG and ZMUDA (1973), THEILE and PRAETORIUS (1973), BERKO et al. (1975), IIJIMA and POTEMRA (1976a, b) and BURROWS et al. (1976). Similar measurements from rockets have been over auroral forms and hence the observed currents have been related to the visual aurora and particles. The number of papers on this subject is rather large and the interested reader is referred to review articles by ARNOLDY (1974), ANDERSON and VONDRAK (1975), EATHER (1975), BOYD (1975) and RUSSELL (1975). Recent papers include CASSERLY (1977), CASSERLY and CLOUTIER (1975), PAZICH and ANDERSON (1975), SESIANO and CLOUTIER (1976) and EDWARDS et al. (1976).

The evidence for the existence of magnetic-field-aligned electric fields comes primarily from two sources: particle pitch angles and spectra observed by rockets and satellites, and the observation of stronger perpendicular electric fields and their effects in the vicinity of auroras. The observations of particle spectra and pitch angles by rockets are numerous and the interested reader is referred to the same review articles as in the previous paragraph. Recent observations include those reported by BOYD and DAVIS (1977), EVANS et al. (1977) and CARLSON and KELLEY (1977). The satellite observations have been dominated by groups at the University of Iowa and Rice University. These are discussed in reviews by FRANK (1975) and the others previously listed. Recent studies include MENG (1976), CRAVEN and FRANK (1976), DOERING et al. (1976), and LUI et al. (1977). Particularly interesting results were reported by MIZERA et al. (1977) in which, from ion, electron and electric field observations, the conclusion was reached that three quarters of the potential drop was above the satellite at 7,300 km and one quarter was below. Evidence for the existence of stronger perpendicular electric fields in the vicinity of aurora can be found in barium-cloud motions (WESCOTT et al., 1975, 1976; JEFFRIES et al., 1975), ion streaming motions (WHALEN et al., 1974), counter streaming motions observed in auroral movies, satellite observations (GURNETT and FRANK, 1973; BURCH et al., 1976a, b; MOZER et al., 1977) and indirectly in the success of the Kelvin-Helmholtz instability in explaining auroral curls (HALLINAN, 1976). Occasional direct observations of parallel electric fields have been reported (MOZER et al., 1977). The altitudes of the parallel electric fields according to the above observations range from 2,000 to 8,000 km. The instantaneous vertical extent is not yet known.

2.   The Current Systems

The arc-associated field-aligned currents must close across field lines both under the arcs, and above the arcs somewhere in the magnetosphere or conjugate ionosphere (see Fig. 1). Thus the arcs must be associated with mechanical forces at these locations which are equal to $J \times B$, and it is reasonable to regard the magnetospheric forces as the cause of the arcs since ionospheric currents are dissipative. Most theories assume that closure below the arcs is horizontally through the E region ionosphere to oppositely directed field-aligned currents. The principal difference between theories is the manner in which currents close above the arcs and hence in the forces involved. The closure mechanisms and theories will be considered individually later in the paper, and in fact only theories presenting a closure mechanism will be discussed here, since the author believes this to be basic to understanding arcs.

The current system is shown schematically in Fig. 1. The arcs are represented as a vacuum diode with from 0.1 to 10 kV potential drop across it. These numbers correspond to the energies of electrons in discrete arcs in the dayside cusp and in nightside discrete arcs. The resistance in parallel with the diode represents anomalous resistance due to wave-particle interactions and the ionospheric resistance is shown as variable to account for the variation of conductivity with precipitation. The generator, shown in the outer magnetosphere, drives the current system, and it is the nature of this generator, as mentioned above, that will be discussed later in the paper as the different theories are discussed. The voltage across it must be at least 0.1 kV to produce dayside arcs and could be as high as the cross-polar cap potential, which is the maximum available in a steady state. A typical value for this is 50 kV.

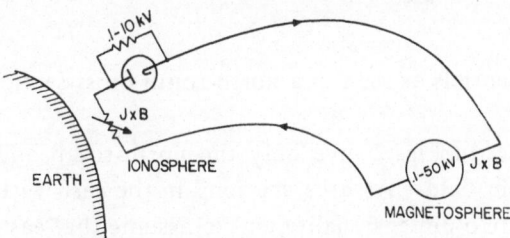

Fig. 1.   Schematic representation of auroral current system.

3.   The Electric Field Configuration and Energy Flow

The electric field models used in this discussion are the ones shown in Fig. 2 and have been developed by CARLQVIST and BOSTROM (1970), BLOCK (1972, 1976) and SWIFT (1970, 1976).

The detailed form, vertical extent and location of the configuration will not be discussed here. Neither will the nature of the associated resistance (double layer, plasma turbulence, magnetic mirroring or ion-cyclotron shocks). This type of model is adopted because: (1) it fits the observations well; (2) it satisfies curl $E=0$; (3) parallel electric fields exist only in a very limited region. The diagrams in Fig. 2

ELECTRIC POTENTIALS

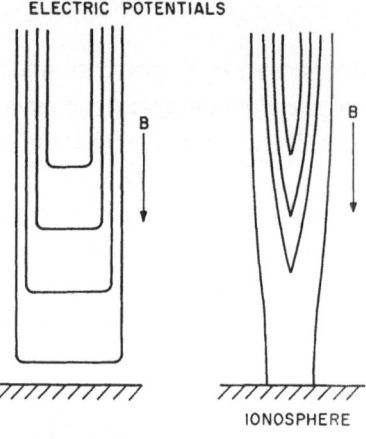

Fig. 2. Two possible configurations of electric equipotential surfaces as seen in a north-south cross-
section of an east-west arc.

IONOSPHERE

ENERGY FLOW

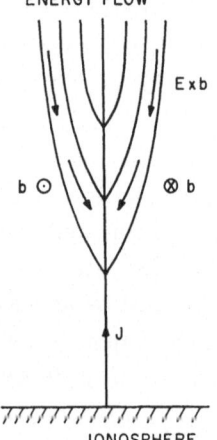

Fig. 3. Energy flow as viewed in a north-south cross-section of an
arc.

IONOSPHERE

show electric equipotentials as seen in a north-south cross-section through an east-
west arc.

The potential lines in Fig. 2 give only the north-south and the vertical components of the electric field. As arcs are long in the east-west direction, it seems
reasonable to assume two-dimensionality and to assume that east-west electric fields
are constant or slowly varying (in space). The limitations on east-west fields will
be explored in a later section of this paper.

There is an electromagnetic energy flux given by the downward component of
the Poynting vector in the two regions of intense electric field on each side of the
arc. This is illustrated in Fig. 3 which is again a north-south cross-section through
an arc with electric potentials similar to those in Fig. 2. The upward current, J,
produces east-west magnetic field components b, which when cross-multiplied by the
electric field gives the Poynting vectors shown. This energy flux is equal to the rate
at which energy is given to the particles. For a sheet-current density I, the induced
magnetic field is $b = \mu_0 I/2$. The downward component of the Poynting vector is then
$E \times b/\mu_0$ which can be integrated over the two regions of intense electric fields to
give IV, which is the power input to the particles, where V is the voltage drop.

From the foregoing it is clear that there is an energy flow downwards from the magnetosphere adjacent to the current sheet in the region of intense electric fields. The source of the energy must be stresses (magnetic or mechanical), acting farther out on these magnetic field lines, subject to the fast convective velocity associated with the E field, and hence doing work. The power input from these forces is of course the product of force and velocity. An equivalent viewpoint is that $J \cdot E$ is negative in the outer magnetosphere at the generator.

In the next section the forces and energy source will be identified for various theories that have been proposed to explain auroral arcs.

4.   Theories of Arcs and Current Closure

Most recent theories agree that the field-aligned currents close below the arcs to oppositely-directed field-aligned currents by horizontal flow in the E-region ionosphere. The force associated with this closure is the 'line-tying' or 'foot dragging' force due to collisions of the convecting E-region ions with the neutral atmosphere. There are some differences between theories as to the distribution of ionospheric conductivity and maintenance mechanisms for the conductivity distribution. Thus ATKINSON (1970) suggests that conductivity variations are maintained by precipitation, whereas OGAWA and SATO (1971) studied the growth of arcs and invoked a cross-field instability to produce irregularities in the E region. This irregularity was further amplified and maintained by precipitation.

However, the biggest difference between theories is in the way the currents close above the arcs and it is these closing currents and their associated stresses and energy source that are discussed at length in this section. Cross-magnetic-field currents postulated to close the systems above the arcs involve four types. Each of these will be considered in turn below.

4.1   Current due to pressure gradients in the plasma (the grad P term in the magneto-hydrodynamic equation)

The corresponding force or stress that provides the energy is repulsion of the plasma by the earth's magnetic field due to the plasma's diamagnetic properties. The convective flow has to have a component in the magnetosphere towards a region of weaker magnetic field on the plasma-rich side of the arc. This is consistent with processes at the ALFVÉN shielding layer (WOLF, 1974). Since, observationally, the ring current plasma attains energies of a few tens of keV, it is a possible source of arcs and could produce kilovolt potential differences along lines. However this mechanism should produce arcs on the morning side in the southern part of the oval since is here that upward currents flow. There is little evidence for arcs at this location and hence pressure gradients are an unlikely explanation of the discrete arcs.

4.2   Polarization currents (the inertial term $\rho d\boldsymbol{v}/dt$ in the magnetohydrodynamic equation)

In ATKINSON (1970) the polarization currents result from accelerations perpendicular to magnetic field lines as the convection velocity changes due to variations in the ionospheric conductivity. The same is true in SATO (1977). In SATO and HOLZER (1973) the currents close above the arcs partly by polarization currents (similar to Atkinson) and partly in the conjugate ionosphere, as will be discussed in the next section. This class of theories has been the most successful to date at explaining multiple parallel arc structure, but, as developed thus far, does not successfully explain the regions of strong electric field adjacent to the arc. That is, a flux tube could not be expected to accelerate as it entered such a region due to inertial forces alone.

In KAN and AKASOFU (1976) the currents again result from accelerations and decelerations perpendicular to the magnetic field lines. However these are produced by downward streaming of plasma parallel to the magnetic field through regions of different perpendicular electric field (convection velocity). Thus in a converging magnetic field with equipotential field lines, the perpendicular velocity increases with height. Similarly the passage of plasma along field lines through the electric field configuration described in the previous section would result in accelerations perpendicular to field lines and hence in currents. The concept in the latter case is that plasma streaming along magnetic field lines goes through an electrostatic shock (SWIFT, 1975, 1976) in which energy is transferred to the electrons from the ions and in which the potential distribution is similar to those in Fig. 2. Hence there is a change in the horizontal convection velocity of the downward moving plasma and polarization currents occur. However, in order to separate the current sheets sufficiently to agree with observation (> 10 km at ionospheric altitudes) the shock must form at large distances from the earth ($10\,R_E$). This does not agree with the observations listed earlier and further it is difficult to produce a shock model at $10\,R_E$ in which momentum balance is satisfied.

4.3   Currents in the conjugate ionosphere

OGAWA and SATO (1971) and SATO and HOLZER (1973) close the currents above the arcs either totally or partly in the conjugate ionosphere. The stresses that provide the energy source for the aurora in this case are due to stored magnetic energy. That is the conjugate points are not those required for a minimum-energy magnetosphere and arcs might arise as the flux tubes flow towards the minimum energy configuration. This situation would be most likely to arise due to a mismatch of magnetic field lines during reconnection.

4.4   Currents at the magnetopause due to the solar-wind-magnetosphere interaction

Currents of this nature are invoked to explain the observed field-aligned currents at the high-latitude edge of the oval (VASYLIUNAS, 1972; WOLF, 1974). This is also the location of the inverted V electron precipitation (FRANK and ACKERSON, 1972) which is believed to result from parallel electric field acceleration. Thus this seems to be a likely cause of at least some of the dusk-side discrete auroras. The acceleration process for electrons in such current systems has been explored by LENNARTSSON (1976) who considered mirroring as a basic cause of the magnetic field-aligned electric field.

Auroral arcs would be expected to occur at boundaries between flux tubes pulled in different directions by the solar-wind interaction since a divergence of the surface currents exists at these locations. Thus the boundary between polar-cap and closed field lines as mentioned above is one obvious location, since the polar-cap field lines are pulled towards the nightside whereas the nightside stresses are such as to drive a flow back towards the dayside. The same mechanism accounts quite well for polar-cap auroras since magnetic flux tubes with different merging histories, due, for example, to a change in the interplanetary magnetic field direction, would be intertwined to varying degrees and subject to different forces for some considerable time after merging.

5. Success of Theories

The success of the various theories in explaining the arcs is summarized in Table 1. The column at the left lists the areas of possible success: explaining the existence of discrete dusk aurora and polar cap aurora, explaining why multiple parallel structure occurs, asking if the model allows the development of several kilovolts of potential difference and if this potential difference occurs at 2,000 to 8,000 km. None of the models are completely successful and it is possible that combinations are needed.

Table 1

	Inertial (Atkinson)	Inertial (Kan and Akasofu)	Conjugate (Sato and Ogawa)	Magnetopause	Pressure
Dusk Aurora	Yes	Yes	Yes	Yes	No
Polar cap aurora	Yes	?	No	Yes	No
Multiple arcs	Yes	No	Yes	No	No
Develop voltage	No	Yes	No	Yes	Yes
$E_{\parallel}$ altitude	?	No	?	Yes	?

6. Energy Conversion

Traditionally, review papers contain a section speculating on areas for future research and that is the intent in this section.

In all the models discussed, except the KAN and AKASOFU (1976) model, the mechanism, as viewed locally at the altitudes of parallel electric fields, represents magnetic (or more precisely electromagnetic) energy being converted to plasma energy. It is true that in the steady-state theories discussed, this energy is replaced at an equal rate by the action of mechanical forces in the outer magnetosphere, but nevertheless as seen locally the energy flow represented by the Poynting vector into the plasma-acceleration region represents local conversion of magnetic to plasma energy. An equivalent way of looking at this conversion process is that the convection velocity at a horizontal surface is in the same direction as the Maxwell shear stress across the surface and hence work is done by the magnetic field.

It can be concluded from these arguments that the mechanism producing arcs is one by which magnetic field energy is converted to plasma energy when the local plasma cannot carry the field-aligned current due to some resistivity-producing mechanism, possibly one of those in the literature. There appears to be no basic reason to limit this mechanism for energy transfer from magnetic fields to plasmas to cases near the ionosphere, or to particles with energies of kiloelectronvolts, or to a single resistivity-producing mechanism. It seems likely that in any plasma with a tangential discontinuity and magnetic field lines not parallel in the two regions on either side of the discontinuity, finite resistivity effects might occur, producing magnetic-field-aligned electric fields to maintain current continuity. This would occur as in the auroral case with relative 'slippage' between different sections of a field line and plasma energization at the expense of magnetic field energy. The 'region of slippage' (region where parallel electric fields exist) is elongated and lies in the current sheet and orthogonal to the bisector of the magnetic fields if the fields are equal in magnitude.

A comparison with the other popular magnetic-energy-release process, magnetic merging, is clearly in order. Both invoke finite resistivity in a small but elongated region which lies in the current sheet of a tangential discontinuity. However the 'region of slippage' is orthogonal to the 'reconnection' or 'neutral line' of merging if the magnetic field strengths are equal. Work needs to be done to see if some combination of the two exists.

Flow patterns for two-dimensional merging are well understood and a similar understanding is desirable for 'slippage.' Since the largest flow component is along auroral arcs it is necessary to be concerned with end effects. First we shall consider some possible flow patterns for auroral arcs, that is, with an ionosphere limiting electric fields on one side, and then some possible flow patterns if 'slippage' occurs in free space.

Individual discrete arcs, though long, do not run completely around the oval and hence do have a finite length. The best way to illustrate the flow is to consider equipotential surfaces and examine, not the cross sections as shown in Fig. 2, but instead a view along magnetic field lines down towards the earth. Figure 4, top diagram, is such a plan view of an arc (with the thickness to length ratio greatly exaggerated) showing electric equipotentials (streamlines of convection) for the portion of flux tubes above the magnetic field-aligned potential drop (solid lines) and for the flux tube feet below the potential drop (dashed lines). The heavy line encloses the region in which field lines are not equipotential. This is for the case where the cross section is the same as Fig. 2 and a weak westward electric field exists everywhere. A potential line such as the one labelled '3' passes above the one labelled '5' in the middle of the arc and hence produces a 2 kV drop (if '3' and '5' in the

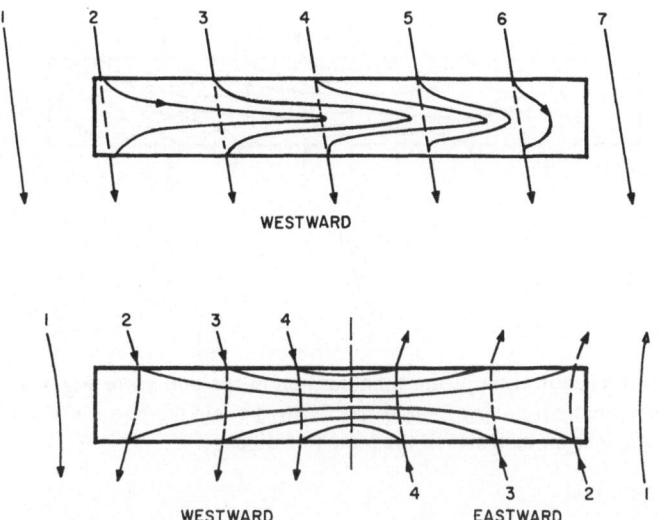

Fig. 4. Plan view of two possible electric equipotential configurations that
are consistent with arc models and measurements. The heavy line is the
arc boundary and solid and dashed lines are equipotentials above and
below the parallel-electric-field region.

figure are in kilovolts) along magnetic field lines at this location. To satisfy Maxwell's
equations, steady state and the condition that magnetic field lines be equipotential
outside the arc, the potentials must match up as shown at the boundary.

A second possible configuration is shown at the bottom of Fig. 4. Again the
cross section has the usual form but the component of electric field along the arc
varies from westward at one end to eastward at the other. It is clear that many
patterns of alternating eastward and westward electric fields are possible. However
these two are the simplest. There seems to be no basic reason to prefer one of these
to the other at the present time, although there are large differences in terms of
plasma transport. Both configurations produce field-aligned potentials and they both
can be fit into the magnetospheric flow. The top one in Fig. 4 would fit well on the
evening side of the polar cap boundary where the flow would be first dragged tailward, and then after merging, sunward. Clearly this is in a direction so as to release
magnetic stress in the magnetosphere consistent with the earlier discussion. In this
example, in the order one tenth of the total convective flow both in the polar cap
towards the nightside and in the oval towards the dayside occurs in the narrow arc
if a potential drop of 5 kV is involved. The second flow in Fig. 4 could well occur
at the boundary of sheared flux tubes.

Finally, possible flow patterns (equipotentials) are drawn for the general non-
ionospheric case. The equipotential configuration at the bottom of Fig. 4 has been
generalized in Fig. 5 at the top to the case where convective flow can occur equally
above and below the 'region of slippage.' Cross sections of the potential surfaces
are shown at appropriate locations along the 'region of slippage.' In Fig. 6 a symmetric solution is shown with the electric field component parallel to the 'region of

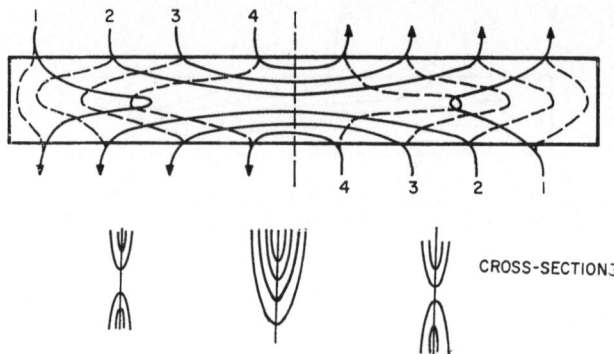

Fig. 5. Generalization of the configuration at the bottom of Fig. 4 to the case remote from any ionosphere. Viewed along the bisector of the field directions on each side of the region of slippage.

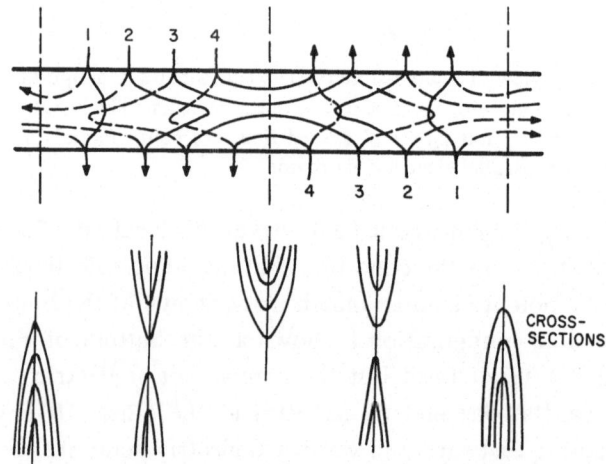

Fig. 6. A symmetric configuration. The basic pattern is repeated indefinitely.

slippage' reversing at the vertical dashed lines. The basic structure of Fig. 6 is assumed to be repeated indefinitely along the 'region of slippage.' Again representative cross sections of the surfaces are shown below the view along the bisector of the magnetic field lines. The solutions shown here are steady state. However there seems to be no reason to believe that time-dependent situations do not exist also.

7. Conclusions

A knowledge of the generator producing the electric currents is essential to understanding auroral arcs, since the energy flow into the acceleration region can be traced back to the generator. Types of currents postulated to close the auroral current systems include: (1) currents due to pressure gradients; (2) polarization currents; (3) currents in the conjugate ionosphere; (4) currents on the magnetopause. The forces driving these currents, which compose the generator, since $J \cdot E$ is negative, are, in order corresponding to the above: (1) diamagnetic repulsion of the plasma by the earth's field; (2) inertial forces as a flux tube gains and loses momentum due to interaction with the ionosphere; (3) magnetic stresses due to conjugate points not being those required for a minimum energy magnetosphere, possibly resulting from a 'mismatch' of field lines during reconnection; (4) discontinuities in the magnetic stresses at the magnetopause arising from either the different histories of flux tubes or topological features such as the cleft. No one of these, as currently presented, fits all aspects of aurora and some combination seems likely.

It appears that the auroral particle acceleration mechanism is one in which magnetic energy is converted into particle energy when a finite resistivity appears at the boundary between sheared flux tubes, that is, in a current sheet. It is not necessary to specify the exact mechanism producing finite resistivity to understand the energy flow. It is suggested that the mechanism might apply to any tangential discontinuity in plasmas in space in that the production of parallel electric fields allows the relative 'slippage' of different segments of a magnetic field line so as to create a lower energy magnetic field configuration with resulting energization of plasma.

Three-dimensional limitations on arcs provide boundary conditions on regions in which magnetic-field-aligned electric fields can exist ('regions of slippage') and two simple flows are shown in Fig. 4 for auroral arcs which are consistent with measurements and generally-accepted models of arcs. Hypothesized flows accompanying 'slippage' in space in a region remote from an ionosphere are shown in Figs. 5 and 6.

REFERENCES

AKASOFU, S.-I., Recent progress in studies of DMSP auroral photographs, *Space Sci. Rev.*, **19**, 169–215, 1976.

ANDERSON, H.R. and R.R. VONDRAK, Observations of Birkeland currents at auroral latitudes, *Rev. Geophys. Space Phys.*, **13**, 243–262, 1975.

ARMSTRONG, J.C. and A.J. ZMUDA, Triaxial magnetic measurements of field-aligned currents at 800 km in the auroral region: Initial results, *J. Geophys. Res.*, **78**, 6802–6807, 1973.

ARNOLDY, R.L., Auroral particle precipitation and Birkeland currents, *Rev. Geophys. Space Phys.*, **12**, 217–231, 1974.

ATKINSON, C., Auroral arcs: Result of the interaction of a dynamic magnetosphere with the ionosphere, *J. Geophys. Res.*, **75**, 4746–4755, 1970.

BERKO, F.W., R.A. HOFFMAN, R.K. BURTON, and R.E. HOLZER, Simultaneous particle and field observations of field-aligned currents, *J. Geophys. Res.*, **80**, 37–46, 1975.

BLOCK, L.P., Potential double layers in the ionosphere, *Cosm. Electrodyn.*, **3**, 349, 1972.

BLOCK, L.P., Double layers, in *Physics of the Flat Plasma in the Magnetosphere*, edited by B. Hulqvist and L. Stenflo, p. 229, Plenum Press, New York, 1976.

BOYD, J.S., Rocket-borne measurements of auroral electrons, *Rev. Geophys. Space Phys.*, **13**, 735–740, 1975.

BOYD, J.S. and T.N. DAVIS, Rocket measurements of electrons in a system of multiple auroral arcs, *J. Geophys. Res.*, **82**, 1197–1205, 1977.

BURCH, J.L., S.A. FIELDS, W.B. HANSON, R.A. HEELIS, R.A. HOFFMAN, and R.W. JANETZKE, Characteristics of auroral electron acceleration regions observed by atmospheric explorer C, *J. Geophys. Res.*, **81**, 2223–2230, 1976a.

BURCH, J.L., W. LENNARTSSON, W.B. HANSON, R.A. HEELIS, J.H. HOFFMAN, and R.A. HOFFMAN,

Properties of spikelike shear flow reversals observed in the auroral plasma by atmosphere explorer C, *J. Geophys. Res.*, **81**, 3886–3896, 1976b.

Burrows, J.R., M.D. Wilson, and I.B. McDiarmid, Simultaneous field aligned current and charged particle measurements in the cleft, in *Magnetospheric Particles and Fields*, edited by B.M. McCormac, pp. 111–124, D. Reidel Publ. Co., Dordrecht, Holland, 1976.

Carlqvist, P. and R. Bostrom, Space charge regions above the aurora, *J. Geophys. Res.*, **75**, 7140–7146, 1970.

Carlson, C.W. and M.C. Kelley, Observation and interpretation of particle and electric field measurements inside and adjacent to an active auroral arc, *J. Geophys. Res.*, **82**, 2349–2360, 1977.

Casserly, R.T., Jr. and P.A. Cloutier, Rocket-based magnetic observations of auroral Birkeland currents in association with a structured auroral arc, *J. Geophys. Res.*, **80**, 2165–2168, 1975.

Casserly, R.T., Jr., Observation of a structured auroral field-aligned current system, *J. Geophys. Res.*, **82**, 155–163, 1977.

Craven, J.D. and L.A. Frank, Electron angular distributions above the dayside auroral oval, *J. Geophys. Res.*, **81**, 1695–1699, 1976.

Doering, J.P., T.A. Potemra, W.K. Peterson, and C.O. Bestrom, Characteristic energy spectra of 1 to 500 eV electrons observed in the high-latitude ionosphere from atmosphere explorer C, *J. Geophys. Res.*, **81**, 5507–5516, 1976.

Eather, R.H., Advances in magnetospheric physics: Aurora, *Rev. Geophys. Space Phys.*, **13**, 925–943, 1975.

Edwards, T., D.A. Bryant, M.J. Smith, U. Fahleson, C.-G. Falthammar, and A. Pedersen, Electric fields and energetic particle precipitation in an auroral arc, in *Magnetospheric Particles and Fields*, edited by B.M. McCormac, pp. 285–289, D. Reidel Publ. Co., Dordrecht, Holland, 1976.

Evans, D.S., N.C. Maynard, J. Trøim, T. Jacobsen, and A. Egeland, Auroral vector electric field and particle comparisons. 2. Electrodynamics of an arc, *J. Geophys. Res.*, **82**, 2235–2249, 1977.

Frank, L.A. and K.L. Ackerson, Local-time survey of plasma at low altitudes over the auroral zones, *J. Geophys. Res.*, **77**, 4116–4127, 1972.

Frank, L.A., Magnetospheric and auroral plasmas: A short survey of progress, *Rev. Geophys. Space Phys.*, **13**, 974–989, 1975.

Gurnett, D.A. and L.A. Frank, Observed relationships between electric fields and auroral particle precipitation, *J. Geophys. Res.*, **78**, 145–170, 1973.

Hallinan, T.J., Auroral spirals. 2. Theory, *J. Geophys. Res.*, **81**, 3959–3965, 1976.

Iijima, T. and T.A. Potemra, The amplitude distribution of field-aligned currents at northern high latitudes observed by Triad, *J. Geophys. Res.*, **81**, 2165–2174, 1976a.

Iijima, T. and T.A. Potemra, Field-aligned currents in the dayside cusp observed by Triad, *J. Geophys. Res.*, **81**, 5971–5979, 1976b.

Jeffries, R.A., W.H. Roach, E.W. Hones, Jr., E.M. Wescott, H.C. Stenbaek-Nielsen, T.N. Davis, and J.D. Winningham, Two plasma injections into the northern magnetospheric cleft, *Geophys. Res. Lett.*, **2**, 285–288, 1975.

Kan, J.R. and S.-I. Akasofu, Energy source and mechanisms for accelerating the electrons and driving the field-aligned currents of the discrete auroral arc, *J. Geophys. Res.*, **81**, 5123–5130, 1976.

Lennartsson, W., On the magnetic mirroring as the basic cause of parallel electric fields, *J. Geophys. Res.*, **81**, 5583–5586, 1976.

Lui, A.T.Y., D. Venkatesan, C.D. Anger, S.-I. Akasofu, W.J. Heikkila, J.D. Winningham, and J.R. Burrows, Simultaneous observations of particle precipitations and auroral emissions by the ISIS 2 satellite in the 19–24 MLT sector, *J. Geophys. Res.*, **82**, 2210–2226, 1977.

Meng, C.-I., Simultaneous observations of low-energy electron precipitation and optical auroral arcs in the evening sector by the DMSP 32 satellite, *J. Geophys. Res.*, **81**, 2771–2785, 1976.

Mizera, P.F., J.F. Fennell, and A. Vampola, Charged particle distributions in the presence of large DC electric fields, preprint, The Aerospace Corporation, 1977.

Mozer, F.S., C.W. Carlson, M.K. Hudson, R.B. Torbet, B. Parady, J. Yatteau, and M.C.

KELLEY, Observations of paired electrostatic shocks in the polar magnetosphere, *Phys. Rev. Lett.*, **38**, 292–295, 1977.

OGAWA, T. and T. SATO, New mechanism of auroral arcs, *Planet. Space Sci.*, **19**, 1393–1412, 1971.

PAZICH, P.M. and H.R. ANDERSON, Rocket measurement of auroral electron fluxes associated with field-aligned currents, *J. Geophys. Res.*, **80**, 2152–2160, 1975.

RUSSELL, C.T., Magnetospheric physics: Magnetic fields, *Rev. Geophys. Space Phys.*, **13**, 952–955, 1975.

SATO, T. and T.E. HOLZER, Quiet auroral arcs and electrodynamic coupling between the ionosphere and the magnetosphere (1), *J. Geophys. Res.*, **78**, 7314–7329, 1973.

SATO, T., Theory of quiet auroral arcs, preprint, University of Tokyo, 1977.

SESIANO, J. and P.A. CLOUTIER, Measurements of field-aligned currents in a multiple auroral arc system, *J. Geophys. Res.*, **81**, 116–122, 1976.

SUGIURA, M., Identifications of the polar cap boundary and the auroral belt in the high-altitude magnetosphere: A model for field-aligned currents, *J. Geophys. Res.*, **80**, 2057–2068, 1975.

SWIFT, D.W., Particle acceleration by electrostatic waves, *J. Geophys. Res.*, **75**, 6324–6328, 1970.

SWIFT, D.W., On the formation of auroral arcs and acceleration of auroral electrons, *J. Geophys. Res.*, **80**, 2096–2108, 1975.

SWIFT, D.W., An equipotential model for auroral arcs, *J. Geophys. Res.*, **81**, 3935–3943, 1976.

THEILE, B. and H.M. PRAETORIUS, Field-aligned currents between 400 and 3,000 km in auroral and polar latitudes, *Planet. Space Sci.*, **21**, 179–187, 1973.

VASYLIUNAS, V.M., The interrelationship of magnetospheric processes, in *Eath's Magnetospheric Processes*, edited by B.M. McCormac, p. 29, D. Reidel Publishing Company, Dordrecht, Holland, 1972.

WESCOTT, E.M., H.C. STENBACK-NIELSEN, T.N. DAVIS, W.B. MURCRAY, H.N. PEEK, and P.J. BOTTOMS, The $L=6.6$ Oosik barium plasma injection experiment and magnetic storm of March 7, 1972, *J. Geophys. Res.*, **80**, 951–967, 1975.

WESCOTT, E.M., H.C. STENBACK-NIELSEN, T.J. HALLINAN, T.N. DAVIS, and H.M. PEEK, The Skylab barium plasma injection experiment 2, Evidence for a double layer, *J. Geophys. Res.*, **81**, 4495–4502, 1976.

WHALEN, B.A., D.W. GREEN, and I.B. MCDIARMID, Observations of ionospheric ion flow and related convective electric fields in and near an auroral arc, *J. Geophys. Res.*, **79**, 2835–2842, 1974.

WOLF, R.A., Calculations of magnetospheric electric fields, in *Magnetospheric Physics*, edited by B.M. McCormac, pp. 167–177, D. Reidel Publishing Company, Dordrecht, Holland, 1974.

ZMUDA, A.J. and J.C. ARMSTRONG, The diurnal flow pattern of field-aligned currents, *J. Geophys. Res.*, **79**, 4611–4619, 1974.

Acceleration Mechanisms for Auroral Electrons

Daniel W. SWIFT

*Geophysical Institute, University of Alaska,
Fairbanks, Alaska 99701, U.S.A.*

(Received February 7, 1978)

Anomalous resistivity and laminar shock mechanisms for the acceleration of electrons responsible for the discrete auroral arcs are reviewed. Ion acoustic or Buneman mode instabilities are shown unlikely to play any significant role in magnetospheric field line resistivity or electron acceleration. Although ion cyclotron turbulence is commonly observed, it is concluded that it is also unlikely to contribute to auroral acceleration. The double layer, oblique and V laminar shock models are reviewed. The relationship among these shock models and the relationship of the shock to the anomalous resistivity models is described. It is pointed out that the double layer model is inconsistent with the highly structured auroral forms. The V-shock model fits the observations so well that it appears that a fundamental understanding of the acceleration process for the discrete aurora has been achieved.

1. Introduction

Ever since the first rockets were fired into the aurora over two decades ago, it has been known that auroral emissions have been caused by energetic electrons (1 to 10 keV) impacting on the upper atmosphere. What has for so long remained a mystery is how these electrons are accelerated—in particular, how these electrons are accelerated in the intricate patterns necessary to produce the highly structured auroral forms like those shown in Fig. 1.

The purpose of this review is to present some of the theoretical predictions and pertinent observational evidence of the more promising electron acceleration mechanisms. Also a theoretical review of anomalous resistivity and laminar shock models is given and followed by a selective review of satellite and ground-based observations with the purpose of identifying the most promising acceleration mechanism. Attention is confined to the mechanisms responsible for the discrete auroral arcs.

2. Theoretical Review of Auroral Acceleration Mechanisms

2.1 Turbulent resistivity

An important class of acceleration mechanisms is generally referred to as turbulent resistivity. For a complete review see PAPADOPOULOS (1977). The essential idea behind the turbulent resistivity models is that a current-carrying plasma consisting of counterstreaming ions and electrons becomes unstable by any one of a variety of

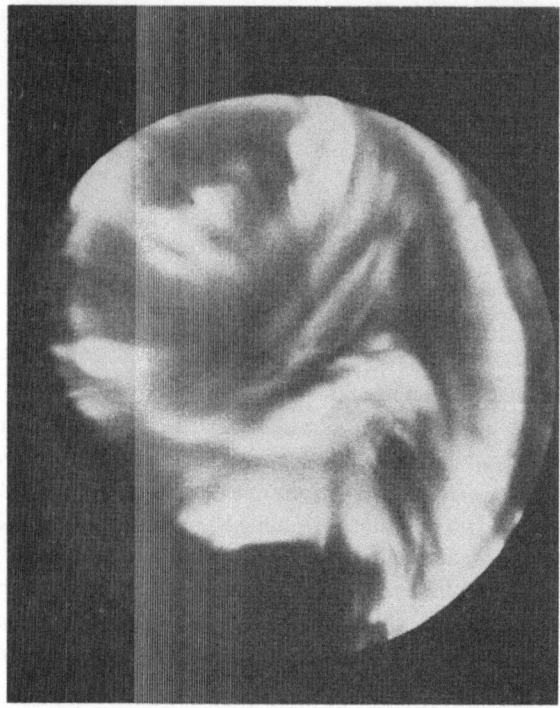

Fig. 1.   A high-speed all-sky camera photograph of an active aurora
illustrating the detailed structure of the discrete aurora.

instabilities. The resulting wave fields scatter the particles, inhibiting their flow thereby giving the plasma a finite resistence to the free flow of electric current parallel to the magnetic field and making it possible to maintain an electric field parallel to the magnetic field. Since $J \cdot E$ is positive in a resistive plasma, electrical energy is going into the energizing of plasma particles.

PAPADOPOULOS (1977) has identified three basic resistivity mechanisms resulting from instabilities excited by the electric currents carried by the plasma, namely the Buneman, ion acoustic and ion cyclotron instabilities. The Buneman instability occurs when the relative drift velocity between ions and electrons exceeds the electron thermal speed, while the ion acoustic instability occurs when the electron temperature greatly exceeds the ion temperature and when the relative drift between ions and electrons exceeds the thermal velocity of the ions at the electron temperature, $V_D > (T_e/M_i)^{1/2}$. When the ion temperature is equal to the electron temperature the two instabilities become indistinguishable. Both instabilities operate in the absence of a magnetic field, or in the presence of a magnetic field the instabilities are essentially one-dimensional with the electric field vector of the wave parallel to the magnetic field.

The ion cyclotron instability, first proposed by KINDEL and KENNEL (1971) has the lowest threshold for excitation when the ion and electron temperatures are comparable. This instability occurs only in a magnetized plasma, and modes can be excited corresponding to each harmonic of the ion gyrofrequency, the $n=1$ mode having the lowest threshold of excitation. The conditions for excitation of the $n=0$ mode are approximately the same as the ion acoustic instability, and the $n=0$ mode reduces to the ion acoustic mode when the direction of propagation becomes parallel to the magnetic field.

The conditions for excitation and the linear growth rates for the current-driven instabilities in a homogeneous plasma are well known, but the effect of these instabilities upon particle populations can only be determined by nonlinear calculations or by numerical simulations. Because of limitations on these methods, there is some uncertainty about the effects, but the general picture that seems to emerge is described below.

If an electric field is applied parallel to the magnetic field in a cold plasma, the electrons and ions will be accelerated in opposite directions until the relative streaming speed exceeds the electron thermal speed. At this point the Buneman instability sets in and results in the heating of the electrons, raising the threshold for instability and quenching the instability. This allows the electric field to further accelerate the electrons until instability sets in again and the cycle is repeated. The result is a streaming population of electrons in which the thermal and streaming speeds are equal and increasing linearly with time, no stable state being achieved in a homogeneous plasma.

The ion acoustic instability on the other hand preferentially heats the ions. If the ion temperature is much less than the electron temperature, a stable resistance can be maintained up to the point where the ion temperature reaches the electron temperature. At this point the instability would take on some of the characteristics of the Buneman instability, and likely both ion and electron temperature would rise together. The result would then likely be ion and electron temperatures and streaming speeds rising linearly with time with the streaming speed comparable to the electron thermal speed. The electron and ion temperature would be nearly equal with no steady state resistance.

However, in a magnetized plasma, the ion cyclotron instability would be excited first. Nonlinear calculations indicate that the instability goes into heating the ions, which raises the threshold drift velocity, bringing the system toward stability and reducing the plasma resistance. The ion cyclotron mechanism also does not seem to lead to a stable plasma resistivity. (The reader interested in further details is referred to PAPADOPOULOS (1977) and references therein.)

The models described above all assume a homogeneous plasma of infinite extent. When the finite size of a system is considered, the effect of the current-driven turbulence could be quite different. If cold populations of electrons and ions were to enter a region of turbulence of finite extent, it might be possible to maintain a steady resistivity because new particles would be continuously entering the system. But the magnitude of the potential that could be maintained across a turbulent region would be comparable to the temperature gain of the particles. Hence it does not appear that significant acceleration potentials could be maintained by anomalous resistivity.

The calculations of anomalous current-driven instabilities have largely neglected effects of variation of the plasma parameters in directions perpendicular to the magnetic field lines. The Buneman and ion acoustic instability models alone are incompatible with such variation. Variations perpendicular to the magnetic field lines must involve coupling to ion or electron cyclotron waves.

There may also be drift wave instabilities, basically utilizing plasma inhomogeneities as their source of free energy, which have enhanced growth rates due to the presence of field-aligned currents. Since these instabilities involve only a few resonant electrons, it is likely that they may be stabilized by the formation of small plateau in the electron distribution function (PAPADOPOULOS, 1977).

Anomalous resistivity models do not seem to lead to electron acceleration or preferential electron heating in a homogeneous magnetized plasma. But, in a finite region of anomalous resistivity maintained by ion cyclotron turbulence, a small fraction of incident electrons might have sufficient energy to escape trapping. These electrons would be freely accelerated by the small parallel component of the electric field. Hence anomalous resistivity models can possibly accelerate electrons with beam characteristics, but the accelerated electrons would contain only a fraction of the $J \cdot E$ energy dissipated.

2.2 Current-driven laminar shock models

The laminar shock models are related to the instabilities described in the previous subsections, except that in the nonlinear state the waves evolve into a solitary wave, Doppler shifted to zero frequency. The potential jump associated with the wave can accelerate electrons in one direction and ions in the other. In a laminar shock, there is no heating of the particles, so the accelerated beam has essentially the same temperature it had before it was accelerated. Thus if the magnitude of the potential jump is large in comparison to the thermal energies of the entering particles, the emerging beams would essentially be monoenergetic. In the laminar models most of the $J \cdot E$ energy would go into the electron beam with a lesser amount in an ion beam accelerated in the opposite direction, so the laminar shocks are far more energy efficient than the turbulent resistivity models.

The best known of the laminar shock models is the double layer described by BLOCK (1972) and CARLQVIST (1972). The double layer is a time-independent nonlinear state of the ion acoustic or Buneman mode current-driven instabilities, depending on the streaming speeds and temperature of the particles entering the shock. The potential jump can have almost any magnitude, so that the plasma resistivity, defined as the ratio of the potential jump to current density, can be arbitrarily large. The thickness of the double layer when the potential jump, $\Delta\phi$, is large in comparison to the thermal energies is the order $(e\Delta\phi/m\omega_p^2)^{1/2}$, where ω_p is the plasma frequency. Double layer solutions corresponding to a wide variety of plasma models can be constructed with relative ease (MONTGOMERY and JOYCE, 1969; DAVIDSON, 1972; KAN, 1975), making the double layer theory much more tractable than the turbulent resistivity theories.

One-dimensional numerical simulations of the double layer have been carried out by GOERTZ and JOYCE (1975) in which they traced the motion of a large number of charge sheets moving in a self-consistent electric field. In this simulation, charge sheets with a Maxwellian distribution of velocities of one sign were introduced from one side of the domain, and charge sheets of the opposite sign were introduced from the other side. The simulation was initiated with a small perturbation in the flow. A solitary potential wave developed and evolved into a quasi-stable potential profile resembling a hyperbolic tangent function; and resulting in a significant potential difference between the two ends of the simulation domain. The most noteworthy feature of the simulation is that the system evolved into a laminar rather than a turbulent state. This simulation suggests that instabilities in a current carrying plasma will evolve into a laminar rather than a turbulent state, provided zero-frequency modes can be excited.

In a magnetized plasma, the double layer theory is only compatible with electric fields parallel to the magnetic field, or equipotential surfaces normal to the magnetic field. Since at high latitudes the magnetic field is nearly vertical, the precipitation caused by double layers would be uniform and of great horizontal extent, which is inconsistent with the highly structured auroral forms.

The oblique shock model represents a generalization of the double layer model to shock normal directions oblique to the magnetic field. SWIFT (1975) showed that as long as the magnetic moment of the incident ions was preserved, the electric potential, ϕ, of the oblique shock satisfied the modified Poisson equation

$$\kappa \frac{d^2}{d\eta^2}\phi + \frac{1}{2}\cdot\frac{d}{d\eta}\kappa\frac{d}{d\eta}\phi = -4\pi e(n_i^* - n_e) \tag{1}$$

where $\eta = z - \alpha x$ is a coordinate normal to the shock plane and α is the tangent of the angle between the shock normal direction and the magnetic field. κ is given by

$$\kappa = 1 + \alpha^2(1 + c^2/V_a^2) \tag{2}$$

where V_a is the local Alfvén velocity. n_e is the electron density, and n_i^* is the ion density computed as if the ions were completely magnetized. The z-coordinate is parallel while x is perpendicular to the magnetic field. The effect of the polarization drift of the ions is included the plasma dielectric c^2/V_a^2.

Because of the large effective dielectric, the thickness of the shock is the order of an ion gyroradius, namely $(e\Delta\phi/m_i\Omega_i^2)^{1/2}$, where Ω_i is the ion gyrofrequency. Because $c^2/V_a^2 \gg 1$, it can be seen that the shock structure is insensitive to the magnetic field direction. When the shock normal becomes parallel to the magnetic field the double layer solutions are recovered. As in the double layer theory, SWIFT (1975) showed that it is possible to construct a wide variety of oblique shock models and that it is possible to support an arbitrarily large potential jump.

This result did not completely solve the problem of accounting for electron precipitation in a spatially confined region because the SWIFT (1975) theory referred to plane shocks of infinite extent. However, SWIFT (1976) was able to take advantage

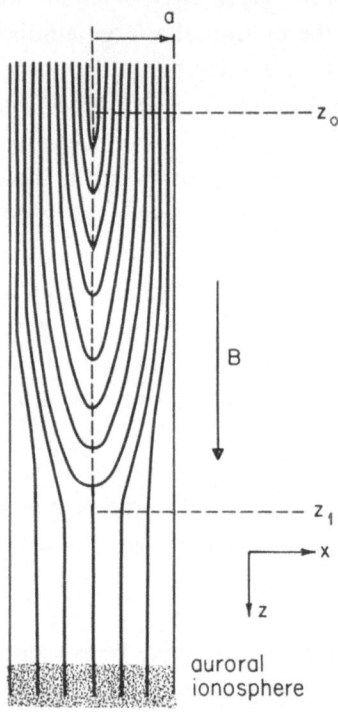

Fig. 2. Equipotentials of the V-shock model. Solid lines represent equipotential surfaces, and the electric field is toward the center axis of the figure. Above the level z_0 and below the level z_1 are the asymptotic regions where equipotentials are parallel to the magnetic field. At $1|x|1 > a$ the electric field perpendicular to the magnetic field is arbitrary, and the field lines are everywhere equipotentials.

of the freedom in shock normal direction and construct a two-dimensional shock model in which the equipotentials are curved surfaces. Thus it was shown possible to construct a shock model which satisfied the nonlinearly coupled Poisson-Vlasov system of equations of the type proposed by SWIFT et al. (1976) in which the equipotential contours assume the form of a V as shown in Fig. 2.

As noted previously, the laminar shock models should lead to essentially monoenergetic ion and electron beams on either side of the shock. The electron beam, especially when it encounters the cold ionospheric plasma is known to be unstable to a variety of instabilities. SWIFT and KAN (1975) and MAGGS (1976) have proposed that the electron beam generate waves on the resonance cone which is responsible for auroral hiss. PERKINS (1968) suggested a beam-plasma instability at the upper hybrid resonance frequency, while BOSWELL (1976) has detected the emission of Bernstein waves by the interaction of an electron beam and plasma in a laboratory device. Hence, instabilities are likely to have a significant effect in broadening the energy spectrum of the beam, so it is unlikely that a simple monoenergetic beam would actually be observed in the ionosphere as a result of acceleration by a laminar shock.

3.   Observations

In this section a selected set of observations is reviewed to decide among the mechanisms reviewed in the previous section which best account for the acceleration of electrons responsible for the discrete auroral arcs like these shown in Fig. 1.

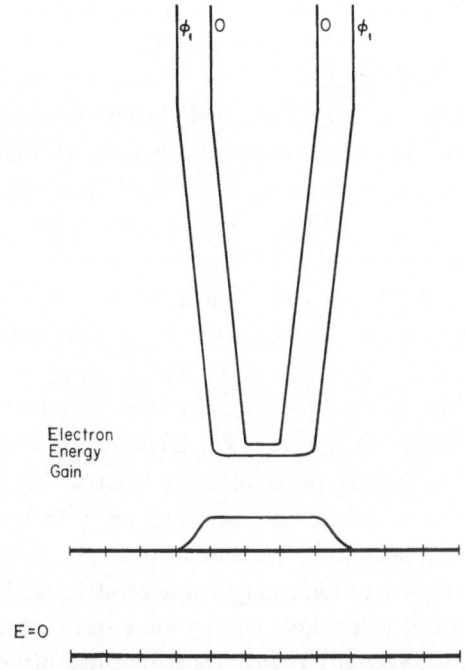

Fig. 3.   A model for auroral precipitation which includes a
double layer pieced together with oblique shocks.

First, as mentioned previously, the discreteness of the arcs rules out the double layer, or the Buneman mode and ion acoustic turbulent resistivity models as stand-alone mechanisms.   But do these mechanisms play a role in auroral acceleration when pieced together with the oblique shock model in the case of the double layer, or with ion cyclotron waves in the case of the turbulent resistivity model?

Figure 3 shows such an example of a laminar shock model.   It can be seen from the figure that a satellite passing under the shock structure would observe an increase in the energies of the precipitated particles, a region of constant energy followed by a region of decreasing energy.   The energy profile would be characterized by a flat plateau in the energy versus time plot.

Of the energy versus time profiles published by FRANK and ACKERSON (1971) none seem to show the predicted plateau.   In fact, the name 'inverted V' given to the precipitation events associated with the discrete aurora suggests the lack of a plateau.   Careful triangulated observations (ROMICK and BELON, 1967) of auroral luminosity with a pair of meridian scanning photometers indicates that the depth of penetration into the atmosphere across an auroral arc increases rapidly and then decreases sharply; no evidence of a plateau in auroral luminosity is seen.   Hence, there seems to be no evidence supporting any role of the double layer model in discrete auroral precipitation.   The same arguments suggest the lack of any role of the ion acoustic or Buneman instabilities in turbulent resistivity models for discrete auroral arcs.

On the other hand, the inverted V-structure of auroral precipitation (FRANK and ACKERSON, 1971) is an explicit feature of the V-shock model of SWIFT et al. (1976) shown in Fig. 2. In fact, the inverted V-events prompted GURNETT (1972) to propose an equipotential model for auroral electric fields which had many similarities to the V-shock model. The energy gain is proportional to the number equipotentials an electron crosses in moving downward along the magnetic field lines. An electron near the center axis of Fig. 2 crosses more equipotentials than an electron to either side of the central axis.

Perhaps the most definitive evidence on the nature of the auroral acceleration mechanism has been provided by S3-3 satellite in a polar orbit with an apogee of about 8,000 km altitude over the north pole. MOZER et al. (1977) report direct observations of dc electric fields with peak magnitudes of 500 mV/m, far stronger than has ever been observed in the ionosphere. The large electric fields are often observed as double structures of opposing perpendicular electric fields accompanied by a component parallel to the magnetic field pointing away from the earth. The high-magnitude electric fields appear to be embedded in a larger region of ion cyclotron turbulence. Within the region of high-magnitude electric field beams of O^+ and H^+ ions of ionospheric origin of a few keV energy have been observed moving up the magnetic field lines (SHELLEY et al., 1976). The high-magnitude electric field structures also occur where the onboard magnetometer indicates field-aligned currents greater than 10^{-6} A/m^2.

Unfortunately the S3-3 observations were not accompanied by simultaneous ground based auroral observations. However, there exist other observations which provide direct links between strong perpendicular electric fields and the visual aurora. HALLINAN and DAVIS (1970) using television imagery of auroral arcs observed rapid auroral ray motions. They interpreted these motions as $E \times B$ drift and deduced perpendicular electric fields a higher than had ever been observed within the ionosphere. Anomalously large electric fields at high altitudes have also been observed with the barium-shaped charge experiments. By tracking the $E \times B$ motion of a field-aligned streak of barium plasma, the electric field along the streak can be determined. WESCOTT et al. (1975, 1976) and JEFFRIES et al. (1975) have observed separate instances of sudden rapid motion of flux tubes containing the barium. In each case the rapid movement coincided with the appearance of aurora on the field lines. In one experiment (WESCOTT et al., 1976) the barium streak above 4,000 km altitude was observed to brighten suddenly accompanied by rapid drifts corresponding to electric fields, referenced to the 100 km level, in excess of 250 mV/m. The barium plasma below 4,000 km did not experience the same large electric fields.

The appearance of aurora has also been associated systematically with upward field-aligned currents. KAMIDE et al. (1976) and KAMIDE and ROSTOKER (1977) have shown that upward field-aligned currents are observed where there are auroral arcs. SESIANO and CLOUTIER (1976), using a rocket-borne magnetometer and ground-based observations, have also shown that regions of visual aurora and upward field-aligned currents coincide.

Returning to the V-shock model of Fig. 2, it can been seen that it is consistent with the presence of strong electric fields on auroral field lines. A field-aligned current is also a necessary condition for the formation of the shock (SWIFT, 1976). The V-shock model also requires the upward acceleration of ions originating from below the shock, which are also observed. Thus the observations seem to provide strong support for the V-shock mechanism.

The observations of MOZER et al. (1977) also show the existence of ion cyclotron turbulence, and HUDSON et al. (1977) have shown that observed levels of ion cyclotron turbulence are sufficient to support a 1 mV/m parallel dc electric field, which over 1,000 km of field-aligned path would be sufficient to accelerate electrons to auroral energies. Field-aligned electric currents are also observed in connection with the ion cyclotron turbulence. The question is whether ion cyclotron turbulence play any significant role in the acceleration of auroral electrons.

All of the observations of high-altitude electric fields and acceleration of ions and electrons, including evidence presented by TEMERIN et al. (1977) and TORBERT and MOZER (1977), associates the acceleration processes with the shock type structures depicted in Fig. 2 rather than with the larger region of ion cyclotron turbulence in which the shocks are imbedded. Further, TORBERT and MOZER (1977) report that electrostatic shocks and auroral arcs have similar occurrence frequencies, similar geographical distribution mapped along the magnetic field lines, similar dependence on magnetic activity and similar thicknesses. On the other hand, ion cyclotron turbulence is observed to occur on field lines occupied by V-shocks as well as in regions between V-shocks. Hence the statistics of ion cyclotron occurrence and particle acceleration appear to be different. Moreover, the effect of ion cyclotron turbulence seems to be that of heating the ions (KELLEY et al., 1975), which would make the ion cyclotron mechanism an inefficient source of electron or ion beam energy. Although not conclusive, the weight of evidence indicates that the V-shock is the only mechanism responsible for accelerating electrons that produce the discrete auroral arcs.

Ion cyclotron turbulence and V-shocks are found in the same regions, so they are likely related. It has previously been mentioned that the V-shocks represent the $n=0$ ion cyclotron mode whereas the $n \geq 1$ modes are associated with the ion cyclotron turbulence. The V-shocks and cyclotron waves also occur on field lines which GURNETT and FRANK (1977) have observed intense plasma wave turbulence extending out to at least $40\,R_E$ in the magnetospheric tail. From the S3-3 satellite data TEMERIN (1977) has identified some of this turbulence as zero-frequency turbulence. Potential differences across the magnetic field lines will be associated with the turbulence.

The picture that seems to emerge is that the potential differences will result in electric currents between the ionosphere, which may be regarded as an approximate equipotential surface, and the turbulent eddies generated in the magnetosphere. The currents will first give rise to $n \geq 1$ ion cyclotron turbulence because that is the most easily excited (KINDEL and KENNEL, 1971). The ion cyclotron turbulence does not significantly inhibit the field-aligned currents so current densities can rise considerably above the $n=1$ instability threshold. In some turbulent eddies, the field-aligned current increases to where the $n=0$ ion cyclotron mode becomes unstable. At this point the V-shocks develop, resulting in the acceleration of electrons and formation of auroral arcs. Hence the auroral arcs arise out of a much broader background of turbulent eddies, field-aligned currents and ion cyclotron turbulence.

In conclusion, among the acceleration mechanisms reviewed, the V-shock model appears to be unique in its fit of the observational evidence. The fact that explicit solutions to the Poisson-Vlasov system of equations for the V-shock model have also been exhibited demonstrates that the V-shock model is consistent with well-known physical laws. It appears that a fundamental understanding of the acceleration mechanism for the discrete aurora has been achieved.

This work was supported by Grant DES 74-19320 from the Atmospheric Science Section of NSF.

REFERENCES

BLOCK, L.P., Potential double layers in the ionosphere, *Cosm. Electrodyn.*, **3**, 349, 1972.

BOSWELL, R.W., VLF hiss generated by supro-thermal electrons, *Geophys. Res. Lett.*, **3**, 705, 1976.

CARLQVIST, P., On the formation of double layers in plasmas, *Cosm. Electrodyn.*, **3**, 377, 1972.

DAVIDSON, R.C., *Methods on Nonlinear Plasma Theory*, p. 72, Academic Press, New York, 1972.

FRANK, L.A. and K.L. ACKERSON, Observations of charged particle precipitation in the auroral zone, *J. Geophys. Res.*, **76**, 3612, 1971.

GOERTZ, C.K. and G. JOYCE, Numerical simulation of the plasma double layer, *Astrophys. Space Sci.*, **32**, 165, 1975.

GURNETT, D.A., Electric field and plasma observations in the magnetosphere, in *Critical Problems of Magnetospheric Physics*, edited by E.R. Dyer, Vol. 123, Natl. Acad. Sci., Washington, D.C., 1972.

GURNETT, D.A. and L.A. FRANK, A region of intense plasma wave turbulence on auroral field lines, *J. Geophys. Res.*, **82**, 1031, 1977.

HALLINAN, T.J. and T.N. DAVIS, Small-scale auroral arc distortion, *Plant. Space Sci.*, **18**, 1735, 1970.

HUDSON, M.R., R.L. LYSAK, and F.S. MOZER, Magnetic field aligned potential drops due to electrostatic ion cyclotron turbulence, Preprint, Space Sciences Laboratory, University of California, Berkeley 94720, June 1977.

JEFFRIES, R.A., W.H. ROACH, E.W. HONES, Jr., E.M. WESCOTT, H.C. STENBAEK-NIELSEN, T.N. DAVIS, and J.D. WINNINGHAM, Two plasma injections into the northern magnetospheric cleft, *Geophys. Res. Lett.*, **3**, 285, 1975.

KAMIDE, Y. and G. ROSTOKER, The spatial relationship of field-aligned currents and auroral electrojets to the distribution of nightside auroras, *J. Geophys. Res.*, **82**, 5589–5608, 1977.

KAMIDE, Y., H.W. KROEHL, G. ROSTOKER, S.-I. AKASOFU, and T.A. POTEMERA, Near simultaneous recording of global auroral features, field-aligned currents and auroral electrojets during magnetospheric substorms, *EOS, Trans. Am. Geophys. Union*, **57**, 993, 1976.

KAN, J.R., Energization of auroral electrons by electrostatic shock waves, *J. Geophys. Res.*, **80**, 2089, 1975.

KELLEY, M.C., E.A. BERING, and F.S. MOZER, Evidence that the electrostatic ion cyclotron instability is saturated by ion heating, *Phys. Fluids*, **18**, 1590, 1975.

KINDEL, J.M. and C.F. KENNEL, Topside current instabilities, *J. Geophys. Res.*, **76**, 3053, 1971.

MAGGS, J.E., Coherent generation of VLF hiss, *J. Geophys. Res.*, **81**, 1707, 1976.

MONTGOMERY, D.C. and G. JOYCE, Shock-like solutions to the electrostatic Vlasov equation, *J. Plasma Phys.*, **3**, 1, 1969.

MOZER, F.S., C.W. CARLSON, M.K. HUDSON, R.B. TORBERT, B. PARODY, J. YATTEAU, and M.C. KELLEY, Observations of paired electrostatic shocks in the polar magnetosphere, *Phys. Rev. Lett.*, **38**, 292, 1977.

PAPADOPOULOS, K., A review of anomalous resistivity for the ionosphere, *Rev. Geophys. Space Phys.*, **15**, 113, 1977.

PERKINS, F.W., Plasma-wave instabilities in the ionosphere over the aurora, *J. Geophys. Res.*, **73**, 6631, 1968.

ROMICK, G.J. and A.E. BELON, The spatial variation of auroral luminosity. II. Determination of volume emission rate profiles, *Planet. Space Sci.*, **15**, 1695, 1967.

SESIANO, J. and P.A. CLOUTIER, Measurements of field-aligned currents in a multiple auroral arc system, *J. Geophys. Res.*, **81**, 116, 1976.

SHELLEY, E.G., R.D. SHARP, and R.G. JOHNSON, Satellite observations of an ionospheric acceleration mechanism, *Geophys. Res. Lett.*, **3**, 654, 1976.

SWIFT, D.W., On the formation of auroral arcs and acceleration of auroral electrons, *J. Geophys. Res.*, **80**, 2096, 1975.

SWIFT, D.W., An equipotential model of auroral arcs. 2. Numerical solutions, *J. Geophys. Res.*, **81**, 3935, 1976.

SWIFT, D.W. and J.R. KAN, A theory of auroral hiss and implications on origin of auroral electrons, *J. Geophys. Res.*, **80**, 985, 1975.

SWIFT, D.W., H.C. STENBAEK-NIELSEN, and T.J. HALLINAN, An equipotential model for auroral arcs, *J. Geophys. Res.*, **81**, 3931, 1976.

TEMERIN, M., The polarization, frequency and wavelengths of high latitude turbulence, *J. Geophys. Res.*, **83**, 2609–2616, 1978.

TEMERIN, M., R.B. TORBERT, and F.S. MOZER, The creation of VLF sources and V-shaped emissions by electrostatic shocks, Preprint, Space Sciences Laboratory, University of California, Berkeley 94720, July 1977.

TORBERT, R.B. and F.S. MOZER, Electrostatic shocks as the source of discrete auroral arcs, *Geophys. Res. Lett.*, **5**, 135–138, 1978.

WESCOTT, E.M., H.C. STENBAEK-NIELSEN, T.N. DAVIS, W.B. MURCRAY, H.M. PEEK, and P.J. BOTTOMS, The Oosik barium plasma injection experiment and magnetic storm of March 7, 1972, *J. Geophys. Res.*, **80**, 951, 1975.

WESCOTT, E.M., H.C. STENBAEK-NIELSEN, T.J. HALLINAN, T.N. DAVIS, and H.M. PEEK, The Skylab injection experiments. 2. Evidence for a double layer, *J. Geophy. Res.*, **81**, 4495, 1976.

Subject Index

AEPS Vol. 1

Special Issue of Journal of Geomagnetism and Geoelectricity (Included in regular issues)

Proceedings of AGU 1976 Fall Annual Meeting, December 1976, San Francisco

ORIGIN OF THERMOREMANENT MAGNETIZATION

Edited by David J. DUNLOP

Contents TRM and Its Variation with Grain Size: A Review (R. DAY)/Single Domain Oxide Particles as a Source of Thermoremanent Magnetization (M.E. EVANS)/Domain Structure of Titanomagnetities and Its Variation with Temperature (H.C. SOFFEL)/The Demagnetization Field of Multidomain Grains (R.T. MERRILL)/The Hunting of the 'Psark' (D.J. DUNLOP)/ On the Origin of Stable Remanence in Pseudo-Single Domain Grains (S.K. BANERJEE)/The Preparation, Characterization and Magnetic Properties of Synthetic Analogues of Some Carriers of the Palaeomagnetic Record (J.B. O'DONOVAN and W. O'REILLY)/Reduction of Hematite to Magnetite under Natural and Laboratory Conditions (P.N. SHIVE and J.F. DIEHL)/Characteristics of First Order Shock Induced Magnetic Transitions in Iron and Discrimination from TRM (P. WASILEWSKI)/The Thermoremanence Hypothesis and the Origin of Magnetization in Iron Meteorites (A. BRECHER and L. ALBRIGHT)/Thermal Overprinting of Natural Remanent Magnetization and K/Ar Ages in Metamorphic Rocks (K.L. BUCHAN, G.W. BERGER, M.O. MCWILLIAMS, D. YORK, and D.J. DUNLOP)/Does TRM Occur in Oceanic Layer 2 Basalts? (J.M. HALL)/The Effects of Alteration on the Natural Remanent Magnetization of Three Ophiolite Complexes: Possible Implications for the Oceanic Crusl (S. LEVI and S.K. BANERJEE)

212 pp. (7×10) 1977 $24.50

AEPS Vol. 2

Supplement Issue of Journal of Physics of the Earth (Not included in regular issues)

Proceedings of the U.S.-Japan Seminar on Theoretical and Experimental Investigations of Earthquake Precursors

EARTHQUAKE PRECURSORS

Edited by C. KISSLINGER and Z. SUZUKI

Contents Earthquake Prediction-Related Research at the Seismological Laboratory, California Institute of Technology, 1974–1976 (J.H. WHITCOMB)/Research on Earthquake Prediction and Related Areas at Columbia University (L.R. SYKES)/Seismic Activities and Crustal Movements the Yamasaki Fault and Surrounding Regions in the Southwest Japan (K. OIKE)/The New Madrid Seismic Zone as a Laboratory for Earthquake Prediction Research (B.J. MITCHELL, W. STAUDER, and C.C. CHENG)/Anomalous Crustal Activity in the Izu Peninsula, Central Honshu (K. TSUMURA)/Recent Seismometrical Works in Japan (S. SUYEHIRO, M. ICHIKAWA, and K. TSUMURA)/Quiet and Violence in Horizontal Movement of the Crust (T. HARDA)/ Anomalous Seismic Activity and Earthquake Prediction (H. SEKIYA)/Seismic Activity in the Northeastern Japan Arc (A. TAKAGI, A. HASEGAWA, and N. UMINO)/Observations of Changes in Seismic Wave Velocity in South Kanto District, South of Tokyo, by the Explosion-Seismic Method (T. KAKIMI and I. HASEGAWA)/Some Precursors Prior to Recent Great Earthquakes along the Nankai Trough (H. SATO)/Possibility of Temporal Variations in Earth Tidal Strain Amplitudes Associated with Major Earthquakes (T. MIKUMO, M. KATO, H. DOI, Y. WADA, T. TANAKA, R. SHICHI, and A. YAMAMOTO)/Gravity Changes Associated with Seismic Activities (Y. HAGIWARA)/Geomagnetism in Relation to Tectonic Activities of the Earth's Crust in Japan (N. Sumitomo) / Precursory and Coseismic Changes in Ground Resistivity (T. Rikitake and Y. Yamazaki) / Geochemistry as a Tool for Earthquake Prediction (H. Wakita) / Recent Laboratory Studies of Earthquake Mechanics and Prediction (W.F. Brace) / Dilatancy of Rocks under General Triaxial Stress States with Special Reference to Earthquake Precursors (K. Mogi) / Possibility of a Great Earthquake in the Tokai District, Central Japan (T. Utsu) / Depth Constraints on Dilatancy Induced Velocity Anomalies (K.W. Winker and A. Nur) / Seismological Precursors to a Magnitude 5 Earthquake in the Central Aleutian Islands (E.R. Engdahl and C. Kisslinger) / Estimation of Future Destructive Earthquakes from Active Faults on Land in Japan (T. Matsuda) / Some Problems in the Prediction of the Nemuro-oki Earthquake (K. Abe) / Responses to Earthquake Prediction in Kawasaki City, Japan in 1974 (H. Ohta and K. Abe) / Socioeconomic and Political Consequences of Earthquake Prediction (J.E. Haas and D.S. Mileti)

304 pp. (7 × 10) 1978 $ 32.50

AEPS Vol. 3

Proceedings of the U.S.-Japan Seminar on Rare Gas Abundance and Isotopic Constraints on the Origin and Evolution of the Earth's Atmosphere

TERRESTRIAL RARE GASES

Edited by E.C. Alexander, Jr and M. Ozima

Contents *EXPERIMENTAL STUDIES* A Mantle Helium Component in Circum-Pacific Volcanic Gases: Hakone, the Marianas, and Mt. Lassen (H. Craig, J.E. Lupton, and Y. Horibe) / Nitrogen to Argon Ratio in Volcanic Gases (S. Matsuo, M. Suzuki, and Y. Mizutani) / Rare Gas Abundance Pattern of Fumarolic Gases in Japanese Volcanic Areas (O. Matsubayashi, S. Matsuo, I. Kaneoka, and M. Ozima) / A Review: Some Recent Advances in Isotope Geochemistry of Light Rare Gases (I.N. Tolstikhin) / Abundances and Isotopic Compositions of Rare Gases in Granites and Thucholites (P.K. Kuroda and R.D. Sherrill) / Rare Gas Isotopic Compositions in Diamonds (N. Takaoka and M. Ozima) / Rare Gases in Mantle-Derived Rocks and Minerals (I. Kaneoka, N. Takaoka, and K. Aoki) / A Comparison of Terrestrial and Meteoritic Noble Gases (O.K. Manuel) / The Composition and History of the Martian Atmosphere (T. Owen) *THEORETICAL STUDIES* Nuclear Components in the Atmosphere (T.J. Bernatowicz and F.A. Podosek) / Trapped Xenon and Cosmic-Ray Effects in Meteorites, in Lunar Sample, and in the Earth's Materials (K. Sakamoto) / Classification and Generation of Terrestrial Rare Gases (K. Saito) / Earth-Atmosphere Evolution Model Based on Ar Isotopic Data (Y. Hamano and M. Ozima) / Terrestrial Potassium and Argon Abundances as Limits to Models of Atomspheric Evolution (D.E. Fisher) / On the Ambient Mantle $^4He/^{40}Ar$ Ratio and the Coherent Model of Degassing of the Earth (D.W. Schwartman) / Earth Degassing Models, and the Heterogenous vs. Homogeneous Mantle (R. Hart and L. Hogan) / Lead Isotope Constraints on the Early History of the Earth (R.D. Russell) / Matter Accretion into the Solar System (S. Hayakawa)

230 pp. (7 × 10) 1978 $ 24.50